T0245327

# Why Penguins Communicate

# Why Penguins Communicate

## The Evolution of Visual and Vocal Signals

**Pierre Jouventin**
Director of Research (retired)
National Center for Scientific Research (CNRS)
Montpellier, France

**F. Stephen Dobson**
Alumni Professor and Curator
Department of Biological Sciences and
Auburn University Museum of Natural History
Auburn University, AL, USA

Academic Press is an imprint of Elsevier
125 London Wall, London EC2Y 5AS, United Kingdom
525 B Street, Suite 1800, San Diego, CA 92101-4495, United States
50 Hampshire Street, 5th Floor, Cambridge, MA 02139, United States
The Boulevard, Langford Lane, Kidlington, Oxford OX5 1GB, United Kingdom



**Library of Congress Cataloging-in-Publication Data**
A catalog record for this book is available from the Library of Congress

**British Library Cataloguing-in-Publication Data**
A catalogue record for this book is available from the British Library

ISBN: 978-0-12-811178-9

For information on all Academic Press publications visit our website at https://www.elsevier.com/books-and-journals

*Publisher:* Sara Tenney
*Acquisition Editor:* Kristi Gomez
*Editorial Project Manager:* Pat Gonzalez
*Senior Production Project Manager:* Priya Kumaraguruparan
*Designer:* Matt Limbert

Typeset by TNQ Books and Journals

*To ecologist, conservation biologist, and teacher—Professor François Bourlière.*

# Contents

# Preface

*Who would believe in penguins, unless he had seen them?*
Conor O'Brien (Across Three Oceans, 1926)

## WHY STUDY THE BEHAVIOR OF PENGUINS?

As we can see painted on the walls of caves during prehistoric times, humans were fascinated by animals, particularly big carnivores and herbivores. It is often compelling to recall these paintings, especially to stimulate the imagination of the general public. Our culture is full of exciting but wrong fables about strong but kind lions, brave or vicious wolves, and clever foxes. We thrill over stories in which animals seem to be, in at least some way, like us. Examples are animals that walk on two feet, such as wild grizzly bears (or less threatening teddy bears), amusing pandas, or funny penguins.

It is more difficult to delve under all these imaginative stories to study the real activities of these animals, and find scientific explanations for their behaviors. That is, to study the actual evolutionary significance of behaviors for reproduction and survival, rather than the dreams and imaginings about such an extraordinary and popular family of animals like the penguins. In this book, we seek to investigate the behavior of penguins without the imaginative bias of anthropomorphism. This is the challenge that confronts us: to study them as penguins and not as humans. We will not lose our wonder at nature with this more realistic approach, because so often the reality is better than fiction. We hope that our book reflects this truism.

Moving beyond an anthropomorphic point of view, we can see that penguins are good models for behavioral studies, because they reconcile the demands of the field and laboratory. The large and clearly visible penguin colonies ensure that we can sample many pairs of breeding birds, and thus get large numbers of observations to typify behaviors and test for environmental contexts. The difficulty of gaining access to the far colonies of breeding penguins is compensated for by their fearlessness toward humans. Even though penguins were relatively easy to observe, however, we still had to measure, mark, and follow individuals to understand their behavioral signals. The known breeding biology of most species gave us a solid working basis for understanding and interpreting their enigmatic ritualized displays and stereotypical songs. The curious displays of some species had already been amply described, but confusion is rife about the ecological and evolutionary context of behavioral displays. So it was necessary to fill in some gaps in previous observations and to standardize the terminology that we used to describe penguin displays.

The family of penguins was particularly opportune for behavioral study because despite their similar lifestyles as marine predators and shore-bound breeders, they display remarkable behavioral diversity. In the same family, you have very different modes of living, with large consequences for ethology and communication. In all genera of penguins but one, they have at least a rudimentary nest, like any other species of bird. But in the two largest species of penguins, there is no nest. Thus, when these penguins return from a feeding trip at sea, they are faced with the difficult problem of finding their mate or chick in a mobile crowd, without the meeting point that a nest provides. As we will see, this particular difficulty has created novel adaptations in the nuptial displays of King and Emperor penguins, both in their optical and acoustical signals.

Why did our research on penguin communication focus on optical and acoustical signals? Different species of animals use a variety of senses for choosing and identifying breeding partners. And such choices are especially important in penguins because a single parent cannot successfully raise and fledge one or two chicks in any of the species. Rather, a breeding pair must cooperate to raise offspring, from incubating one or two eggs to brooding and feeding chicks until they fledge into the sea. Thus, we can look to the senses that appear best developed in penguin species, to understand how breeding partners communicate. Touch and taste lack evidence of relevance. Olfaction appears to be important for such seabirds as petrels: for foraging, nest site identification, and even partner recognition (Jouventin and Robin, 1984; Jouventin et al., 2007a). But there is no evidence of use of olfaction in mate choice or individual recognition in penguins, and olfactory parts of their brains are not well developed (in contrast to petrels; Bang, 1960; Cobb, 1960). Thus, we focus on optical and acoustical signals, transferred through visual and vocal systems of signal production. Both of these signals (visual ornaments and vocal signatures) are overt and used in stereotyped displays. We have studied these signals, described them, and moreover conducted experiments in the field, in an attempt to understand them.

For a study of the evolution of visual and vocal signals, as well as their role in communication, penguins seemed eminently suitable for several reasons. They are relatively tame and consequently easy to observe in the wild where natural selection occurs, using a large number of individuals to typify populations and species. Penguins are a special family of birds in which all the species are unable to fly, rendering them consequently fairly easy to catch and manipulate. They are similar to humans in using visual and vocal signals, so that we can see and hear most of their signals, record and manipulate them experimentally, and ask questions about their meaning to the animals themselves. They are typical birds in having stereotyped behavior patterns, and consequently exhibit consistent responses during experiments that focus on understanding the evolutionary causes of behaviors.

Penguins are more variable than might be supposed for such an apparently homogeneous zoological group, due to the different adaptations to a diversity of environments from polar to equatorial lands and waters. They are diverse in their modes of breeding and have communication patterns that require, as we will see, visual signals for mate choice and vocal signatures for identification of the mating partner. Finally, penguins are monogamous and sexually monomorphic animals, and thus have optical signals that are similar in both sexes. Before our researches, however, mutual mate choice and sexual selection on optical signals of these birds had been poorly studied. The biological significance of bird colors had received extensive experimental study, but almost entirely in species where males are brightly colored and females by contrast are dull and environmentally cryptic.

When we see a photograph of a seabird colony, it seems as though it would be impossible to find a particular partner or chick in such a huge crowd and in penguins only by the voice. Early ornithologists were surprised that seabirds could nevertheless find their mate and chick quickly, and our leitmotif in this book will be that we consider the location of the nest as a rendezvous site for reproduction (Jouventin, 1982; Bried et al., 1999; Dobson and Jouventin, 2003a). So our research task was to first discover how penguins identify their partner, both in terms of mate choice and for meeting the partner while raising offspring, when penguins use a nest as a fixed location. Second, we proceeded to ask the same questions about the species that have no nest. Adults of the two largest penguin species have no fixed location for incubating an egg and raising a dependent chick. The question of what happens when no nest is present is even more curious and compelling because these are the only two species of birds that do not have at least a rudimentary nest, burrow, chink in a rocky surface, or cup of pebbles in which to care for their valuable and vulnerable egg and young chick.

Early accounts of penguin behavior drew unconscious analogies and even fanciful conclusions about the similarity of penguin colonies and human crowds. It was not until the 1950s that scientific and objective behavioral studies of penguin colonies began in earnest. Research on the behavior of penguins is often limited by the difficulty of observations at distant field sites. It is easier to observe penguins in captivity, though these studies are nevertheless rare and not particularly instructive about ecological context. During the three decades from the 1990s to the present, ethological research on penguins has been relatively sparce, perhaps because it seemed old-fashioned and too descriptive. During this time, studies focused almost exclusively on the life and feeding of penguins at sea, something that had previously been little explored. This new fascination was due to the use of new technologies such as electronic recorders that continuously monitor performance at sea. From these sorts of data, the life of penguins in their habitat at sea could be studied, when previously it had only been possible to study penguins on land. Consequently, links with climate changes and conservation status of the species were developed during this period, with the production of some hundreds of scientific articles. A new science that mixed ornithology and oceanography appeared. Penguins were fitted with depth–temperature–salinity recorders, as well as satellite transmitters, to reveal their positions and movements at sea. The focus on the marine environment was necessary, but also was surprising in its exclusivity. In contrast, the behavior of penguins on land is relatively easy to observe and subject to experiments, as we describe in this book. In the penguins' terrestrial environment, mate choice, incubation, and chick rearing provide excellent opportunities for recording behaviors associated with the details of reproduction in relation to nesting ecology. Also, during the breeding period, study of communication is particularly fruitful because penguins are in fact extraordinary models for the study of visual and vocal signals in a constrained social environment, such as a noisy crowd.

Scientific books on the fascinating behaviors of penguins are rare. Using his doctoral dissertation, Pierre Jouventin (1982) published an initial monograph on penguin behavior 35 years ago. He delved into the visual and vocal signals of penguins, including all the information that was known at the time. His monograph is now long out of print. These early and largely descriptive studies gave way to careful comparative research and many experimental studies on acoustical and optical signals. During the past 17 years, Jouventin and Stephen Dobson collaborated on studies of penguins on islands surrounding Antarctica, experimenting on the biological significance of visual ornaments of penguins, such as brightly colored feather patches. Jouventin also collaborated with colleagues, especially Thierry Aubin, on acoustic signals of penguins, producing a fairly complete study of vocal signatures (i.e., for individual recognition). Finally, Jouventin (1982) suggested that behaviors such as nuptial displays and song characteristics might produce reproductive isolation of populations of Rockhopper and Gentoo penguins, perhaps leading to speciation. Recent molecular work has confirmed genetic divergence of populations of these species, supporting the importance of communication behaviors to speciation.

In the present book, we first introduce who the penguins are and explain why they provide an extraordinary model for behavioral study. We divided them into two groups with an important difference in their behavioral ecology. Small to medium sized penguins have a nest in a fixed location, be it in a burrow, among rocks, or on open ground. Two species of large nest-less penguins, however, are unique among birds, walking about with their single egg and young chick on their feet. This basic difference in the breeding ecology of penguins creates remarkable differences in the way that penguins use visual and vocal signals to communicate. We describe nuptial displays, colored ornamental visual signals, and vocal signals embedded in the songs of penguins, illustrated by numerous drawings and photographs. Next, we summarize the many experiments conducted on optical and vocal signals. Finally, we explain how natural selection may have brought about the visual and vocal signals as evolutionary strategies.

We describe how the ethological mechanisms of sexual isolation provide a key to speciation, taxonomy, and ecology of this group, specialized for cold marine waters. And we describe new penguin species, once defined by their nuptial displays and now confirmed with DNA analyses. We discovered that optical signals are used primarily for mate choice before pairing, and acoustical signals are used to identify the partner (only vocally) during incubation and when brooding and raising the chick(s). We concluded from our experiments that, nesting penguins (and by extension, all nesting birds) are helped by a rendezvous site, and that they add to this location information both a relatively simple vocal signature and species-specific head markings. Large penguins, on the other hand, lacking information from the location of a fixed nest, exhibit particularly complex songs and visual signal information from their feather and beak colors. The larger penguins complement their vocal and visual signal systems with the use of their double voice, and UV and pigment colors. They thus exhibit extremely sophisticated and highly evolved traits for communication. So the nesting and non-nesting penguins provide natural experiments that reveal the evolution of sophisticated and complex behaviors in response to challenges in their ecological and social environments.

# Acknowledgments

First of all, we have to thank the French Polar Institute, Institut Français pour la Recherche et la Technologie Polaire (IFRTP), which gave us support for our numerous field trips to Adélieland and the Archipelagos of Kerguelen, Crozet, and Amsterdam. Most of our sojourns were complex to organize, because we worked in penguins colonies that were far from scientific bases. Consequently, our long-term behavioral ecology studies would have been impossible without the logistic support of the IFRTP, an institute of the Centre National de la Recherche Scientifique (CNRS) of France. CNRS is an invaluable public organization for research, and both authors and many postdoctoral fellows and students who worked with us benefited from CNRS support. We also benefited from the support of Auburn University, including sabbaticals for field work and writing. The US National Science Foundation provided financial support.

We made more than 20 campaigns in the field, including our studies of the behavior of penguins. These activities were managed by the French Polar Institute (I.F.R.T.P.), as a research program on the Ecology of Birds and Mammals in the French Austral and Antarctic Territories, led by P. Jouventin from the CNRS. We visited remote islands of the Southern Indian Ocean via births and good fellowship of the crew on the Marion Durfesne, the remarkable French marine research vessel. District chiefs of Amsterdam, Crozet, and Kerguelen provided full and friendly cooperation. Our programs were under the supervision of the scientific committee of IFRTP and of its ethical committee.

Thanks to Oceanopolis, a marine park located in Brest (France), which permitted our feeding experiments to find the origin of Gentoo penguin beak coloration. We also thank the National Museum of Natural History in Paris for access to penguin skins that revealed that only the two large penguins show UV reflectance from the beak.

Any work like ours relies on many other scientists that provided a basis of natural history and evolutionary theory, and we thank them for their efforts. We of course depended on observations and experiments in the field that relied on the help and collaboration of many volunteers, students, post-doctoral fellows, and colleagues. We do not have space to thank them all, but most of their names appear as coauthors on the studies that we cite. Joel Bried read and commented on the manuscript, suggesting many improvements. Julia D. Kjelgaard read and improved the composition of several versions of the manuscript. Penguin postures were drawn a long time ago by C. Gabouriot. Most of the other figures were patiently prepared by Line Ruchon-Jouventin. All photographs are by both authors and I. Keddar, except for Fig. 2.5, sent by J.T. Darby (Otago Museum, Dunedin, New Zealand).

Finally, we thank the excellent staff of Academic Press for many forms of help. Kristi Gomez provided initial encouragement for our project of putting together 50 years of curiosity and study into a book. Pat Gonzalez of Academic Press guided us through the publication process, answering our many questions with infinite clarity and patience. Priya Kumaraguruparan provided excellent help with proofs and production.

Chapter 1

# The Evolutionary Biology of Penguins

## DISCOVERY AND EVOLUTION

Penguins are fascinating and outlandish birds, famous for walking on two feet and wearing a tuxedo. They live in the southern hemisphere, so for many years they were virtually unknown to early European scholars. Penguins were first sighted and thus "discovered," during expedition voyages commanded by Dias, Vasco da Gama, and Magellan in the early 1500s. These early sightings of penguins occurred in both South America (Punta Tombo, Patagonia) and South Africa (Mossel Bay, Cape of Good Hope). It was the time of the first great global expeditions by European explorers. These expeditions resulted in the discovery of new lands, the acquisition of new wealth (from commerce in spices), and finally provided the evidence that the earth is round. In fact, as for many plants and animals, native peoples (such as Hottentot, Negroid, South American Indian, Australian aborigines, and Polynesians) had previously known about penguins for several thousand years. The discovery of penguins is thus meant in a typically ethnocentric way, thinking as we often do that something was discovered when a European first saw it and perhaps wrote a book about it. For penguins, we fortunately have early reports from journal notes of the first Europeans to see these incredible birds.

Ferdinand Magellan was Portuguese by birth and a marine officer under the crown. He renounced his citizenship when the King of Portugal refused to give him what amounted to a veteran's bonus. He then convinced the Spanish king, Carlos V, that he had found a way to the East Indies and their riches. Magellan fitted out five ships and his expedition set sail from Seville on August 10, 1519. The expedition carried the first around-the-world tourist, Antonio Pigafetta, a native of Vicenza in Italy. While Magellan was killed in an engagement with natives, Pigafetta survived and continued on to circle the globe, returning to Seville on September 8, 1522. The single ship that returned contained only 18 men of the original 275 and was commanded by Sebastian del Cano. Pigafetta returned to Italy with the diary of his travels, and he subsequently wrote a book about the adventure (Simpson, 1976).

**Why Penguins Communicate. http://dx.doi.org/10.1016/B978-0-12-811178-9.00001-9**

The original of Pigafetta's book is lost, but a translation is now in Paris at the French National Library. Translating from old to modern French: "Nous trouvâmes deux iles pleines d'oies et d'oisons...nous en chargeâmes les cinq navires en une heure... Ces oisons sont noirs et ont les plumes par tout le corps d'une même grandeur et façon, et ils ne volent point et ils vivent de poisson. Et ils étaient si gras qu'on ne les plumait pas mais on les écorchait et ils ont le bec comme un corbeau." And translating from modern French to English: "We found two islands full of geese and goslings...we loaded all the ships with them in one hour... All these goslings are black and have feathers over their whole body of the same size and fashion, and they do not fly, and they live on fish. And they were so fat that we did not pluck them but skinned them and they have a beak like a crow's." These birds were of course penguins. They were so unusual that Pigafetta did not know what name to give them and finally settled on "geese." Paradoxically for a naturalist, but understandable for an explorer, the first Europeans to see these nonflying birds were most interested in the fact that they constituted food that could be cooked on board and were thus an easy source of fresh meat for sailors (Fig. 1.1). Later, in 1699, William Dampier, an explorer–buccaneer, reported that penguins were really abundant and easy to kill, but their meat was mediocre. This sentiment was echoed in 1775 during an expedition by the famous English explorer James Cook (1785).

Around the turn of the 20th century, a debate began about the origin of birds, and whether they were descended from reptiles. One theory (Menzbier, 1887; Lowe, 1933) considered that the first birds were adapted to aquatic life and later emerged onto land. Penguins were thought to be the most ancient

**FIGURE 1.1** Etching of a ship's crew killing and capturing penguins for a fresh food resource at sea. *By Theodore de Bry (1633); from the Bodleian Library.*

birds, the first birds, because they swim with flippers, rather than wings, and cannot fly. The large Emperor penguin that breeds in Antarctica was supposed to be the first penguin, and thus its egg was the place to find the ancestor from which all of the rest of the bird species evolved. It was thought that the egg should therefore contain an embryonic lizard. Edward Wilson was a famous ornithologist, artist, and explorer, and many Antarctic birds are named for him. He avidly wanted to find and examine some eggs of the Emperor penguin, so that he could prove that the ancestor of birds was a reptile. During the Antarctic winter of 1911, he led an expedition to collect eggs from Emperor penguins on their breeding grounds at Cape Crozier, where chicks could be observed at the end of the Antarctic winter. But to collect eggs, it was necessary to arrive in the middle of the Antarctic winter and thus the heart of the coldest period. One of his companions, Cherry-Garrard (1922) described this expedition, during which they nearly died, in his memoir, *The Worst Journey in the World*. This 60-mile journey was made in almost total darkness, with temperatures reaching as low as −57°C. Having successfully collected three eggs and desperately exhausted, they returned to Cape Evans, 5 weeks after setting off. But when anatomists of The British Museum open these eggs, they found only bird embryos. Still unlucky, Wilson ended his life by freezing to death with Scott in his quest for the South Pole. They reached the South Pole, but 5 weeks after Amundsen had already been there.

In fact, penguins are a very old group of birds, their ancestors appearing more than 70 million years ago in the southern hemisphere, where oceans predominate (Mayr et al., 2017). Penguin species were much more numerous in the past, and fossil remains suggest that some extinct penguins grew to more than 1.5 m high (Fig. 1.2). Modern penguin genera appeared more recently, and they occupied their present geographical zone about 10 million years ago. Currently, paleontologists and taxonomists do not think that the first penguins came from the polar zones, but began their evolution in temperate seas where six different genera evolved (Fig. 1.3):

- the several temperate Little penguins (genus *Eudyptula*);
- the single species of Yellow-eyed penguin, *Megadyptes antipodes*, which breeds only in New Zealand, and it is considered close to crested penguins.
- the diversified group of six crested penguins of the genus *Eudyptes*, which breed from the temperate to the sub-Antarctic zone; the number of species recently enlarged to 7–8 because Rockhopper penguin populations (*Eudyptes chrysocome*) exhibited large differences, first in displays (Jouventin, 1982), and then in DNA markers (Banks et al., 2006; Jouventin et al., 2006; de Dinechin et al., 2009), yielding at least two species.
- another group of 3–4 species (genus *Pygoscelis*), with two polar species (the Adélie penguin, *Pygoscelis adeliae*; the Chinstrap penguin, *Pygoscelis antarctica*) and another species (the Gentoo penguin, *Pygoscelis papua*, likely several species according again to behavioral and DNA

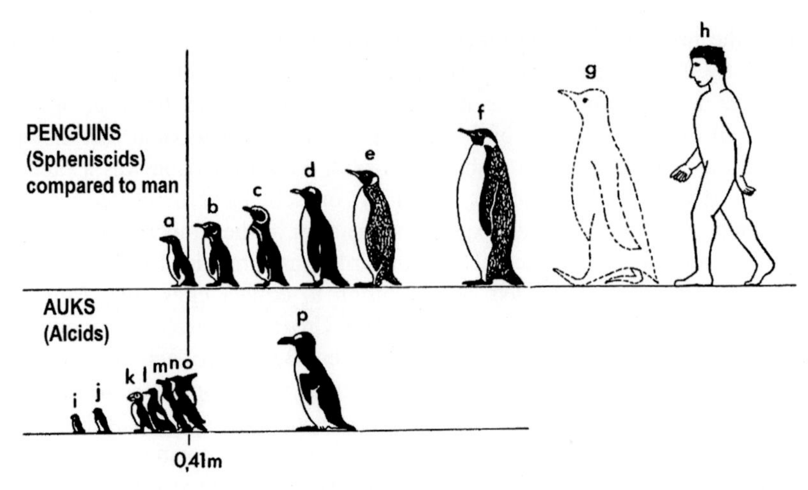


**FIGURE 1.2** Comparison of size between auks, penguins, and man, including the largest known extinct penguin. (A) Little penguin; (B) Galapagos penguin; (C) Magellanic penguin; (D) Gentoo penguin; (E) King penguin; (F) Emperor penguin; (G) the largest extinct fossil penguin from the Miocene; (H) *Homo sapiens*; (I) Least auklet; (J) Little auk; (K) Atlantic puffin; (L) Black guillemot; (M) Razorbill; (N) Common murre; (O) Thick-billed murre; (P) extinct Great auk. *Modified from* Sparks, Soper, 1967. Penguins. David and Charles, Newton Abbot, Devon.

evidence; Jouventin, 1982; de Dinechin et al., 2012; Vianna et al., 2017) that breeds both in the sub-Antarctic and the Antarctic;

- the remaining two genera of specialized penguins, the banded penguins along warm coasts (genus *Spheniscus*), with four species in South Africa and South America, and one species at the Equator, in the Galapagos Islands,
- and the last genus of large penguins, living on cold coasts (*Aptenodytes*, with two species, the King penguin, *Aptenodytes patagonicus*, breeding on sub-Antarctic islands; and the Emperor penguin, *Aptenodytes forsteri*, breeding on sea ice along the Antarctic coast).

Among these genera, each species has a unique distinctive appearance of the head, some with black and white markings, and some with yellow-colored feathers (Fig. 1.4).

Several types of data, including DNA and morphology, indicate that penguins descended from flying seabirds close to the petrels (Procellariiformes, the tube-nosed seabirds), the main group of seabirds (Hackett et al., 2008; Wang and Clarke, 2014). The Procellariiformes contain not only such exemplary flying species as the albatrosses but also some very poor flyers and good divers, like the diving-petrels. The Little penguin, as the name suggests, is small in size. Little penguins do not have clear black and white plumage, as in most penguins, but rather they are blue to gray on the head and back and grading to white on the breast and belly. They nest in a burrow and have

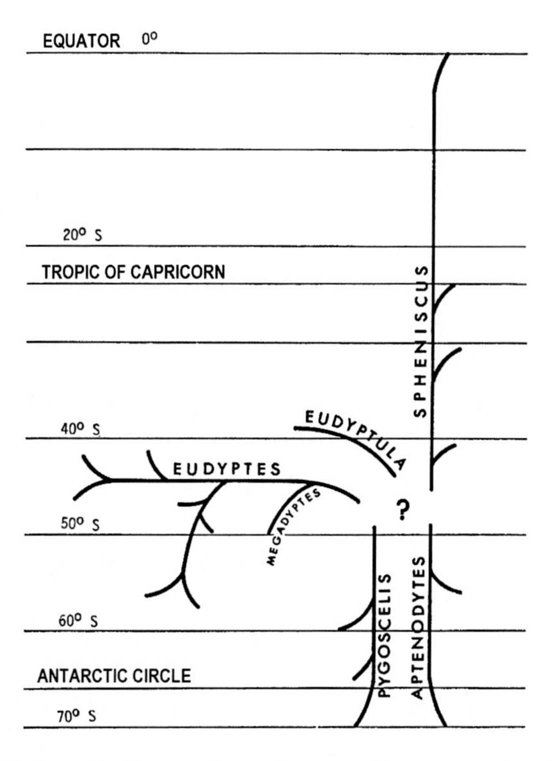


**FIGURE 1.3** Phylogenetic diagram of penguin genera from an ancestor that occurred in temperate seas (the "?" shows the latitudinal location of the ancestral species) and the relationship of the extant genera to latitude. *Modified from* Tollu, 1988. Les Manchots, Écologie et Vie Sociale, Le Rocher, Paris.

nocturnal habits. Little penguins share many features with diving-petrels. They are similar in appearance and habits, and so we might suppose that they share a common origin, the root of petrels and penguins. But we have to be careful to test such hypotheses using DNA markers, because auks, from the Alcid phylogenetic group in the northern hemisphere are far from penguins and petrels on the family tree of birds, despite their similarity in appearance to the seabird species of the southern hemisphere. Like penguins and diving-petrels, Alcids are well adapted to diving. These three groups converged ecologically and morphologically due to the same constraining way of life of fishing underwater (Fig. 1.5). The recently extinct Great auk (*Pinguinus impennis*) was a flightless diving predator and very similar to the two largest species of penguins. Contrary to the 1900 theory, penguins lost their flying abilities secondarily to become the best divers among birds. Thus, in spite of their ancient history, the advanced diving abilities of penguins are a specialization, and the ancient root of birds lies elsewhere.

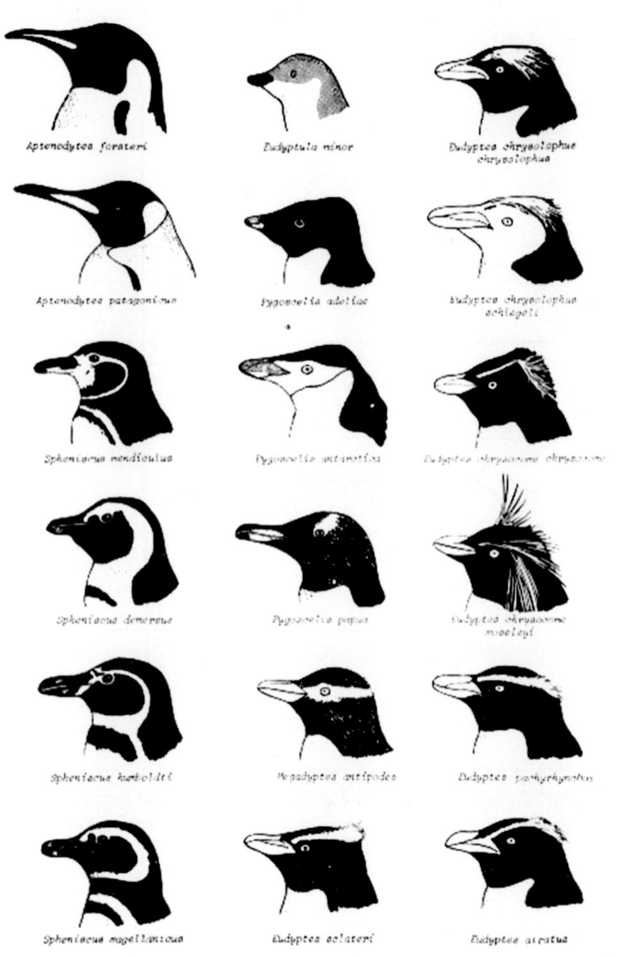


**FIGURE 1.4** Six genera of penguins, showing markings and feathers of the head and upper breast. Two species of large penguins (genus *Aptenodytes*), four banded penguins (genus *Spheniscus*), Little penguins (genus *Eudyptula*), three of the genus *Pygoscelis*, the yellow-eyed penguin (genus *Megadyptes*), and seven crested penguins (genus *Eudyptes*). The actual number of current species is now increased according to recent phylogenetic analyses (see Chapter 6). *From Jouventin, P., 1982. Visual and vocal signals in penguins, their evolution and adaptive characters. Advances in Ethology,* ***24****, 1–149.*

## WHAT IS A PENGUIN?

Penguins are flightless seabirds of the bird order Sphenisciformes and the single family Spheniscidae. Why are they so popular in comic strips, novels, or movies? Like monkeys or bears, but different from reptiles and insects, they invite anthropomorphism. We unconsciously perceive them as thinking as we

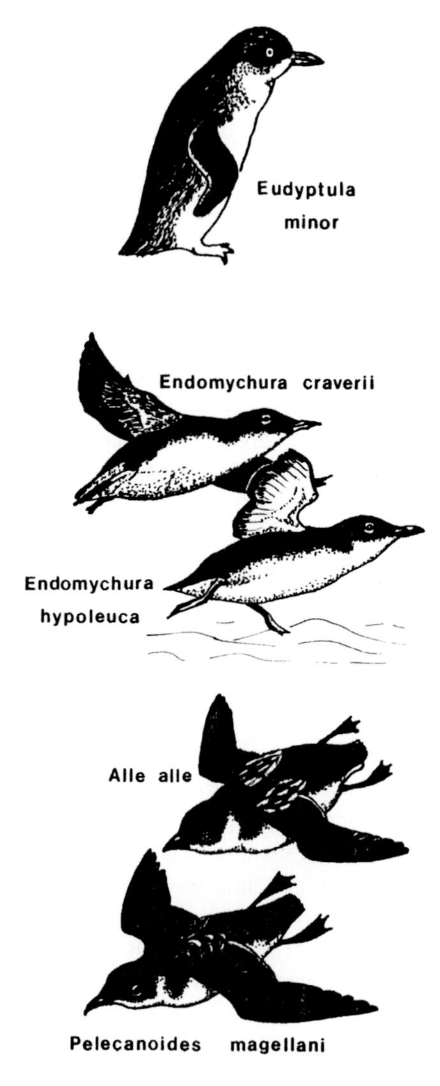


**FIGURE 1.5** These five diving seabirds look similar, but they belong to three different groups. At the top is the Little penguin (*Eudyptula minor*), living only in the Southern Hemisphere. At the bottom is a diving petrel (*Pelecanoides magellani*), also a Southern Hemisphere species and in a group related to penguins. The three birds in the middle are auks (*Alcidae*), related to gulls, and living only in the Northern Hemisphere.

do, because on land they walk upright on two feet. Penguins are funny for people to watch as they waddle about on land. But if we see them in the sea or in a marine park, we learn that they are much more active and talented in moving about underwater. They are specialized for sea life and not for terrestrial habitats. On land, their short legs help them to jump from one rock to another; or to go tobogganing on the snow while propelling and steering with their feet. They do not walk rapidly, but they swim very well using their feet as rudders. Consequently, their legs have migrated to a more posterior attachment on the body than in other birds, due to the evolution of the entire family to adaptation for an underwater life. The shortness of the lower leg bone, the tarsometatarsus, makes the leg unusually short and contributes to their waddling walk, so similar to that of some humans.

Like any bird, penguins breathe air and are unable to stay under water more than a few minutes; they regularly porpoise to breathe when at sea. Their wings have evolved into strong flippers. In the ocean, they swim using an "underwater flight," the general movement of flippers being not so different from wings of a flying bird. Due to the constraints of underwater propulsion in a more advanced mode than other diving birds, penguin wings are reduced and transformed into flippers that are shorter than wings. These flippers are flexible, but not folding, almost like the fixed wings of a plane, and with each beat they function as propellers. Their underwater locomotion is impressive, the individual flippers acting like oars that facilitate rapid changes in direction.

The modified feathers of penguins are imbricated, like the tiles of a roof, insuring a waterproof plumage. The feathers are short and hard, and particularly shortened on the flippers, especially on the leading edge. They enclose underlying down feathers and air, increasing thermal isolation of the body from the relatively cold temperatures of seawater. As seabirds in Antarctic and temperate seas, they are particularly well adapted to colder environments. In fact, a greater problem for penguins is living in warmer places where cold currents occur, such as South Africa or the equatorial Galapagos Islands, where once on land they must escape the heat. Consequently, penguins that live in warmer environments nest in cool caves or in burrows close to the beach. These penguins open their beaks to ventilate and pant. In these paradoxical species that are adapted to warmer environments, the difference between the temperature of the sea and land can be extreme. Thus, their greatest challenge is when they are breeding on land, where temperatures are relatively high. In these species, the areas around the beak and eyes lack feathers, with pink naked skin that radiates and thus eliminates extra heat. When at sea, however, plumage serves as a diving suit for penguins, similar to human diving suits. If some of the penguin's feathers are injured or lost, it constitutes a hole in the suit by which the water can enter and heat escape. Penguins in good condition are so well insulated that they have an incubation patch free of feathers, the brood patch, that can be folded against a layer of feathers for swimming, or opened for incubation on land by putting their eggs or chicks against the warm skin.

Penguins inhabit cold waters, and the biggest problem for most of the species is to keep warm, always at sea and at times on land, especially in the

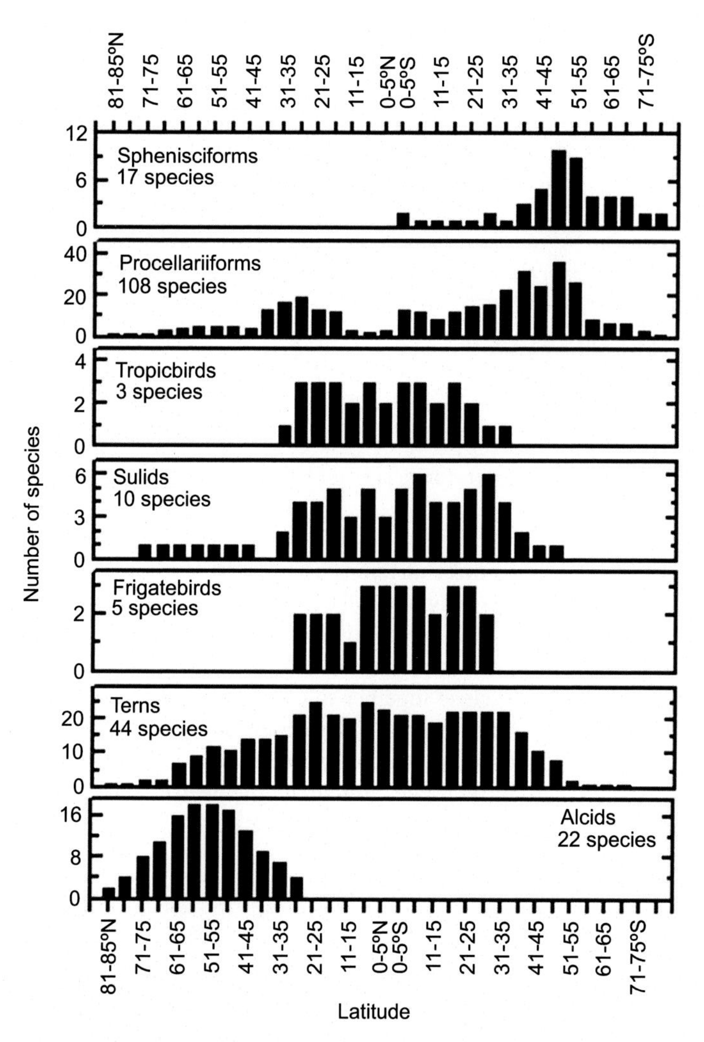


**FIGURE 1.6** Breeding distribution of all seabirds with respect to latitude. Note that penguins (*top bar*) live exclusively in the Southern Hemisphere, while auks (*bottom bar*) only in the Northern Hemisphere. *From Bried, J., Jouventin, P., 2002. Site and mate-choice in seabirds: an evolutionary approach to the nesting ecology. In: Schreiber, Burger (Eds.), Biology of Marine Birds. CRC Press, New York, pp. 263–305.*

far south. This is why penguins are preadapted to a polar way of life. Comparing the distribution of the penguin family to those of other seabirds, we see that penguins are the most polar family of birds, even compared to the auks of the northern hemisphere (Fig. 1.6). No other species has been able to match the incredible way of life of the Emperor penguin (Fig. 1.7). Emperor penguins

**FIGURE 1.7** The large Emperor penguin breeds on the sea ice during the Antarctic winter.

breed during the Antarctic winter, often at temperatures of −40°C, far below zero, when the cooling power of temperatures is increased by blizzards, and "wind-chill" temperatures may reach the equivalent of −200°C. To resist the strongest winds in the world, sometimes exceeding 300 km/h, adults of this species use a unique behavioral solution. They form a crowd like a scrum of rugby players, huddling close together to create a warm shelter against blizzards. The lee Emperor penguins in the scrum are sheltered from the high winds by the bodies of their windward peers (Fig. 1.8). Those along the windward edge constantly rotate around to the leeward side, causing the huddle to move across the ice during the blizzard.

**FIGURE 1.8** The huddle of Emperor penguins on the right shelters many individuals from the cold and wind. The isolated bird in the center of the left side is singing (head down and with a flipper band) to find its mate in this crowd of brooders. Around it, several brooders are walking with their single egg on feet, with the egg caught up behind a skin fold of the stomach.

Indeed, penguins have extraordinary behavioral and physiological adaptations for dealing with cold temperatures, such as those experienced in the waters of their deepest dives. But penguins are birds, and the species cannot be without air more than a few minutes. The Emperor penguin has the longest dives, reaching depths of more than several hundred meters. Their eyes are more adapted to the water than to aerial vision, and they can see even in the dark depths of the ocean by opening their pupils widely. On land, they have poor stereoscopic vision, having to move their heads constantly to see well. Because of the difference between the indices of refraction between air and water, they have a myopic view and have to be close to see accurately (Sivak, 1976). On the other hand, their sense of hearing is particularly good, and much better than human hearing, as we found in our experiments (see Chapters 5 and 6).

## HOW MANY PENGUINS AND WHERE DO THEY LIVE?

The number of recognized penguin species depends on vagaries of the science of systematics. Some forms are considered true species, and others as subspecies or even populations, and of course different scientists place penguins into these categories slightly differently (Table 1.1). The International Ornithological Commission recognizes 18 species, and BirdLife International lists their conservation status (http://www.worldbirdnames.org/ioc-lists/master-list-2/). As more phylogenetic studies using DNA markers become available, the number of species will likely increase. For example, using molecular techniques to confirm the status of species that were suspected 40 years earlier from behavior or appearance, we have provided evidence for new species in Rockhopper and Gentoo penguins (Jouventin, 1978, 1982; Jouventin et al., 2006; de Dinechin et al., 2009, 2012), as we will see in Chapter 6. Depending on how systematic studies translate into taxonomy, there are now about 20 species of penguins. These species can be differentiated by morphology, but as we will see, by their visual and vocal signals, both used in nuptial displays and for the crucial evolutionary process of breeding. Indeed, in this book, our goal is to explain how and moreover why penguins communicate, that is, how these behaviors contribute to efficient reproduction and maintain the cohesion of adaptations in the species.

Penguins live and breed from the coasts of Antarctica to the Equator. So they do not always appear as cold-adapted as we might expect. Rather, they are specialized for life in cold waters, but they have necessarily adapted for many other habitats as well, especially relatively warmer terrestrial breeding habitats. In the Southern Hemisphere, where all penguins live, 70% of the surface of the earth is covered by water. Thus, there is limited land to stop the ocean currents from circling the globe. This situation differs from the Northern Hemisphere, where most of the continents are present to block the movements of oceans. Consequently, in the austral ocean the thermocline of seawater

**TABLE 1.1** Scientific and English Equivalent Names of Penguin Species. Weights and Lengths

| Latin Name | English Name | Mean Mass and Length |
|---|---|---|
| *Pygoscelis adeliae* | Adélie penguin | 5 kg, 70 cm |
| *Pygoscelis antarctica* | Chinstrap penguin | 4.5 kg, 70 cm |
| *Pygoscelis papua papua* | Northern Gentoo penguin | 6.2 kg, 80 cm |
| *Pygoscelis papua ellsworthi* | Southern Gentoo penguin | 5.5 kg, 70 cm |
| *Eudyptes chrysolophus chrysolophus* | Macaroni penguin | 4.5 kg, 70 cm |
| *Eudyptes chrysolophus schlegeli* | Schlegel penguin (Royal penguin) | 4.5 kg, 70 cm |
| *Eudyptes chrysocome chrysocome* | Southern Rockhopper penguin | 2.5 kg, 55 cm |
| *Eudyptes chrysocome moseleyi* | Northern Rockhopper penguin | 2.5 kg, 55 cm |
| *Eudyptes sclateri* | Erect-crested penguin | 4 kg, 65 cm |
| *Eudyptes pachyrhynchus* | Fiordland penguin (Thick-billed penguin) | 3.5 kg, 60 cm |
| *Eudyptes robustus* | Snares penguin | 3 kg, 60 cm |
| *Megadyptes antipodes* | Yellow-eyed penguin | 5 kg, 70 cm |
| *Eudyptula minor minor* | Southern blue penguin | 1.2 kg, 38 cm |
| *Eudyptula minor chathamensis* | Chatham island blue penguin | 1.2 kg, 38 cm |
| *Eudyptula minor albosignata* | White-flippered penguin | 1.2 kg, 38 cm |
| *Eudyptula minor variabilis* | Cook-strait blue penguin | 1.2 kg, 38 cm |
| *Eudyptula minor iredalei* | Northern blue penguin | 1.2 kg, 38 cm |
| *Eudyptula minor novaehollandiae* | Fairy penguin | 1.2 kg, 38 cm |
| *Spheniscus mendiculus* | Galapagos penguin | 5.5 kg, 55 cm |
| *Spheniscus demersus* | Jackass penguin (Black-footed penguin) | 2.8 kg, 70 cm |
| *Spheniscus humboldti* | Humboldt penguin | 3.5 kg, 65 cm |
| *Spheniscus magellanicus* | Magellanic penguin | 3.5 kg, 70 cm |
| *Aptenodytes forsteri* | Emperor penguin | 28 kg, 110 cm |
| *Aptenodytes patagonicus* | King penguin | 15 kg, 90 cm |

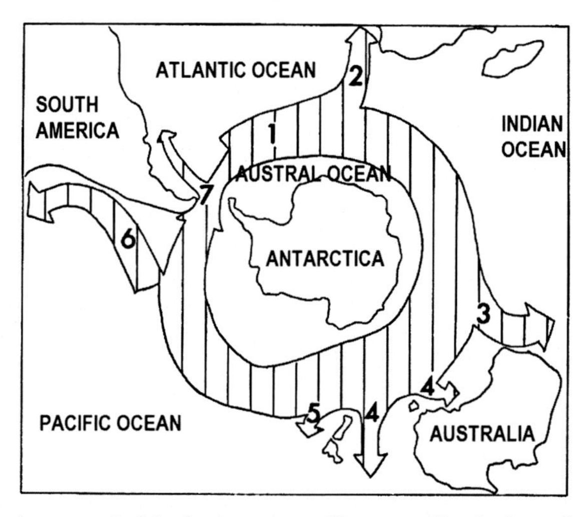


**FIGURE 1.9** As a target, both land and water are colder approaching the Antarctic continent, and the cold currents of water running North along the other continents and islands explain the presence of penguins (even as far as the equator) because these cold currents are rich in food. Numbered currents: (1) main Circumpolar current, (2) Benguela current, (3) west Australian current, (4) Tasmanian current, (5) Southland current, (6) Humboldt current, and (7) Falkland current. *Modified from* Sparks, Soper, 1967. Penguins. David and Charles, Newton Abbot, Devon.

temperatures is not far from the appearance of a target of concentric circles. A strong and extensive circumpolar current due to the rotation of the Earth and Coriolis forces keeps cold waters circulating around the globe (Fig. 1.9). Near the coast of Antarctica, the polar current is barely interrupted by the Antarctic Peninsula. But to the north, a major current of circulation in the Pacific Ocean is the Humboldt current, cold water that runs up the west coast of South America and extends as far north as the Galapagos Islands. This and other currents, with their numerous marine inhabitants, provide important habitat resources, especially food resources, for all of the penguin species. Seabirds live in a marine environment where they can move about the oceans freely, but the surface of the oceans lacks topographical relief that would otherwise divide up the populations of penguins. Nonetheless, masses of water circle the southern globe, and they vary in both salinity and temperature, forming ecological environments where marine organisms, seabirds, and particularly penguins occupy different ecological niches. Penguins are nearly out of competition with the flying seabirds because penguins fish at greater depths, but each penguin species competes with other penguins for shared marine resources. Consequently, all penguins, including the equatorial species, feed in cold waters where marine life is abundant. According to the ecological rule of niche separation, however, each penguin species occupies a unique ecological zone, so that competition with the other species is minimized.

The ocean is not as flat as it looks. Waters of different temperature and salinity are continuously in motion due to the rotation of the globe and contours of the ocean floors. Oceans are a 3-dimensional environment, much like the air that circulates around the planet. Upwelling from the ocean depths brings cold waters to the surface. These waters are particularly rich because they carry deep nutrients to the surface and thus to marine animals. This explains why these currents, flowing along the South American pacific coast and South-west African coast, are well known for their marine and fish productivities. Consequently, these regions have immense colonies of seabirds and penguins that depend on this wealth of marine food. Since the first satellite tracking of a bird (Jouventin and Weimerskirch, 1990) and of a penguin (Jouventin et al., 1994), an increasing number of studies have used satellite transmitters to study the life of penguins and other seabirds at sea (Fig. 1.10). For example, we observed sophisticated feeding patterns in King penguins (Fig. 1.11) that changed with the swimming distance to the "polar front zone," where the main upwelling occurs (Fig. 1.12). When this mobile marine boundary between two masses of water is not too far from the breeding colonies, King penguins do "direct trips" to this source of abundant fish. When it is too far, they do "circular trips" nearer to their breeding colonies that are less costly energetically, but through poorer fishing grounds (Jouventin et al., 1994; Guinet et al., 1997).

The first radio-tracking of King penguins at sea revealed that during the season of abundant resources, they go directly to the zone of modified Antarctic waters that lie roughly 500 km from the Crozet Islands. In these waters they find myctophid fishes, their favorite prey, at depths between 100 and 300 m (Fig. 1.13). Such technologies have produced a more accurate description of both the marine faunae and the seabirds that prey on them. Thus, knowledge of both ornithology and oceanography has improved because of the

**FIGURE 1.10** Equipping the first transmitters on a King penguin (between the legs of the student) for tracking at sea.

**FIGURE 1.11** A King penguin rearing its chick on its feet and without a nest.

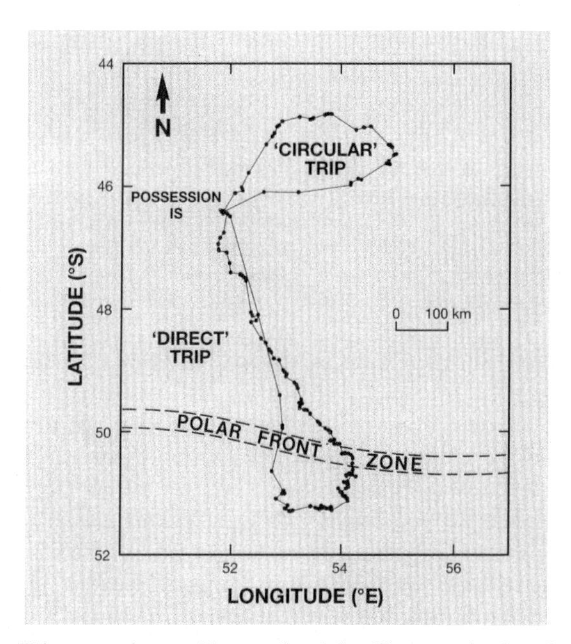


**FIGURE 1.12** Using transmitter tracking, we found that King penguins from Possession Island (Crozet Archipelago) have two strategies at sea. "Circular trip" shows a short trip near the breeding colony. "Direct trip" shows a run to the polar front zone 500 km from the breeding colonies, where the richest food resources occur. *From Jouventin, P., Capdeville, D., Cuenot-Chaillet, F., Boiteau, C., 1994. Exploitation of pelagic resources by a nonflying seabird – satellite tracking of the King penguin throughout the breeding cycle. Mar. Ecol. Prog. Ser. 106, 11–19.*

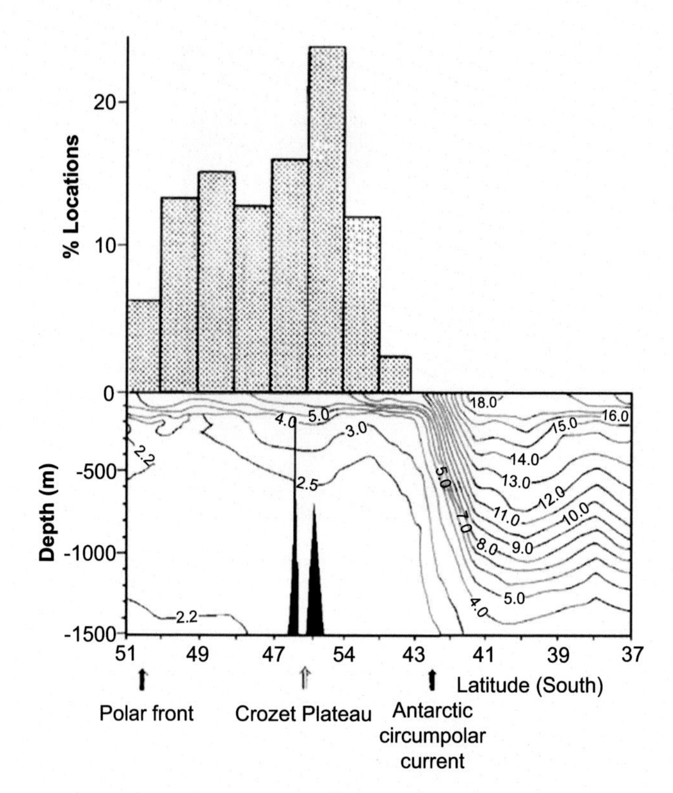


**FIGURE 1.13** In the first study that tracked King penguins, we found that they feed from the Crozet Plateau, where Possession Island is located, to the polar front zone (where the greatest densities of food fish were) and 100–300 m below the surface of the ocean. Numbers are water temperatures at different depths. *From Jouventin, P., Capdeville, D., Cuenot-Chaillet, F., Boiteau, C., 1994. Exploitation of pelagic resources by a nonflying seabird – satellite tracking of the King penguin throughout the breeding cycle. Mar. Ecol. Prog. Ser. 106, 11–19.*

fruitful encounter of two sciences. Many hundreds of marine birds and mammals have been equipped with satellite transmitters, and electronic recorders of depth, temperature, salinity, $CO_2$, and other measures. These transmitters aid the understanding of life at sea, especially the movement behaviors of these mysterious species, as well as the physical properties of oceans. The oceans have a major role in absorbing the anthropogenically increased $CO_2$ in the earth's atmosphere but this also acidifies the seas. In addition, the aerial heat, which influences terrestrial weather, produces in part the well-known oceanographic phenomenon of El Nino. So penguins and marine mammals equipped with satellite transmitters now contribute to the understanding of processes in the austral ocean.

As predicted according to the same natural laws that are better known for terrestrial species, marine masses of water constitute ecological boundaries for marine predators and their prey, even though the physical frontiers at sea are not easily visible to the casual observer. On a map of the geographical distribution of penguins, we can first see that their distribution on land is in relation to their marine distribution, because their feeding grounds are as close as possible to their breeding grounds. Secondly, each species occupies a circle of breeding localities around the Antarctic continent because it is the pole of cold temperatures on land as well as at sea. Consequently, each species is adapted to a marine ecological niche that is defined by sea temperature. In turn, sea temperatures in the Antactic are organized into concentric circles, similar to a target, around the South Pole (Mackintosh, 1960). Several cold currents start from the Antarctic and in the case of the west coast of South America, one cold current continues until the Equator. Similar currents occur in southwestern Africa and around Australia and New Zealand (Fig. 1.9). The breeding distribution of penguins follows these rich cold currents. Penguins forage in these waters and form immense breeding colonies wherever they can find land that is isolated from terrestrial predators.

No one really knows exactly how seabirds, including penguins, are able to orient themselves in the open sea without visual landmarks. We only observe that each year they find their way to an island or coast where their nesting area is found. Some field experiments on orientation were conducted with Adélie penguins. A combination of several systems of orientation was suggested, including both astronomical and visual cues (Emlen and Penney, 1964; Penney and Emlen, 1967). We observed a natural experiment in which Emperor penguins "lost their way" in getting back to their traditional breeding location (Jouventin, 1971a). Magnetism and sky reckoning (the sun during the day and stars at night) were suspected at sea, followed by visual sighting when land was in view. In the year of this natural experiment, Emperor penguins that were returning to breed on the sea ice were lured away from their traditional breeding location by the unusual movement of a tongue of a glacier (Fig. 1.14). So the penguins had to lay their eggs where they were, but they were not numerous enough to form large huddles and thus resist the Antarctic cold. During brooding, some penguins were able to get around the glacier and back to other Emperor penguins and the large traditional huddle. They did this by walking 3 km with their egg on their feet. The remaining isolated penguins failed to keep their eggs alive. The fact that some of the penguins walked around a tongue of the glacier (with a 40 m high cliff of ice) and successfully entered the main colony, which they could neither see nor hear, shows remarkable homing ability. The brains of these penguins are relatively small, but their ability to orient by either the earth's magnetic field, or by the sun and stars, allowed them to accomplish a task that would be beyond human observers without electronic help.

In another example, at the beginning of each austral summer, male Adélie penguins return from the open sea to their nest of the previous year. But when they return, the colony area is invariably completely covered by snow. Males sing above the previous nesting place, atop the snow, and they display to

**FIGURE 1.14** A mistake in bird orientation. The *black star* is the French base on an island in Adélie land (on the coast of the Antarctic continent). The *white asterisk* (#5) is the traditional location of the Emperor penguin colony on the frozen sea ice during the winter. In 1968–69, a large part of the colony mistakenly settled on area #1 and then moved progressively to area #2, #3, and then #4. Subsequently, some of the brooders returned (moving with their egg on their feet) to the traditional location of the colony at the *white asterisk* (#5). *From Jouventin, P., 1971b. Comportement et structure sociale chez le Manchot Empereur. Revue d'Ecologie Terre et Vie 25, 510–586.*

retrieve their previous female partner and to fend off other males. The snow melts as summer advances and marked males can be seen exactly on the same nest location as in either the following or the previous year, another remarkable feat of orientation, and at a relatively small scale.

## PENGUINS AS DIVERS AND BREEDERS

Penguins swim faster than any of the other waterbirds. Under water, they make abrupt movements, twists, and turns, swimming at about 10 km/h, depending on the species. Penguins are often considered maladapted or at an evolutionary "dead end," because they do not fly like other birds. In fact, they lost the ability to fly in a distant ancestor, and they became completely specialized for diving. Several species of birds hunt prey by diving, such as shags, gannets, and auks. But in penguins, the diving specialization was carried to an extreme. Natural selection seems to have taken one of two opposed directions. As an engineering problem, but recast for birds, it is impossible to combine a glider and a submarine. The penguin family opted without compromise for the diving way of life, contrary to the usual aerial way of life for just about all other birds. So, to exploit the third dimension of the ocean and to avoid competition with other seabirds, penguins evolved into the best divers among birds. At the same time, they definitively lost the capacity to fly (for an exception in other birds and especially Alcids, the Great auk was also a flightless diver; see Fig. 1.2). Small and medium sized penguins can dive up to 100 m deep, scores of times each day, and the largest species, the Emperor penguin, is able to fish at depths of more than 400 m (Table 1.2). While penguins are the best divers among birds (as opposed to being called the worst flyers, as is often concluded in comparison to other birds), they are outdone among vertebrate species by mammals. Mammals such as seals and especially whales are able to dive thousands of meters, and stay submerged without breathing for much longer periods of time. An elephant seal can dive for the better part of the day, at depths of more than a thousand meters, and for up to an hour at a time. Even after the introduction of radio-tracking studies, it was difficult to follow this air-breathing mammal because it spends so much of its life underwater.

Penguins weigh from as little as 1 kg for aptly named "Little penguin" to 20–40 kg for the largest extant species (Table 1.1), the Emperor penguin. Of course, penguin mass changes greatly depending on whether one measures it before or after fasting. During breeding, a penguin's life is a succession of

**TABLE 1.2 Maximum Diving Depths and Maximum Dive Durations**

| Species of Penguin | Maximum Diving Depth (m) | Duration of Longest Dive (min) |
|---|---|---|
| Adélie | 180 | 5.90 |
| Chinstrap | 179 | 3.65 |
| Gentoo | 212 | 11.28 |
| Macaroni | 163 | 6.30 |
| Schlegel | 226 | 7.50 |
| Northern Rockhopper | 168 | 3.22 |
| Southern Rockhopper | 113 | 11.00 |
| Yellow-eyed | 56 | – |
| Little Blue | 69 | 1.50 |
| Galapagos | 32 | 3.10 |
| Jackass | 130 | 4.58 |
| Humboldt | 54 | 2.75 |
| Magellanic | 97 | 4.60 |
| Emperor | 564 | 27.60 |
| King | 343 | 8.70 |

From: Ropert-Coudert and Kato, 2012. The Penguiness Book. World Wide Web Electronic Publication (http://penguinessbook.scarmarbin.be/), version 2.0.

foraging trips at sea and fasting on land. Their body is torpedo shaped and extremely hydrodynamic. To fly, a bird has to be light. But penguins are freed from this constraint and thus can accumulate body stores of heavy fat. They are thus protected from the cold marine environment, where they fish in oceanic currents. Water is a well-known heat sink and draws a lot of energy from a submerged penguin. In fact, the cooling power of water is 20 times the cooling power of air. When a man enters the sea near the poles, he can survive for only about 15 minutes because the water temperature is 40°C below his body temperature. But a penguin in the same habitat is in a neutral environment and is able to stay as long as needed in cold Antarctic (or temperate) waters. Swimming and especially diving also produce warmth from exercise. As well, it turns out that the very fat that makes a penguin too heavy to fly provides strong insulation against heat loss while swimming. What is more, stored fat reserves allow penguins to fast on land for several weeks or months, giving more flexibility to their breeding cycle, so that they can incubate or brood for long periods without feeding.

Penguins live in the sea, but they must come on land to reproduce. A lot of the research on penguins has focused on their extraordinary capacity to fast and to resist cold on land. Breeding strategies in the same harsh environments can greatly differ among species. During the Antarctic summer when temperatures rise and become positive, the Adélie penguin rears one or two chicks. They build a stone nest to keep their eggs above the level of the ice. Parents alternate fishing and brooding eggs or chicks, even when the colony is in a blizzard and when nests are covered by snow (Fig. 1.15). Contrast this breeding cycle with that of the Emperor penguins. During the Antarctic winter when temperatures fall to −40°C, Emperor penguins continue to breed, even when they are confronted by blizzards blowing more than 300 km/h. At the same time, the open sea on which they depend for food can be scores of kilometers away. Moreover, males can stay for 4 months without food, drinking only a little snow. When the Emperor chick hatches, the male may have lost up to 50% of its mass, but he continues to wait for his female partner. He produces a sort of "milk" from his esophagus to feed the newly hatched chick. The Emperor penguin incubates and broods not on land, as do other birds, but on flat sea ice, where it can huddles to keep warm with others of its species (Fig. 1.8).

The length of the breeding cycle of an animal is generally linked to its size and the phenology of penguins is diverse, with breeders staying on land a little more than 4 months in the Adélie penguin, while another species, the King penguin, regularly needs more than 1 year. The songbirds in your garden reproduce faster than ostriches. It simply takes more time to make an ostrich than a wren. Nevertheless the Emperor penguin, weighing about 30 kg, twice the King penguin, has to contract its reproductive cycle into less than 9 months. It cannot be longer because this is the length of time that the sea ice will last. Consequently, juvenile Emperor penguins leave the breeding colony at about half the size of an adult and stand on ice floes and drift away from the

**FIGURE 1.15** In Adélie penguins, breeding pairs have nests of pebbles that are relatively dispersed. Even in the blizzard and under the snow as here in Antarctica, breeding partners relieve each other for egg and chick care, one partner tending the nest while the other feeds at sea.

breeding site (pieces of the sea ice, break off and float away; see the whole breeding cycle in Fig. 1.16).

Another impressive example occurs in the King penguin, that increases the duration of its breeding cycle on islands of the Austral ocean. When food is scarce during the austral winter, parents stay at sea and forage while the chicks fast on land or are poorly fed for several months. Parents must return in the spring and "fill" the chick back up with food, so that it can molt from the juvenile plumage, suitable only on land, into a swimming and diving plumage similar to that of the adults. With the new plumage, the chicks can "fledge" into the sea. Consequently, the population of this large sub-Antarctic penguin breeds all the year round, but at best pairs can produce a surviving chick only twice in 3 years, and even this is an extremely rare event (Weimerskirch et al., 1992). Most commonly, King penguin pairs produce a single chick every 2 years (Jouventin and Lagarde, 1995; Jiguet and Jouventin, 1999). Indeed, the breeding frequency in this species is controlled more by the individual body condition of each adult than by the yearly seasonal cycle of sub-Antarctic weather (Jouventin and Mauget, 1996). In the sub-Antarctic, temperature variation is much lower than in Antarctica. The breeding cycles of these two penguin species are very different, despite their large body size and living in cold oceans (Fig. 1.17).

Penguins feed not only on fish but also on crustaceans and other marine animals such as squid. Their diet varies according to the species, the season, and the place. A penguin is able to dive as many as 20 times in a single day, and their impact on the marine environment can be exceptionally high. For example, the largest colony of King penguins in the world (Fig. 1.18; Ile aux Cochons, Crozet Archipelago, Indian Ocean) numbers about 300,000 pairs,

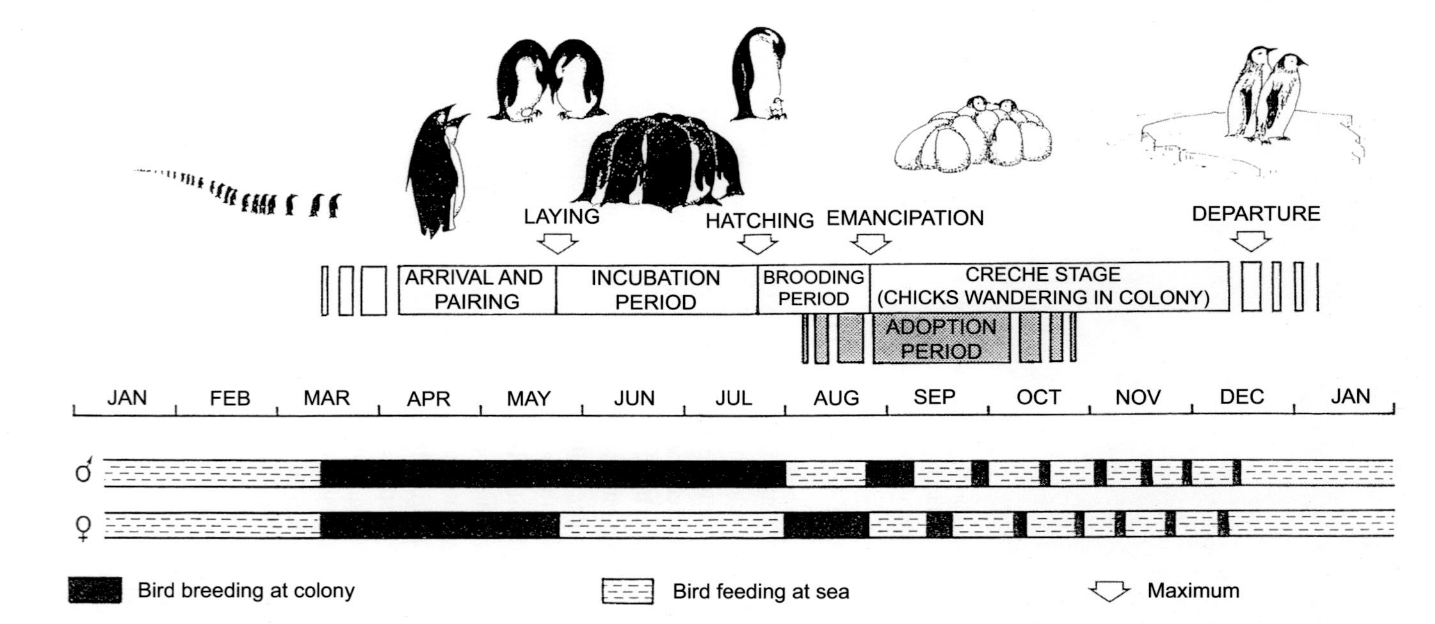


**FIGURE 1.16** The breeding cycle of the Emperor penguin is unique because it occurs on sea ice during the winter, with only the male incubating, while the female is at sea (see the bottom for the timing of male and female presence in the breeding colony). The breeding cycle of this large penguin has consequently to follow the annual cycle of ice presence of about 9 months. Chicks have to molt into swimming plumage on ice floes as the sea ice is breaking up. *From Jouventin, P., Barbraud, C., Rubin, M., 1995. Adoption in the Emperor penguin,* Aptenodytes forsteri. *Anim. Behav. 50, 1023–1029.*

**FIGURE 1.17** Comparing the very different breeding cycles of the two largest penguin species of the genus *Aptenodytes*. The Emperor penguin breeds on the Antarctic coast every year and contracts its reproductive time to fit with the cycle of sea ice. The King penguin breeds on sub-Antarctic beaches over a period of about 14 months. *From Sparks, Soper, 1967. Penguins. David and Charles, Newton Abbot, Devon.*

**FIGURE 1.18** The biggest colony of King penguins in the world at about 300,000 pairs (Cochons Island, Crozet Archipelago, Southern Indian Ocean). The huge colony reveals the difficulty of finding a breeding partner when returning from a foraging trip at sea. The colony also illustrates the high productivity of the austral ocean that feeds here as many as a million birds, each weighing about 12 kg.

close to 1 million birds. Each King penguin weighs about 12 kg, so, to sustain this many birds, the colony takes in more than half a million tons of fish and cephalopods per year. This amounts to about twice the total annual fishing volume of France (Jouventin et al., 1994).

The oceanic foraging range of penguins can be very large, and this range influences their reproductive strategies. Even though penguins are great swimmers, they are not as mobile as flying seabirds. Even large penguins, like the King penguin, travel much shorter distances to feed and collect food for the young than a large albatross (Wandering albatrosses, gliding or "sailing" on the quasi-permanent winds at these latitudes, fly several 1000 km from the nest for similar foraging trips, Jouventin and Weimerskirch, 1990). Of course, a large species like the King penguin swims much farther than a medium sized species like the Gentoo penguin. Their dispersive behavior also reflects the opposing ways that these two species exploit food resources from the sea, the former being broadly pelagic and the latter coastal. The size of a penguin influences its speed in the water and consequently the foraging range. The situation is similar to that of a sailing vessel, where maximum rate of travel through the water is proportional to the length of the waterline of the boat.

Among the guild of sub-Antarctic seabirds, the number of eggs and chicks per breeding adult pair is less for true pelagic foragers than for coastal birds due to the high cost of long feeding trips (Dobson and Jouventin, 2007, 2010, 2015). Of course, this pattern is further augmented by body size, as it is a well-known pattern among birds that the number of offspring declines with increasing body mass (Bennett and Owens, 2002). Among the penguins, feeding trips at sea require constant swimming that is energetically costly. Coastal and medium-sized species, such as the Gentoo penguin, lay two eggs and breed every year, while the strongly pelagic and large King penguin lays a single egg and generally rears only one chick every 2 years (Weimerskirch et al., 1992; Olsson, 1996; Bried and Jouventin, 2002). Indeed, when the foraging distance from the shore increases from coastal to pelagic seabirds, fecundity decreases, and the cost of chick rearing increases. This demographic strategy of seabirds explains why coastal penguins lay two eggs and pelagic penguins lay only one egg (Fig. 1.19).

## PENGUINS AS SOCIAL ANIMALS

Whether on land or at sea, penguins are exceptionally social animals: when walking or swimming, they are most often in groups. When young penguins are close to fledging into the sea, they wait in a dense crowd and make their first dive into the water as a group. While the advantage of this behavior may seem to us as though they are waiting for their friends before embarking into the new and unfamiliar aquatic habitat, it appears to provide distinct survival advantages. Giant petrels or leopard seals lay in wait for young fledgling penguins and will eat them as soon as they hit the water. When they dive together, these predators are lucky to catch one bird in the group. Without the innate herd mentality of the young penguins, the risk of predation would increase as the predators pick them off one by one.

Penguins are so social that first explorers who discovered Emperor penguins, breeding in huddles on the sea ice in the heart of the Antarctic winter, supposed that these birds reared their chicks collectively. Initially, the ornithologist Edward Wilson (1907) observed that adult Emperor penguins were fighting to adopt a lost chick or were able to kidnap a chick from the feet of another brooding adult. He concluded that this species became communistic while colonizing the Antarctic: they huddle to resist to the cold, mate and brood collectively, and feed their chicks together. Later, Prévost (1961) showed from banding of individual penguins that partners were faithful to their mate of the year, and that each pair reared its single chick. Finally, Jouventin et al. (1995) found from more extensive observations of Emperor penguins that chicks were fed by failed breeders as well as their parents, particularly when the winter was hard and several parents had lost their chicks. These failing parents continued to be motivated by parental hormones to care for chicks, even if they were not their own (Lormée et al., 1999; Angelier et al., 2006). Here the initial story of scientific discovery was a sequence of relative truths and, as we see often in science, with the interpretation initially misguided. But, as we see when further examining their behavior, the eventual story of penguin life is both more complicated and much more interesting than one can imagine.

Penguin colonies or rookeries are usually impressively large, often numbering several thousand or even hundreds of thousand pairs. The colony, however, can only be as large as the coastal site and the marine resources

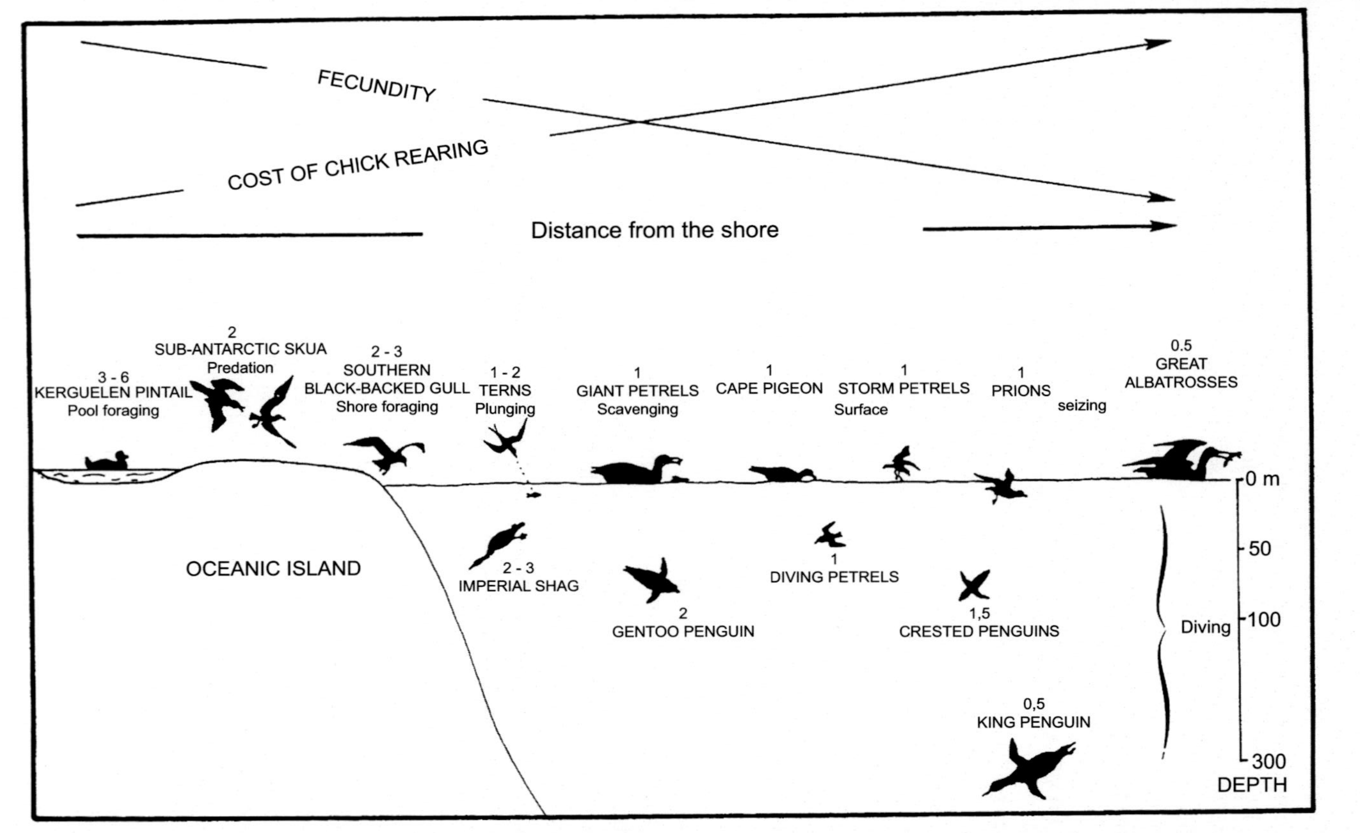


**FIGURE 1.19** Relationships in a sub-Antarctic avian community between foraging range and life-history traits such as the clutch size (the number above the named species). The key factor in all seabirds is the distance between the feeding and breeding grounds. Foraging trips, both flying and moreover diving, represent an energetic cost that prevents the most pelagic seabirds from rearing more than one chick per year (or even every other year). For penguins breeding on the same sub-Antarctic island, coastal species such as Gentoo penguins lay two eggs annually, pelagic King penguins lay one egg every other year (0.5 eggs/year), and in the intermediate feeding zone, Crested penguins produce both a normal and a small egg annually (i.e., 1.5 eggs/year). Thus, as distance of the feeding zone from the shore increases, fecundity decreases for species of penguins, just as it does for other seabirds because the cost of chick rearing increases. *Modified from Jouventin, P., Mougin, J.L., 1981. Les stratégies adaptatives des oiseaux de mer. Revue d'Ecologie Terre & Vie 35, 217–272.*

permit. Penguins are awkward on land and thus are easy for scientists to approach and capture by hand. Before human arrival on their islands, penguins had few or no real enemies on land. Unlike most terrestrial species, they were not under mortality selection from predators. Consequently, they are usually not afraid of humans. They assemble on special locations like sea ice, cliffs, rocky slopes, or sandy beaches. In the southern hemisphere, these habitats usually occur on islands that lack terrestrial predators. These "islands" not only occur far from continents (for example, the Galapagos and sub-Antarctic Islands), but also along continental coasts that are surrounded by a landward belt of forbidding frozen or deserted habitat, such as the Antarctic continent or the Valdes Peninsula in Argentina. These breeding sites on land are close to rich feeding zones at sea, to reduce the cost of movements between the terrestrial colony and foraging zones of the ocean.

A colony of penguins is usually composed, as for any of the seabirds, of a mosaic of families, each consisting of a breeding pair and their current young in the nest, or in the case of the two largest species, on their feet. Due to this unusual incubation and brooding habit, the two largest penguins produce only one egg per breeding attempt. Other penguin species lay two eggs. In some of the species of crested penguins, eggs are strongly dimorphic in size, with the smaller egg laid first. The nest can be built with stones, branches, or seaweed, and it isolates eggs from the wet soil. It can be on open land, under bushes, in a burrow, or in a cave. The nest can even be absent, as in the two largest penguins that incubate their egg on their feet. For the penguins, even those without a nest, the extent of their territory or personal space is limited by the distance that they can reach with their beak. The territory thus constitutes a small circle around the incubator or brooder.

Just as penguins have to occupy different feeding niches at sea due to competition for food resources, they have to differentiate their nesting habits on land. On a sub-Antarctic island, you can find four species of penguins in separate breeding colonies. These colonies occupy different nesting habitats, and other seabird species occupy additional habitats. A typical sub-Antarctic island (Fig. 1.20) has both penguins, petrels, and albatrosses breeding: on cliffsides, for sooty and mollymawk albatrosses, or on leas for the Wandering albatross (*Diomedea exulans*). Penguin colonies can also be found on leas, such as the Gentoo penguin. Excrement around the nests of Gentoo penguins kills the vegetation and leaves a mucky habitat, so the colony moves slightly from year to year. The King penguin nests on sandy or pebble beaches. The Macaroni penguin (*Eudyptes chrysolophus*) nests above or atop masses of fallen rocks, while the smaller Rockhopper penguin (*Eudyptes chrysocome*) can often be found nesting low down or even below the same masses of fallen rocks, sometimes in grass tussocks along the base of cliffs.

In spite of crowded conditions in their colonies, successful breeding in all species of penguins requires paired males and females that are faithful to their mate for at least a complete breeding season. Together with their partner, they

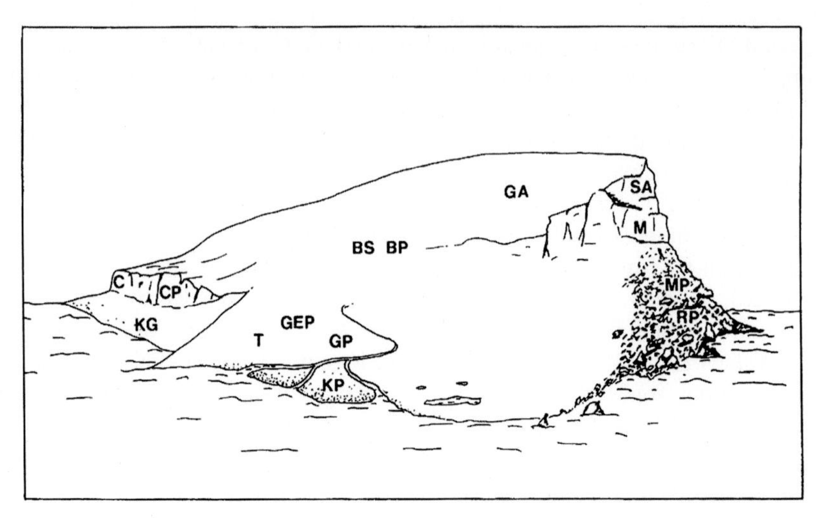


**FIGURE 1.20** Idealized partitioning of breeding habitats in a seabird community on a sub-Antarctic island where four species of penguins breed. *BP*, small petrels dig burrows; *BS*, Brown skuas on grassy slopes and plateaus; *C*, cormorants; and *CP*, Cape pigeons on coastal cliffs; *GA*, great albatrosses on grassy plateaus; *GEP*, Gentoo penguins; and *GP*, Giant petrels on grassy slopes; *KG*, Kelp gulls nest on the shore; *KP*, King penguins on sandy beaches and estuaries; *M*, mollymawks nest on steep slopes; *MP*, Macaroni penguins in boulders on the surface; *RP*, Rockhopper penguins under boulders below the Macaroni penguins; *SA*, sooty albatrosses on ledges in cliffs; and *T*, terns on graveled coastal plateaus. *From Bried, J., Jouventin, P., 2002. Site and mate-choice in seabirds: an evolutionary approach to the nesting ecology. In: Schreiber, Burger. (Eds.), Biology of Marine Birds. CRC Press, New York, pp. 263–305.*

cooperate in the feeding of their own chick(s), returning from foraging trips at sea to relieve the incubating or brooding partner and to feed their young chick(s). But such mate fidelity varies strongly according to the species and sometimes by the locality (Table 1.3). The two large nonnesting penguins have very low fidelity to their previous mate from year to year. The territorial penguins, however, often exhibit four times higher fidelity to the mating partner over the years. The nest provides a landmark that greatly assists reuniting with the past year's mate, as it serves as a rendezvous location. The advantage for communication of a fixed nesting location will be a major theme in this book (Bried et al., 1999; Dobson and Jouventin, 2003a).

Despite the variable degree of mate-switching from year to year in penguins, they are generally thought of as monogamous because they have to work closely with their partner and thus be faithful in the same year, to breed successfully. Seabirds are monogamous in this way, primarily because they cannot mix breeding and feeding grounds (Lack, 1968). One member of the pair is unable to reproduce without the help of the partner, both for incubating the egg and for raising chicks through brooding and fledging, tasks that require close attention of both pair members. Because they have eggs to incubate, 90% of birds are socially monogamous (Bennett and Owens, 2002). This is

**TABLE 1.3** Nest and Mate Fidelity in Penguins. Survival Rates Are Given Only for Those Species for Which We Provided Fidelity Rates

| Taxon[a] | Locality | % Nest Fidelity[b] | % Mate Fidelity[b] | Adult Survival | References[c] |
|---|---|---|---|---|---|
| **Sphenisciforms** | | | | | |
| *Aptenodytes patagonicus* | Iles Crozet | 39.4 (site fidelity) | 22.4 | 0.952 | 1, 2, 3 |
| *A. patagonicus* | South Georgia | – | 18.8 | – | 4 |
| *Aptenodytes forsteri* | Terre Adélie | – | 14.5 | 0.91 | 2, 5 |
| *Pygoscelis adeliae* | Cape Crozier | 59.4 | 18–50 | 0.696 | 6, 6, 7 |
| *P. adeliae* | Cape Bird | 98.2 | 56.5 | 0.736 | 8, 8, 8 |
| *P. adeliae* | Wilkes Land | 76.8 | 84 | 0.77 | 9, 9, 9 |
| *Pygoscelis papua papua* | Iles Crozet | – | 76 | 0.865 | 10, 7 |
| *P. p. papua* | South Georgia | 93 | 90.2 | c.0.8 | 11, 11, 12 |
| *Pygoscelis papua ellsworthi* | King George Is. | 61.4 | 90 | – | 13, 13, 13 |
| *Pygoscelis antarctica* | King George Is. | 87.9 | 82 | – | 13, 13, 13 |
| *Eudyptes chrysolophus chrysolophus* | South Georgia | 83 | 90.8 | – | 11, 11 |
| *Eudyptes chrysolophus schlegeli* | Macquarie Is. | – | 80 | 0.86 | 14, 15 |
| *Eudyptes chrysocome chrysocome filholi* | Iles Kerguelen | 53 | 78.6 | – | 16, 16 |
| *Eudyptes chrysocome moseleyi* | Amsterdam Is. | 34.9 | 46.3 | 0.84 | 16, 16, 17 |
| *Megadyptes antipodes* | New Zealand | 30 | 82 | 0.87/0.86 | 19, 20, 21 |

*Continued*

**TABLE 1.3** Nest and Mate Fidelity in Penguins. Survival Rates Are Given Only for Those Species for Which We Provided Fidelity Rates—cont'd

| Taxon[a] | Locality | % Nest Fidelity[b] | % Mate Fidelity[b] | Adult Survival | References[c] |
|---|---|---|---|---|---|
| *Spheniscus mendiculus* | Galapagos Is. | – | >89 | 0.844 | 22, 14 |
| *Spheniscus demersus* | South Africa | 59.8 | 62.1 | 0.617 | 23, 23, 23 |
| *Spheniscus magellanicus* | Punta Tombo | 80/70 | 90.4 | 0.85 | 24, 14, 14 |
| *Eudyptula minor* | Philip Is. | 43.9 | 82 | 0.858 | 25, 25, 25 |

[a] *Only adult individuals were considered. Data from populations known to live in unstable environments, or not to be in equilibrium, were excluded.*
[b] *Studies involving less than 25 individual years or 25 pair years for site fidelity and mate fidelity, respectively, were excluded. Only adult individuals (i.e., known to have bred in the past) were considered. Site fidelity rates were calculated as 1 minus (number of site changes/number of adult years). Mate fidelity was calculated as 1 minus the probability of divorce (Black, 1996). When two values separated by a slash (/) are given for the same parameter (e.g., 80/70), the former is for males, the latter for females.*
[c] *Numbers refer to the source of nest fidelity, mate fidelity, adult survival rate, respectively, and when data were available. Although some of these sources did not express fidelity rates in the same manner as ours, they provided the data that enabled us to calculate them as described above. (1) Barrat (1976), (2) Bried et al. (1999), (3) Weimerskirch et al. (1992), (4) Olsson (1998), (5) Jouventin and Weimerskirch (1991), (6) Ainley et al. (1983), (7) Ainley and DeMaster (1980), (8) Davis (1988), (9) Penney (1968), (10) Bost and Jouventin (1991), (11) Williams and Rodwell (1992), (12) Croxall and Rothery (1995), (13) Trivelpiece and Trivelpiece (1990), (14) Williams (1995), (15) Carrick (1972), (16) J. Bried (unpubl. data), (17) Guinard et al. (1998), (18) Marchant and Higgins (1990), (19) Richdale (1949), (20) Richdale (1947), (21) Jouventin and Mougin (1981), (22) Boersma (1976), (23) LaCock et al. (1987), (24) Scolaro (1990), (25) Reilly and Cullen (1981).*

From Bried, J., Jouventin, P., 2002. Site and mate-choice in seabirds: an evolutionary approach to the nesting ecology. In: Schreiber, Burger. (Eds.), Biology of Marine Birds. CRC Press, New York, pp. 263–305.

particularly true of penguins, since there is a great distance between their nest and foraging sites. Thus, the sexes alternate during cooperative incubation at the nest and feeding trips at sea.

Mammalian species provide a nice contrast to this situation because the sexes have completely different breeding roles. Female mammals usually carry and suckle their young alone, and males take no part in rearing young in most of the species. According to Darwin's (1871) sexual selection theory, the most polygynous species are usually strongly dimorphic in body size, particularly in mammals such as elephant seals or red deer, where males are socially dominant over the smaller females. In birds, this social structure is uncommon due to the constraint of incubation. Incubation often produces an obligatory sharing of egg care between mates, a single bird being unable to rear the young. In spite of their relatively low number of offspring, this constraint occurs particularly in seabirds. Nonetheless, even in highly monogamous birds, there is usually at least some sexual dimorphism due to promiscuous mating. Pairs are socially monogamous, but extra-pair copulations can occur. In these cases, males copulate when they are paired, and they try to sneak copulations with the females that are paired to other males.

In seabirds, where monogamy is strong and cooperation essential to reproduction, just about all of the species appear monomorphic. That is, males are similar to females in appearance, and the sexes are usually difficult or impossible to distinguish for human observers. This is surprising because so many songbird species have extremely colorful males, and females are drab by comparison (reviewed by Hill and McGraw, 2006). The most detailed studies of sexual selection are those of song birds, where color dimorphism is especially strong. Because the sexes are so alike in appearance in seabird species, and particularly for penguins, seabirds provide an excellent opportunity to further our understanding of how sexual selection works to produce not only similarity of the sexes but in some species similarity of beautifully bright ornamental color patterns. We conducted experiments to specifically address the curious similarity in coloration or markings of the sexes in most of the penguin species. In Chapter 3, we focus on this color similarity of the sexes in King penguins.

## THREE BREEDING PATTERNS

Seabirds feed at sea and reproduce on land. Thus, they are constrained by feeding trips while reproducing. As a result, they breed rapidly, and then return to the open sea. Similar to the Adélie penguin, most penguins are classical nesting seabirds, breeding in huge colonies because space is limited on islands or coasts that are free from terrestrial predators. As we will see, the nest constitutes a meeting point for the pair and then between parents and chicks (Figs. 1.21 and 5.1). Coming to the nest, the parent identifies their partner or chicks vocally because confusion is possible in such huge and crowded colonies (Fig. 1.22). But in the penguin family, we have also two unusual species of large

FIGURE 1.21 In Adéle penguins, each pair builds a nest with stones on the Antarctic coast. Here, the second chick is hatching, the egg being open, and the first chick is behind, warm under the parent.

size, and they incubate their single egg and later brood their chick on their feet, both breeding on flat areas. The smaller species of the two, the King penguin, breeds on sandy beaches (Fig. 1.23). Like Adélie penguins, they use vocal signatures for individual recognition, as does any seabird. But the problem of being identified only by the voice is much more complicated for King penguins

FIGURE 1.22 Adélie penguin colonies are coastal and huge with several hundred or thousands of nests made of small stones and each nest is occupied by a pair. Mates are 100% faithful during the breeding cycle and about 80% (according to the locality) from one year to the next.

**FIGURE 1.23** In a colony of King penguins, there is no nest and only a few landmarks. Mobile parents incubate their egg and brood the young chick on their feet. Finding the breeding partner or chick is accomplished vocally using a sophisticated song to both locate and identify the singer. Mates are faithful within a breeding cycle, but only 20% of pairs reunited from one year to the next.

because they lack the meeting point of the nest. They have few or no landmarks that can help them find their breeding partner. This problem only gets worse for the Emperor penguin, as brooders move about on the sea ice and feed their single chick on their feet. Later, when the chick is semiindependent and waits in a crowded crèche for either parent to return with food, the only way that the parents can identify the chick is by the voice (Fig. 1.24).

We compared the very different breeding strategies of these three penguins, the small Adélie penguin, a classical nesting seabird, and the two large

**FIGURE 1.24** Emperor penguins breed on the sea ice and move about continuously. Returning from a fishing trip, an Emperor penguin regurgitates to its single chick after recognition of the vocal signature confirms the reunion.

penguins. The King penguin breeds throughout the year, on the beaches of sub-Antarctic islands. The Emperor penguin breeds during the winter, on the sea ice that surrounds the Antarctic continent. In animals, there is an association between body size and the duration of the breeding cycle. The small size of the Adélie penguin helps to explain the shortness of its breeding cycle, compared to the two much larger penguin species. Moreover, the Antarctic weather strongly constrains both Adélie and Emperor penguins that breed along the Antarctic coast. But these two species breed at opposite times of the year. The smaller Adélie penguin is present during the austral summer when the sea ice is absent, and there are open waters for feeding close to the colony. The larger Emperor penguin, on the other hand, breeds in winter, when the sea ice occurs for scores to hundreds of kilometers off the Antarctic shore. Emperor penguins must walk to the open sea to fish. Faced with the harsh Antarctic climatic constraints, these two breeding cycles are relatively short but vary according to the penguins' size. Breeding lasts the 4.5 months of spring/summer for the small Adélie penguin and must end before the sea ice cuts off access to the sea for the chicks. And breeding lasts for the 9 months of the long Antarctic winter for the Emperor penguin, because it is essential for this large penguin to breed on a flat place, so that it can move about with the chick on the parent's feet. Partners in Adélie and King penguins share the incubation tasks between sexes, as they do in any seabird. But remarkably, male Emperor penguins complete virtually all incubation of the egg alone, before being relieved by his female partner (Fig. 1.25). And as mentioned before, the male exudes an esophageal "milk" to feed the chick if the female is delayed upon her return from the open sea. The female has often walked tens of kilometers, returning to brood and regurgitate food for the chick.

The long breeding cycle of the Emperor penguin is in fact short for such a large bird. Emperor penguins weigh up to around 40 kg. But they stay on the sea ice for not more than about 9 months. At the end of the 9 months, chicks leave the colony on floes of sea ice that are breaking up and floating away from the coast. The chicks molt and then fledge when they about half the size of an adult (Figs. 1.16 and 1.26). The other large species of the genus *Aptenodytes*, the King penguin, weighs less than half its Antarctic relative, but lives in the sub-Antarctic zone where there is little food during the 4 months of winter. King penguins have a much longer breeding cycle, lasting about 14 months. Consequently, they are unable to breed every year like the two previous penguins. The breeding frequency of the King penguins is constrained mainly by the body condition of each breeder, but successful individuals might expect to breed once every 2 years (Fig. 1.27).

To return to individual recognition and the territories of birds, we see in Fig. 1.28 that the nest of any seabird, such as the Adélie penguin, is the central meeting place from mating to chick emancipation. The territory of the large penguins is much reduced or absent and is necessarily mobile. This is a relatively permanent place that shifts just a few meters from laying to thermal

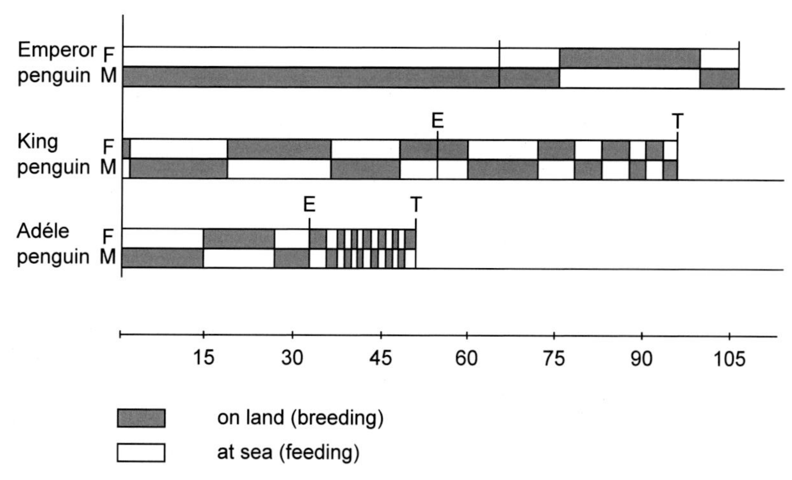


**FIGURE 1.25** Schematic showing egg and chick attendance for males and females. The scale at the bottom of the figure shows the number of days from the beginning of the breeding season. E indicates the mean hatching date, and T the end of the brooding period. The subsequent feeding periods of the chick(s) are not shown. *Clear and colored bars* shows periods of parental egg and chick care versus foraging trips at sea. Male Emperor penguins do all the incubation of the egg. Then parents alternate feeding the chicks, as do King penguins and Adélie penguins during the whole breeding cycle, incubation included. *From Isenmann, P., Jouventin, P., 1970. Eco-ethologie du Manchot empereur (*Aptenodytes forsteri*) et comparaison avec le Manchot adelie (*Pygoscelis adeliae*) et le Manchot royal (*Aptenodytes patagonica*). L'Oiseau et La RFO 40, 136–159.*

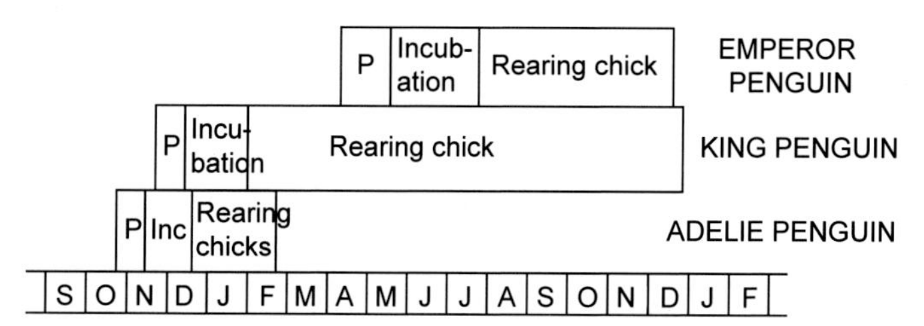


**FIGURE 1.26** The Emperor penguin breeding cycle is shorter than that of the King penguin, despite the former species being about two times larger. The Emperor penguin breeding cycle has to fit with the 8–9 months sea ice duration. The Adélie penguin is much smaller than Emperor or King penguins, so it contracts its breeding cycle to 4.5 months of the brief Antarctic spring/summer. "P" represents the laying period. *From Isenmann, P., Jouventin, P., 1970. Eco-ethologie du Manchot empereur (*Aptenodytes forsteri*) et comparaison avec le Manchot adelie (*Pygoscelis adeliae*) et le Manchot royal (*Aptenodytes patagonica*). L'Oiseau et La RFO 40, 136–159.*

emancipation of the chick for King penguins. This contrasts with more extensive adult and then chick mobility during the whole breeding cycle in Emperor penguins. So, if we classify the two large species as "nonnesting penguins," the Emperor penguin really is without a territory. The King penguin also has no nest, but a small territory that provides "personal space," and occurs only during

**FIGURE 1.27** For a long time, it was impossible to understand the breeding cycle of King penguins. They seem to be constantly breeding, with mating pairs, eggs, young chicks, molting chicks, and adults looking for breeding partners spread over much of the year. The solution was to examine breeding cycles of individual birds, and recognize that late breeders almost always failed and returned with more success the following spring. The diagram shows the breeding sequence of an idealized individual beginning with spring molt in October, followed by pairing, laying, and chick rearing until the following December, with a late breeding attempt to follow. Failure can occur at any time. Successful breeders produce a chick every 2 years, and several reproductive stages overlap. *From Jouventin, P., Mauget, R., 1996. The endocrine basis of the reproductive cycle in the King penguin (*Aptenodytes patagonicus*). J. Zool. 238, 665–678.*

incubation and brooding. This territory moves, however, if there is flooding, a storm, or other disturbance. The territory disappears completely from the Adélie to the Emperor penguin, but the King penguin is intermediate. Why have some penguins lost their nesting territory, such a useful place for protecting pairs and their chicks from disturbances? Breeding grounds are a limited resource for seabirds. By abandoning the nest, the large penguin species colonize immense areas, such as beaches for the King penguin and sea ice for the Emperor penguin.

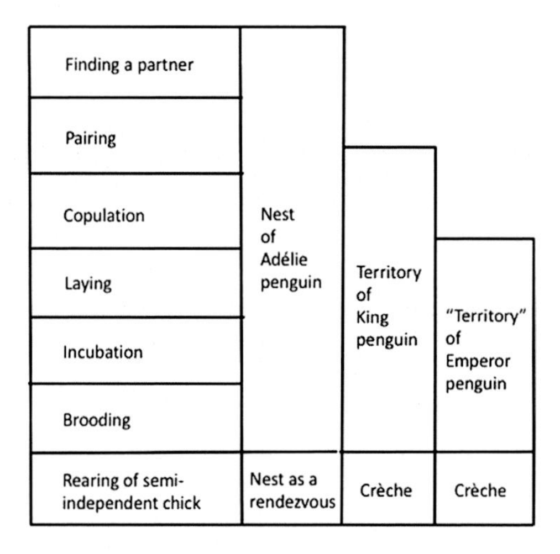


**FIGURE 1.28** For Adélie penguins, the nest territory is used during the whole breeding cycle, first to mate, to lay and incubate eggs, then to rear small chicks and finally as a meeting point to feed emancipated chicks. In King penguins, each carrying a single egg then chick on the feet, the territory is limited to the beak range of a standing parent, and only during the laying–brooding stages. Flooding, fighting, or other disturbances cause movements during breeding. In Emperor penguins, there is no nest or semipermanent brooder-position because they move continuously. The mobile parents only defend their egg and then chick. *From Isenmann, P., Jouventin, P., 1970. Eco-ethologie du Manchot empereur (*Aptenodytes forsteri*) et comparaison avec le Manchot adelie (*Pygoscelis adeliae*) et le Manchot royal (*Aptenodytes patagonica*). L'Oiseau et La RFO 40, 136–159.*

But these flat areas are not sheltered, and only large penguins that have sufficient fat reserves can stand in the colony for long periods.

## SPECIAL PENGUINS

Penguins are very good models for the study of behavior, because the species are very different, in spite of their common appearance. As previously explained, penguin species exhibit extreme variety in their nesting habits, body size, evolutionary origins, and ways of life, leading to wide variations in behaviors. One extraordinary genus includes two exceptional species of penguins that are large in body size and breed without a nest. So, penguins include territorial species with nests that are easy to find topographically, and also two species that are confronted with the exceptional challenge of finding a mating partner and their chick in a mobile crowd of several thousand breeding pairs (first studied in the Emperor penguin by Prévost, 1961; Jouventin et al., 1979; in the King penguin by Derenne et al., 1979). Because of these challenges, Emperor and King penguins have exceptionally sophisticated characteristics of behavioral communication. Consequently, we focused much of our research, including experiments, on the species of the genus *Aptenodytes*, the King and Emperor penguins.

The fact that penguins are able to find their partner or chick by the voice alone (Jouventin, 1982) gave us the opportunity to discover and to test some of the most original acoustical adaptations in animals (Aubin and Jouventin, 1998, 2002a,b; Jouventin et al., 1999a; Aubin et al., 2000; Jouventin and Aubin, 2000; Searby and Jouventin, 2005). For optical signals, these two large species are again the most sophisticated of the penguins. Thus, King and Emperor penguins also gave us the opportunity to experimentally and objectively study the way in which complex colored signals permit various kinds of signaling, such as mate quality, given their harsh, unusual, and constraining environmental living conditions (Jouventin et al., 2005, 2008; Dobson et al., 2008, 2011; Pincemy et al., 2009; Nolan et al., 2010; Keddar et al., 2013, 2015a,b,c; Schull et al., 2016; Viblanc et al., 2016).

Before beginning the description and experiments of optical and vocal signals, we need to describe the ecology of these two special penguins, to learn from the "natural experiments" that these species provide, and to understand the context of their behavioral challenges. For the 16−18 species of penguins that have a nest, an adult can return from foraging at sea and find their partner at a stable and consistent location where the nest and egg or chick also occur. In this, they are like any other seabird and any other bird species. When this return from the sea is accomplished, the penguin parent has to verify that the partner and/or the chick at the nest is the right one. This is accomplished by voicing a song, that is, giving a vocal signal. If, in the next year, the penguin repeats this pattern, it will likely recover both the nest site and the previous partner. In comparison, large penguins breed on flat areas, sand for the King penguin and sea ice for Emperor penguins. When breeding, these species can move around with the egg or young chick on their feet, and thus a returning partner does not have a fixed location for a rendezvous (Jouventin, 1982; Dobson and Jouventin, 2003a). Due to the flat nature of the habitat and dense crowds of breeding individuals, they are likely not assisted by topographic landmarks (Bried et al., 1999). Yet, extensive research reveals that they successfully and faithfully locate their mating partner and young chick. How do they do it?

When incubating and brooding, the sub-Antarctic King penguin initially returns to a specific general location or "zone" in the colony (Dobson and Jouventin, 2003a). If there is a storm or flooding from a nearby river or the sea, the parent caring for the egg or chick may be forced to move about the beach (Fig. 1.29; Bried and Jouventin, 2001). When breeding, Antarctic Emperor penguins have to move about continuously on the sea ice due to frequent blizzards, during which they huddle for warmth. Incubators and brooders cannot stand alone against the Antarctic cold, nor can their egg survive more than a few minutes without the parent's warm brood patch. If the wind blows for several days, many brooders leave the cold windward side of the huddle to go on the lee side, step by step; but this also exposes a new "leading edge" to the cold. The huddle itself moves, not as a whole, but as a result of the short

**FIGURE 1.29** A colony of King penguins (Possession Island, Crozet archipelago) flooded during a winter storm. Chicks can be washed away to the sea.

treks of hundreds or thousands of individual brooders. The first observers of this mini-migration (Prévost, 1961) supposed that the penguins moved because they felt movements of the sea ice beneath them and wanted to feel safe on a flat surface. But in time it became clear that temperature and wind speed produced the movements (Jouventin, 1971b; Gilbert et al., 2006). In fact, in other places than Adélie Land, where the phenomenon was first studied, the same movements were noted (see video with these slow movements accelerated: https://www.youtube.com/watch?v=OL7O5O7U4Gs). When the wind stops, Emperor penguins stop their huddle and become more widely spread over a larger area, returning to their main breeding site on the sea ice (Fig. 1.30). Later, when the chicks are semiindependent and mobile, parents encounter a new problem: finding their chick in a crowded crèche so that they can feed them.

The reproductive cycle of the Emperor penguin seems to be the most complicated and difficult of all the penguins. Emperor penguins breed once a year, thus compacting the timing of the breeding cycle (Fig. 1.16). On the other hand, King penguins breed successfully over a 2-year period, in a reproductive cycle that almost fits into a single year (Fig. 1.17). Clearly, the breeding cycle of King penguins is more complicated in both its timing and the difficulty of reproduction. In a colony of this species, breeding begins in November, at the beginning of the austral summer. During the subsequent months, several reproductive events occur continuously. At the same time, some parents are feeding their large young, chicks as well as adults are molting, unpaired adults are singing to find a mate, pairs are copulating, females are laying, and males are incubating. The inception of the breeding cycle is not well synchronized but extends for at least 4 months. This breeding cycle is consequently much more complex to understand than for an annual species, where each bird may be synchronized to a similar reproductive stage. In King penguins, each individual bird has its own timing (Jouventin and

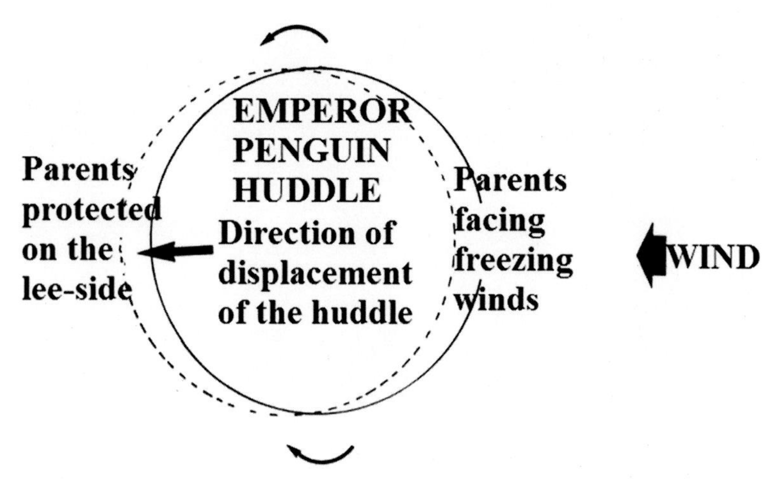


**FIGURE 1.30** Schematic of the breeding huddle of Emperor penguins that forms during the Antarctic winter, especially for warmth during blizzards. Incubating penguins move from the windward side of the huddle to the leeward side to avoid the bracing wind and cold of the leading edge. This creates a continuous flow of individuals through the huddle, moving on the sea ice. When the wind blows for several days, the huddle moves about 100 m. When it stops, brooders come back to their traditional colony area. *From Jouventin, P., 1971a. Incubation et elevage itinerants chez les manchot empereurs de Pointe Geologie (Terre Adelie). Revue du Comportement Animal 5, 189–206.*

Lagarde, 1995). If parents fail during mating or laying or brooding or rearing, they go on a foraging trip, fishing far from the breeding colony, feed to recover their fat, and return to the colony to mate and commence breeding again. One failing parent is enough to stop the breeding cycle of a pair because they absolutely must have the care of two partners to rear a chick. In this, penguins are like most seabirds in that breeding is necessarily strongly cooperative. If they are successful, a King penguin pair need about 14 months to rear the chick through its molt, before fledging and subsequent departure into the sea. During the winter, however, parents have to stop feeding the chick because the water masses with abundant fish move toward Antarctica and thus occur too far from the breeding grounds for parents to return. During this period, the chick must fast for 4 months on the beach during the sub-Antarctic winter, suffer flooding and storms (as we saw in Fig. 1.29), and live solely on its fat reserves (Jiguet and Jouventin, 1999). The duration of the breeding cycle and moreover breeding frequency are so highly variable in this species that individuals are permanently in competition with each other and selective constraints are very strong, first to find a high-quality mate, and second to rear the single chick in a very constraining environment. Consequently, King penguins have the most sophisticated behavioral adaptations of the penguins for solving such difficult problems. Because of this, King penguins were the best model for our studies of mate choice.

## PENGUINS AS MEAT AND FUEL

Even before the "discovery" of penguins by Europeans, they were threatened by humans. We know from bone fragments that penguins were cooked in the historical villages of the Maoris of New Zealand. While Maoris were eating penguins, they were eating and actually exterminated Moas, a family of large nonflying terrestrial birds. One of the first Europeans to see the Maoris was Joseph Banks, naturalist of the James Cook expedition in 1769 (Beaglehole, 1963). Banks reported that penguins were their usual meal. Around the same time, in 1779, Colonel Robert Jacob Gordon was exploring the Orange River of South Africa, and he drew natives dressed with penguin skins. Accumulations of subfossil bones of Jackass penguins were found in Lüderitz, Namibia, close to prehistoric camps. Natives that are now extinct in Patagonia, Argentina, that Charles Darwin met during his voyage on the *Beagle* were often seen hunting Magellanic penguins and fur seals. Until recently, the collecting of eggs was usual in many places, such as the Valdes Peninsula of Argentina, Falkland Islands, and Great Britain's island of Tristan da Cunha (Fig. 1.31).

Early human exploitation of penguins was nothing compared that of sealers and whalers that occupied sub-Antarctic islands as outposts during their hunts. Wood was lacking on these islands, and the sailors exterminated whole penguin colonies, using the birds as a combustible energy source for melting down the fat of elephant seals. A penguin contains a lot of oil, so, after they were killed, the birds were crushed in a penguin squeezer, and then dried and used as a form of organic coal. During the early 1800s, whaling ships filled their holds with barrels of penguin eggs in fur seal oil. Guano was exploited on the Pacific coast of South America (Chile, Peru) for making agricultural fertilizers. The removal of guano exposed rocky surfaces that were less appropriate

**FIGURE 1.31** Ladies collecting (with wheel barrow) Rockhopper penguin eggs on the Falkland Islands in the Southern Atlantic Ocean. (Port Stanley Museum, South Georgia Island).

nesting habitats for burrowing penguins. These days, threats to penguins come from oil slicks near breeding colonies that are close to shipping lanes. In addition, industrial fishing by Russian, Japanese, and Spanish trawler factories is increasing, mainly in international waters of the southern oceans where overfishing, particularly of krill, is permitted. Penguins are frequent bycatch. Penguins are also threatened when they breed close to human settlements with dogs, for instance on the Otago peninsula of New Zealand. Penguins of New Zealand, now protected, were first collected as food by Maoris and then by Europeans, who killed great numbers of birds. Species such as the Yellow-eyed penguin became quite frightened of humans, even though the species is still quite tame on islands nearby the mainland. When big American cities substituted mineral oil for penguin oil as an energy source for lights, most large industrial massacres stopped. After about a century, penguin populations recovered, even on islands where they had been exterminated. The recovery of penguin colonies was due to their ability to colonize new sites from the sea, similar to recovering rookeries of fur seals.

The use of penguins as meat by explorers was easy because these awkward birds were easy to catch on land and keep alive on shipboard. This was also the likely cause of the massive killing of great auks in the northern hemisphere. The Great auk, the first bird to be called a penguin, was extremely adapted to diving and was unable to fly. However, Great auks were not closely related to modern penguins. Great auks became extinct in 1844 because sailors had for several centuries caught them on islets of the Atlantic Ocean and put them on board their vessels as live food reserves. Great auks had already been preyed on by prehistoric men, as we can see from drawings of them on the walls of a French Mediterranean cave (Fig. 1.32). Eighty cm high, 5 kg in weigh, and with a flipper length of only 15 cm, it was the biggest species of the Alcid family of birds, and the only species in the family that could not fly. The other species in this family both fly and dive but are not so well adapted to diving as penguins. Alcids are more closely related to gulls than petrels, these latter species being the group (viz., the bird order Procellariiformes, petrels, and albatrosses) most closely related to penguins. Alcids show an evolutionary convergence with penguins, especially in their ecological niche as marine predators. The Great auk was the Alcid most like a penguin, both in habits and appearance (Fig. 1.2). But they were geographically opposite to penguins, living only in the Northern Hemisphere where men and mammal predators were present. In this hemisphere, mammalian predators were able to reach numerous polar islands where Alcids breed, swimming like polar bears, or by treks over the sea ice during winter. Consequently, Alcids usually breed on cliffs and chicks fledge as early as possible to escape land-based predators.

In summary, penguins cannot fly but are excellent divers, in fact the best divers among seabirds. They exploit the depth of oceans where they are

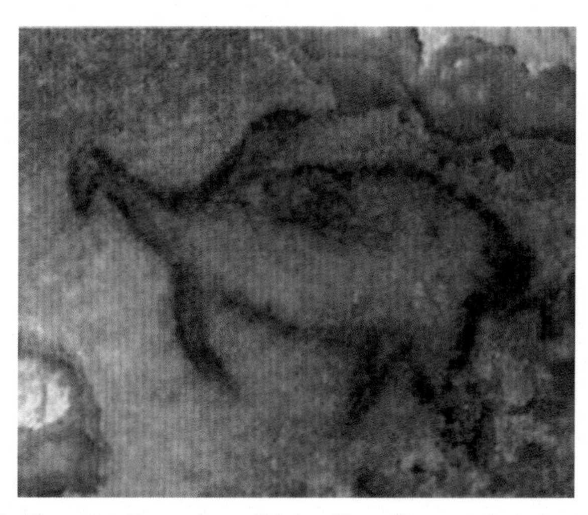

**FIGURE 1.32** The extinct Great auk was flightless like modern penguins and was present in the Mediterranean Sea 19–27,000 years ago, at the end of the last glaciation. Prehistoric men likely ate the Great auk and painted this image on the wall of Grotte Cosquer, a cave (now 37 m under the sea) close to Marseille, France.

without avian competitors. Living in the oceans of the southern hemisphere, they use fat to resist the cold waters of polar climes. A penguin is not primarily a polar bird, rather a couple of species have secondarily invaded the Antarctic continent. Like other seabirds, their life is a succession of feeding at sea and fasting on land. They need to have a mating partner that will share roughly equally in substantial parenting duties, and they are all faithful for at least a year at a time for cooperative breeding. The penguin family not only has species with the common bird habit of nesting, but also two large species that incubate and brood on their feet. When breeding, penguins identify each other, adult breeding partners and their chicks, only by their voice. Thus, study of their special vocal signatures is especially appropriate, given their extreme breeding conditions. In nesting penguins, the choice of a mate is made from the presence and location of a nest and the appearance of potential partners. The appearance of potential partners involves signals, in most of the species head markings and colored ornaments. In the two largest species, the absence of a nest provides an additional challenge to successful breeding that influences the complexity both visual and vocal communication. Studies of the biological meaning of colors in animals have mainly examined birds where the male is more colorful. For our studies of the monomorphic penguins, with males and females virtually indistinguishable in appearance, we tried to discover why males and females look so similar and so ornamented. In Chapter 2, we begin with a description of these similar visual signals and the associated postures that put them on display.

Chapter 2

# Description of Visual Signals

## STUDY LOCALITIES

Early comparative behavioral research on penguins was based on observations by a single researcher, usually working at a single locality and for a limited amount of time. We have observed the sexual and agonistic behavior of 15 species and subspecies in the six genera of penguins, as defined by Pycraft (1898) and Peters (1931). Field observations were often made at a distance using binoculars, so that individual penguins usually behaved in an undisturbed manner. Observations were collected during stays of several days or weeks in captive and wild situations: the zoological gardens of Edinburgh, Basel, Duisburg, Antwerp, and Amsterdam (reproduction in captivity of *Pygoscelis papua*, *Aptenodytes patagonicus*, *Eudyptes chrysocome*, *Spheniscus magellanicus*, *Spheniscus humboldti*, *Spheniscus demersus*); in South Africa at Dassen Island near Cape Town (reproductive behavior of *S. demersus*); in Australia at Phillip Island near Melbourne and in the Sydney Zoo (reproductive behavior of *Eudyptula minor novaehollandiae*); in New Zealand near Dunedin (reproduction of *Megadyptes antipodes* and *Eudyptula minor minor*); at Galapagos Islands (breeding biology of *Spheniscus mendiculus*); at the Valdes Peninsula in Argentina (reproductive behavior of *S. magellanicus*); at St. Paul and Amsterdam Islands (reproduction of *Eudyptes moseleyi*); at the Kerguelen Islands (reproduction of *Eudyptes chrysocome* and *E. chrysolophus*); in Adélie Land (on migrating *Eudyptes chrysolophus schlegeli* and on *Pygoscelis antarctica*); and in the Orkney and Shetland Islands of the United Kingdom, and Grimsey Island of Iceland (for comparisons with vocal signals of Alcidae species).

More detailed and long-term studies were made: in Adélie Land from November 1968 to January 1970 and from December 1975 to March 1976 (breeding cycles of *Aptenodytes forsteri* and *Pygoscelis adeliae*); at the Crozet Islands from December 1971 to March 1972, from December 1976 to February 1977, and from October 1979 to February 1980 (reproduction of *P. papua*, *A. patagonicus*, *E. chrysocome*, *E. chrysolophus*). Over the next 20 years, regular field seasons in Antarctic and Sub-Antarctic French field stations were devoted to experimental studies in acoustics (by P. Jouventin and T. Aubin). Subsequently, field trips over the next 17 years were used in experimental study of

**Why Penguins Communicate. http://dx.doi.org/10.1016/B978-0-12-811178-9.00002-0**

optical signals (the two authors of this book working together) on the Sub-Antarctic Archipelagos of Crozet and Kerguelen. Most sojourns in the field were spent in huts near relatively undisturbed penguin colonies that were several hours walk from the main island bases.

## FRAGMENTARY ETHOLOGY

In ethology, it is much easier to describe "how" a behavior happens than to understand "why" the behavior occurs in the first place. "How" questions are simply descriptions of what a particular animal does, or in general what a species does, and the developmental, genetic, physiological, or ecological "proximate" underpinnings of a behavior. When asking "why" a behavior occurs, we ask about the ultimate cause of our observations, by which we mean the adaptation of an animal to its environment through the process of natural selection. Asking "why" questions is the purview of behavioral ecology. We can often infer the biological function or "utility" of an animal behavior by examining its ecological context under natural conditions in the field. The general idea is that behavioral traits are often "solutions" to the problem of surviving and reproducing, given an environmental context. The behaviors that are favored by natural selection are expected to result in maximizing the offspring contribution that successful individuals make to future generations. The utility of a behavior is in furthering the processes of survival and reproduction, and natural selection over time should favor the contribution of offspring from individuals that exhibit behaviors that contribute most to these processes. This is termed fitness maximization, and the most successful individuals are those that succeed in maximizing fitness by producing the greatest number of offspring that themselves survive and, most importantly, reproduce.

Testing the ecological conditions that favored the evolution of observed behaviors requires long-term observations of the species in nature, with the goal of understanding how the environment may have molded behavioral actions over evolutionary time. Unfortunately, fossils seldom preserve behavioral traces. The lack of fossil evidence forces us to compare different species to understand the variations in behavior of modern species, a classical procedure called the comparative method. Early applications of the comparative method to surveys of small numbers of species led to circular reasoning and "adaptive stories" that were not scientifically testable. With the study of whole families with numerous and substantially different species, such as penguins, such problems can be overcome. For a study of the problematic role of behavior in evolution, penguins seem eminently suitable. The difficulty of study in the field, as opposed to the laboratory, is access to penguin breeding sites, often on remote and isolated islands. The compensation for such difficulty is the relative fearlessness of penguins to humans that allows close approach for behavioral study. The known breeding biology of most species gives us a solid working basis for understanding their behaviors. Indeed, the curious displays of several species have already been amply described. But confusion in the scientific literature on penguins is rife, and in this section, before interpreting their behaviors, we describe their ritualized behaviors and introduce standardized terminology, while filling in gaps in previous observations.

Although penguins have always interested anatomists and embryologists, broader natural history data have been fragmentary. Even though pioneering behavioral ecologists such as Richdale, a New Zealand teacher, began their work as early as 1945, studies of the breeding cycles of penguins have been hampered by the physical difficulty of gaining access to penguin colonies and by the logistical needs of scientists in nature on the small remote islands of austral seas. Scientific descriptions of penguin behavior are more recent; first confined to short paragraphs in monographs, the quantification of penguin natural history and behavioral analysis has increased in the last few decades of research. As in initial field studies of primates, penguins invited anthropomorphic digressions, and early accounts drew rash and often unconscious anthropomorphic analogies between penguin colonies and human societies.

Only in the 1950s did systematic and objective behavioral studies appear. Comparative ethological studies of this group of birds have been relatively rare. A synthesis of penguin comparative behavior was ventured upon by Roberts, as early as 1940. For its age and heuristic value, his study should be rehabilitated. In “Sexual Behaviour in Penguins,” Richdale (1951) made a more serious attempt to tackle the subject of penguin behavioral variation. However, of the three species Richdale observed, only the Yellow-eyed penguin was studied for their behavioral traits, rendering comparison impossible. Isenmann and Jouventin (1970) published a short essay that described environment-behavior relationships, comparing the effects of territory-holding on social organization in the Adélie penguin, the King penguin, and the Emperor penguin. Finally, Jouventin (1982) published the first extensive comparative study of behaviors of the penguin species. Our current goal is to actualize and expand this previous information and add nearly 40 years of experiments and genetic studies of penguins to the mix.

With nesting areas scattered over the oceanic islands of the southern hemisphere, a single scientist can observe only one or two species during the reproductive season. The vocabulary that a researcher uses will naturally be specific to the species observed, and different scientists may give different names to homologous behaviors, either in the same species or in two related species. Or the name assigned to a behavior in one species may also be given to a similar-looking posture in another species, but the behaviors may be completely different, not only in their ecological context, but perhaps in the evolution of the behavioral trait (that is, a behavioral analogy, rather than a homology). When comparisons are made among species, only photographs or brief descriptions may be at an author’s disposal. Even when excellent behaviorists make detailed observations, difficulties arise in evaluating the subject or topic under study, leading all too often to misconceptions and even incomprehension.

Examples are the behavioral analyses of the Rockhopper penguin by Warham (1963, pp. 244–252) and of the Macaroni penguin by Downes et al. (1959, pp. 59–66). From their accounts, one might conclude that the ethogram of the Rockhopper penguin is rich and the Macaroni penguin poor. However, this divergence is in the descriptions only, for the diversity and elaborations of postures are very similar in the wild. Then again, ornithologists that are untrained in ethological methods of observation may not be able to distinguish similar or identical behaviors between species from distinctly unique ritualized behaviors. An example of this problem is the difference between the normal walking on land and elaborated nuptial walking (the curious "waddling gait") in the Emperor and King penguin. Finally, the behaviors of some genera have as yet to be completely described or have been relatively neglected.

As is natural for behavioral studies, pioneering descriptive studies came first, often brief and with only preliminary descriptions, followed by more complete descriptive studies, and then experimental studies. For example, the behavior of Little penguins was briefly described long ago (Warham, 1958; Kinsky, 1960). A complete behavioral study of the genus *Eudyptes* was published later, revealing that different species and subspecies exhibit very similar behaviors (Warham, 1975). The genus *Spheniscus*, having four apparently very closely related species, differentiated on a morphological and behavioral basis, was almost completely unknown until 1979 (Eggleton and Siegfried, 1979). The Yellow-eyed penguin was described in detail by Richdale (1951, 1957), a pioneer in penguin biology. Without training in ethology, Richdale, however, did not attribute precise meanings to the behaviors he described, and he often failed to distinguish ritualized behaviors, such as mating displays, from more common behavioral sequences such as those used in fishing or walking. Behaviorally, both species of the genus *Aptenodytes* are nowadays well known. A monograph was published by Stonehouse in 1960 on the King penguin and in 1953 on the Emperor penguin. The remarkable biology and behavior of the Emperor penguin was further documented by Prévost (1961), Isenmann (1971), and Jouventin (1971a,b, 1972a,b, 1975; Jouventin et al., 1979; Guillotin and Jouventin, 1979).

As for the genus *Pygoscelis*, a short publication (van Zinderen Bakker, 1971) described the behavior of the Gentoo penguin. In a comparative study, Roberts (1940) briefly reported on the Chinstrap penguin, seemingly very closely related to the Adélie penguin. The behavior of the latter species has been better studied than all the other penguins put together (Sapin-Jaloustre, 1960; Sladen, 1958; Penney, 1968; Ainley, 1975; Spurr, 1975; Jouventin and Roux, 1979).

## MODELS FOR VISUAL SIGNALS

Penguins are the family of birds that are most strongly adapted to life in the water. They seem fairly homogeneous when we examine their morphology, particularly because all have a torpedo shape and wings that have been evolutionarily modified into flippers for swimming. But in fact, different genera and even species of penguins in the same genus are very different in their way of life at sea. Moreover, the reproductive behavior of penguin species on land is highly diverse. The challenge for any seabird, and particularly for those unable to fly, is to breed where there are no land-based predators and where there is a rich feeding zone nearby. Oceanic islands, especially when uninhabited by humans, are perfect breeding locations for oceanic birds that must have land (or solid ice) to stand on for nesting or nonnesting reproduction. But in the Southern Hemisphere, islands are rare and often small, so space is extremely limited for breeding birds that can be numerous where seas are productive. Consequently, most seabirds and all penguins are colonial where suitable islands occur.

About 90% of birds are socially monogamous (Bennett and Owens, 2002), and virtually all seabirds are faithful to their partner during the breeding cycle because both partners have to alternate between land and sea (Lack, 1968). While one partner incubates or broods the offspring, the other partner forages in the ocean. Thus, both partners must shift back and forth from different tasks: relieving their partner for care of eggs or chicks on land, foraging at sea to recharge their energy reserves, and returning with food for their chicks during brooding. A single mate is simply unable to rear the young alone. The social structure of most seabirds and penguins in dense breeding colonies is based on a mosaic of nesting territories occupied by pairs with their eggs or chicks. The distance that a brooding penguin can reach with its beak determines the size of the territory that it can defend. This optimal use of the space limits the number of pairs that can fill an appropriate cliff, beach, or ice field as breeding habitat, but nonetheless a colony can hold hundreds, thousands, or even millions of pairs.

Marine animals that spend most of their time swimming underwater are counter-colored, light underneath and dark on the back. Penguins are no exception to this nearly universal pattern. It allows diving penguins to be mimetic with the surface when seen by prey or predators from below in the water column and also mimetic with the bottom when see from above. On the other hand, each species has a different color or marking pattern of the head. Some aspects of these markings are examined carefully during the nuptial display by potential breeding partners. A bird that arrives at the colony and searches for a breeding partner has to avoid immature individuals that are not ready for breeding and members of the opposite sex that are in poor condition. Or, the bird will be unable to successfully rear a chick and will likely lose a breeding year.

The potential breeder must also avoid possible partners from other, perhaps closely related species. Migrant individuals occur in the open ocean, and hybrids between closely related species are possible but may not be so well adapted to the local water mass or the ecological situation of the terrestrial breeding colony. As we will see, all types of poor potential mates can be detected by a careful visual inspection during the nuptial display. We conducted experiments to reveal the characteristics that penguins use to avoid selection of a poor partner.

All penguins live in the Southern Hemisphere where the Antarctic continent occupies the geographic center and has the coldest temperatures of the hemisphere. The water flows of the Austral Ocean decrease in temperature going from the equator to the pole, with broad current bands at different temperatures. These bands form a pattern of radiating circles that resemble a bullseye target around the Antarctic continent. Each penguin is adapted to a water mass of particular characteristics (e.g., warmer or cooler) and breeds on an oceanic island inside their marine "zone." Even while swimming at sea, when only the upper part of the body is visible, the head pattern of each species indicates the identity of the species. Thus, head markings give clues, not only to ornithologists about the different species, but to penguins about which potential partners are most likely to facilitate and cooperate in successful breeding. This prevents interspecies breeding, and matings with immature individuals or those in poor condition, as we will see in Chapters 3 and 6.

## AGONISTIC AND APPEASEMENT SIGNALS

### Nonritualized Postures

Displacement activities are behaviors that do not appear to be relevant to the environmental context in which they occur. These behaviors often indicate stress, and they can be seen amid natural patterns of individuals interacting with their environments. But they also can be easily induced by human manipulation. For example, an observer can induce displacement behaviors by blocking the path of a penguin. Most displacement behaviors of penguins involve preening. For example, 54.8% (of 750) of nonritualized behaviors in the Emperor penguin and 54.6% (of 150) in the King penguin are preening behaviors (Jouventin, 1982). Some nonritualized attitudes convey information to congeners, owing to the colonial life of these birds. Penguins flap their flippers when alarmed or uneasy, and this movement spreads from one individual to the next. So, flapping the flippers works like an alarm signal. In a group of 50 Emperor penguins, for example, the number of flapping penguins may rise from 3 to 25 when a human appears 50 m away (Jouventin, 1982). The flapping birds then turn towards the intruder and thrust their beaks forward to strike them as they pass, or more often, they try to waddle away. In the attack posture of these large penguins, the bird faces the enemy, flippers raised ready to strike, neck and beak stretched out to poke and stab. Directed at a congener, a Skua (*Catharacta maccormicki*), a Giant petrel (*Macronectes giganteus*), or sometimes toward a human observer, this nonritualized intention movement serves its purpose well.

## Genera *Eudyptula, Megadyptes,* and *Eudyptes*

Aggressive and appeasement behaviors are common and well differentiated in crested penguins (genus *Eudyptes*). For example, incubating Macaroni penguins crouch on the nest, flippers half-raised, head stretched out toward the enemy with their beak open and ready to nip and pinch. While on the nest, they turn their heads from time to time while staring at an approaching intruder. At the opposite, social interactions are scarce in Yellow-eyed penguins (genus *Megadyptes*), since this species does not nest in dense colonies. The different Little penguins of Australia and New Zealand (genus *Eudyptula*) are closely related morphologically as well as behaviorally. They are strictly nocturnal and their agonistic signals make good use of their white ventral area. An aggressive penguin sings while approaching an adversary, then stands erect, raises its flippers and appears as a luminous spot in the dark. Towering over its enemy, it snaps at it with its beak and often forces a retreat.

## Genus *Spheniscus*

Observations of agonistic behaviors of Jackass penguins differ between Roberts (1940, p. 219, Fig. 16) and Jouventin (1982). The latter considered the function of "twisting" or rotating the head around an axis formed by the beak pointed at an intruder as agonistic and territorial, and not purely sexual as Roberts (1940) claimed. The movement is made by penguins in their burrows when an intruder is approaching, even if the burrow is occupied by only one partner. It is most probably the same movement that is described above in crested penguins, a behavior pattern common to many avian species: i.e., flippers half-spread and body close to the ground, the bird attempts to exhibit a maximum of surface to impress its enemy by its size. The protruding head and open beak bar the entrance to the burrow.

## Genus *Pygoscelis*

Numerous very elaborate agonistic behavior patterns are seen in the Pygoscelid penguins, with the exception of Gentoo penguins. This latter species is quite peaceful and has rudimentary agonistic behavior patterns. A behavior common to all penguin species that nest in dense colonies has been termed the "tête" by Richdale (1951, p. 34) and the "tête à tête" by van Zinderen Bakker Jr. (1971). Two neighbors take a firm stand on their nests and face each other, head forward and beak open. Each sporadically tries to grab the adversary's beak with its own and twist it. In fact, this behavior might be considered a variant of that described in Macaroni penguins. It is also similar to the curious attitude observed by Roberts (1940) and Jouventin (1982) of banded penguins (genus *Spheniscus*), in which the display is more ritualized. However, this behavior is not exhibited between incubating neighbors, since the banded penguins nest in burrows.

The agonistic behavior of the Chinstrap penguin, a second member of the Pygoscelid genus, is poorly described. Roberts (1940) was able to observe this species, but concentrated on its sexual behavior and provided no information on agonistic behavior patterns. However, the latter might be similar to behavioral accounts of the Adélie penguin, as these two species were classified in the same group by Roberts on the basis of their behavior and other criteria.

The most varied agonistic attitudes, deserving more detailed description, are found in the Adélie penguin (Fig. 2.1). Its behavior was more or less completely described long ago by Sapin-Jaloustre and Bourlière (1952), Sladen (1958), Sapin-Jaloustre (1960), and subsequently further detailed by Penney (1968), Ainley (1975), and Spurr (1975). We will use Spurr's behavioral classification. Depending on the enemy's distance, a sequence of agonistic actions unfolds and may end in a real attack. In the "beak-to-axilla" attitude, the penguin puts its beak under one flipper and grunts while whirling its head. This is often used to signal to distant congeners. It is all the more ritualized in that it does not recall a fighting movement and is not necessarily directed toward an adversary (unlike the "horizontal head–circling motion" of the Emperor penguin, which is a ritualized fighting movement). In the "sideways-stare" the bird, either lying or standing, turns its head sideways and stares with one eye at the intruder; if the intruder moves closer still, the "alternate stare" follows: the bird slowly turns its head from side to side staring with first one eye, then the other facing the opponent (Fig. 2.2). These two patterns correspond to the "twisting" of banded penguins and a similar movement in Little penguins. It can be divided into three elements: the bird stands firmly on its nest, maximizing its size, flaps its flippers, and points its

**FIGURE 2.1** A pair of Adélie penguins is displaying on their nest, first for mating and then when they meet during incubation and brooding. The mutual display comprises both body movements and vocal recognition.

**FIGURE 2.2** The aggressive Adélie penguin on its nest turns the head facing an intruder, the white sclera of the eye (top of the eye) showing and the feathers on the nape of the neck erected. This common agonistic posture was named "the oblique stare" by Penney (1968).

beak at the intruder, as it moves its head to evaluate the enemy's distance. If the intruder comes very close to the bird, the beak is stretched still further forward, the crest is erected, and the pupil of the eye is lowered to reveal the white (the "point") marking. The penguin prepares to attack by opening its beak ("gape"), and charges the adversary with flippers half-opened ("charge"). Finally, it pinches the adversary ("peck"), pushes it in the chest, and strikes at it with its flippers.

## Genus *Aptenodytes*

Both King and Emperor penguins perform other types of agonistic behavior in addition to flipper strikes and beak stabbing. Because these species lack a nest, they exhibit unique breeding behaviors. Paradoxically, the most ritualized agonistic behaviors are found in the peaceful Emperor penguin. In many territorial species, territory owners often threaten intruding neighbors with the beak, an efficient intention movement of blows. The Emperor penguin, on the other hand, shows a stereotyped "horizontal head−circling motion" ("balayage latéral de la tête" of Jouventin, 1971b): the bird throws its head back, then moves it forward in lateral half-circling motions, grunting. In most cases, the beak-swinging movement is not directed toward a particular individual, and can even be executed by solitary birds. It almost never ends with a blow of the beak. This highly ritualized movement is common during crucial periods of the breeding cycle, when the animal tends to isolate itself from the crowd, for example during pair formation, copulation, egg-laying, and egg or chick exchange. It occurs rarely in King penguins, primarily during brooding in a fixed place, when flooding and storms force them to move (Stonehouse, 1960,

p. 31). However, Emperor penguins are very mobile, even at the beginning of their breeding cycle.

## Conclusions on Agonistic Behaviors

Sexual displays frequently involve agonistic behavior patterns. For example, in the Adélie penguin, the "beak-to-axilla" posture is not only observed after threats against intruders but also after a typical sexual behavioral sequence, the "ecstatic display." The association of these two apparently functionally different attitudes recalls a similar pattern in the Emperor penguin, in which a sexual attitude ("mutual display") is also often followed by an agonistic one ("horizontal head-circling motion"). According to Penney (1968, p. 101), 49 out of 96 (51%) "ecstatic displays" were followed by the "beak-to-axilla" in the Adélie penguin. In the Emperor penguin, the "horizontal head-circling motion" follows courtship song about 52% of the time (out of 108 cases, Jouventin, 1982). Although the similarity of these percentages may be due to chance, they show that about half the time an agonistic attitude follows a sexual one. This expression in two species of different genera confirms the strong relationship between sexuality and aggression. It is thus understandable that two clear-sighted observers of penguin ethology were led astray. Sladen (1958) mistook the "beak-to-axilla" of the Adélie penguin, and Roberts (1940) mistook the "twisting" attitude of the Jackass penguin and the "aposematic display" of the Adélie for sexual behavioral patterns, but these are aggressive and territorial postures.

In the whole penguin family, behavior patterns can be classified by their degree of increasing aggressiveness. First, displacement behaviors showing conflict, such as headshaking followed by swallowing. Second, warning postures such as flipper flapping (which can be integrated into more complex attitudes such as the "alternate stare"). Third, behaviors based on a lateral movement of the head, revealing a low level of aggressiveness and acting as an alarm signal regarding intruders (such as the "horizontal head-circling motion" in the Emperor penguin and the "beak-to-axilla" in the Adélie penguin). Fourth, behaviors based on the rotation of the head around the beak, showing a mean level of aggressiveness, and constituting a threat signal just before an attack (such as the "alternate stare" of the Adélie penguin and the "twisting" of the banded penguins). And fifth, variable combat techniques depending on position and size of the birds. These last behaviors can be subdivided into three groups: (1) frontal attacks made by standing birds, including the "direct stare display" of Adélie penguins (according to Penney, 1968), and "attack posture" of the Emperor penguin ("posture d'attaque" from Jouventin, 1971b); (2) frontal attacks made by birds lying on their bellies, usually on the nest ("tete" of Yellow-eyed penguins, "tête à tête" of Gentoo penguins); and (3) attacks from the rear, in which the dominant bird grabs the back feathers of the adversary with its beak while striking it with its flippers (Adélie and Jackass penguins).

Table 2.1 summarizes these descriptions of the agonistic behavior of penguins. We tried to show equivalents between systems used by different authors.

## Appeasement Signals

In most cases, an appeasement behavior is defined as the opposite of an agonistic one and is often literally opposite (Darwin, 1872). An excellent example is provided by the nocturnal Little penguins, in which the erect agonistic posture reveals the white underside of the flippers and body, whereas when in appeasement the bird bends over, head and flippers lowered, completely hiding its white zones, and thus becoming indistinguishable from the dark background. A characteristic posture common to all penguins is assumed when the birds sneak through the colony's brooders, avoiding beak and flipper blows on their way. It is frequent in Adélie, King, and Emperor penguins, which nest in tight formations. The bird walks rapidly, flippers outstretched in the large *Aptenodytes* penguins ("defence posture" = "posture de defense," Jouventin, 1971b) and on the tips of its feet in Adélie penguins ("slender walk" of Spurr, 1975, after Warham, 1963, although this behavior differs in context in Rockhopper penguins). The head is held high. This rapid walking is little or not at all ritualized. While rapid walking cannot be confused with agonistic behavior, neither is it the exact opposite.

In the crested penguins, alternating with rapid walks as described above, a new behavior appears, which seems less relevant to rapid traversing a colony and thus has a signal value. On its way to the nest, the bird slows down occasionally, climbs on a rock to locate its position, then adopts a curious walk in which the body is bent as if bowing, flippers parallel and pointed forward. The term "submissive attitude" was used by Downes et al. (1959) to describe this highly ritualized posture in Macaroni penguins, whereas Richdale (1941) called it "sneering attitude" and Warham (1963) "slender walk" in Rockhopper penguins. It is present in all species of crested penguins (Warham, 1975). In Yellow-eyed penguins this submissive behavior might be what Richdale (1951) called the "sheepish look." The context in which it occurs shows that it functions to inhibit the territory-owner's aggressiveness when crossing a territory. On arrival at the nest, this posture is followed in the crested penguins by what Warham (1975) called the "shoulders-hunched attitude," which is considered merely a ritualized variant of the submissive attitude (Jouventin, 1982). Thus, the transition between these postures and those used in a sexual context is quite imperceptible.

# COURTSHIP DISPLAYS

## Genus *Spheniscus*

For banded penguin species, a scientific description of courtship displays developed over a period of about 50 years. Kearton (1930) wrote a book with

**TABLE 2.1 Agonistic Postures. Only Species Observed Personally, or Whose Behavior Has Been Well Described, Are Compared**

| Species | References | 1 | 2 | 3 | 4 | 5 | 6 | 7 |
|---|---|---|---|---|---|---|---|---|
| *Eudyptula minor* | Jouventin (1982) | | | Threat with whistle | Erect posture | Attack posture | | Appeasing posture |
| *Megadyptes antipodes* | Richdale (1951) | Shake, excited shake | ? | ? | Glare? | Open yell? | Téte | Sheepish look |
| *Eudyptes chrysocome* | Warham (1963) | Headshake | Wing-shivering | | Mild threat? | | | Slender Walk |
| *Eudyptes chrysolophus* | Downes et al. (1959) | | | | | | | Submissive attitude |
| *Spheniscus demersus* | Jouventin (1982) | Beak shaking | Wing-flapping | | Twisting | Attack posture | | |
| *Spheniscus magellanicus* | Roberts (1940) | | | | Twisting | | | |
| *Pygoscelis papua* | van Zinderen Bakker (1971) | Headshake | Arms-waving | | | | Téte à téte | |
| *Pygoscelis adeliae* | Spurr (1975) | | Flicking | Bill-to-axilla<br>Sideways stare | Alternate stare<br>Point<br>Gape<br>Charge | Peck | | Slender walk |
| *Aptenodytes patagonicus* | Stonehouse (1960) | Head-flagging | | | | Direct stare display | | |
| *Aptenodytes forsteri* | Jouventin (1971a,b) | Beak-shaking | Wing flapping | Horizontal head circling motion | | Attack posture | | Defense posture |

Postures classed in progressive stages of aggressiveness: 1. displacement activities showing conflict, such as beak-shaking, found in all species; 2. warning postures, e.g., wing-flapping, present in all species; 3. low-level aggressive postures such as the "horizontal head—circling" of the Emperor penguin; 4. mean-level aggressiveness, such as "twisting" in the Jackass penguin; 5. frontal attack by a standing bird; 6. frontal attack by a bird lying down; 7. appeasement postures. Names for these 7 postures, as assigned by the authors quoted, are inserted. Dark gray = common behavior not described by author, light gray = rare behavior not described by author.

From Jouventin, P., 1982. Visual and vocal signals in penguins, their evolution and adaptive characters. Advances in Ethology, 24, 1–149.

some remarkable photographs of the Jackass penguin, in which the penguin was represented as a hero. Roberts (1940) mentioned this species briefly in his review of courtship behaviors. Jouventin (1982) observed Magellanic and Jackass penguins both in the wild and in zoos, and the description of their behavior may be extended to the whole genus, given what can be seen during reproduction in the wild for the Galapagos penguin and in captivity of another congeneric species (Humboldt penguins). These penguins have complex displays (Fig. 2.3). They nest in holes, mostly burrows dug with their feet but alternatively in rabbit burrows. Courtship displays usually occur close to the burrow. A penguin looking for a mate stands up, spreads its flippers a bit, lifts its beak skyward and utters a song (the "ecstatic attitude," of Roberts, 1940). As a bird of the opposite sex approaches, the singer stands still in an awkward position, flippers pointed forward, body stretched and head bent, resembling the appeasement posture described for crested penguins. The newly arrived bird then adopts a similar characteristic posture and circles the motionless singer once or more, a behavior pattern termed "La Ronde" ("turn-around stare") by Jouventin (1982). Sometimes the singer begins turning around

**FIGURE 2.3** Courtship displays of Jackass penguins (*Spheniscus demersus*).

the circling bird, each staring at the other's chest. After a while the pair walks away in an "automaton" gait (="l'automate" for Jouventin, 1982), interrupted by pauses when the birds stand still and contemplate each other, recalling the "face-to-face" display of the Emperor penguin. The male then puts his beak on the female's neck and presses her to the ground. When mounted by the male, the female raises her tail, revealing the cloaca. During intromission, her beak is held parallel to the ground while his is perpendicular, enabling them to rub beaks frantically against each other while similarly vibrating their flippers. After copulation, a twig-gathering session follows, and the bottom of the burrow is eventually covered with twigs. When approaching the burrow with a twig, the nesting bird assumes a "bowing" posture, head bent forward, beak and flippers pointing downward. At times, partners "mutually preen" their necks and heads or perform the "mutual display" inside the burrow on the nest. Partners sing duets in the postures described above, flippers raised and body bent.

Eggleton and Siegfried (1979) also published a detailed description of the Jackass penguin's courtship displays. The correspondence between their nomenclature and ours is the following: their "alternate stare" = our "twisting"; "ecstatic" = "ecstatic display"; "oblique stare bow" ("intermediate bow" and "sideways-stare") = "face to face" and "waddling gait"; "extreme bow" (and "bending-down bow") and "vibratory headshake" = "bowing"; "allo-preening" = "mutual preening"; and "mutual ecstatic" = "mutual display."

## Genus *Pygoscelis*

The genus of Pygoscelid penguins contains two species showing very complex courtship displays (Adélie and Chinstrap penguins) and one with more rudimentary displays (Gentoo penguin). Adélie penguins have the richest repertoire of optical signals of all penguin species (Fig. 2.4). A single male stands on a pile of pebbles, which constitutes its nest, stretches out raising head and beak vertically and, vibrating its chest, it claps its beak repeatedly, while synchronously flapping its flippers perpendicular to the body (Fig. E "ecstatic display" Wilson, 1907; "position extatique," in Sapin-Jaloustre and Bourlière, 1952, p. 11). A receptive female approaching the nest bends forward, its beak 30–60 degrees from the horizontal, and slowly exhibits the side of its head to the male ("bowing," Sladen, 1958; "oblique stare bow," Penney, 1968). The male adopts the same posture but its feathers are ruffled and its pupils lowered, which, as we have seen, is a sign of aggressiveness. The two bow to each other ("bowing," Roberts, 1940; Sladen, 1958). Then the male sits in the nest and scrapes around the nest bowl with his feet, rolling the pebbles about, and then gives way to the female. This is what Spurr (1975) called "nest scraping" and "rearranging." The female settles on the nest, the male lowers its head, vibrates its flippers, and jumps on the back of the female, as she raises her beak and

**FIGURE 2.4** Courtship displays of Adélie penguins (*Pygoscelis adeliae*).

tail. They copulate, rubbing their beaks together. Male and female can often be found together on their territory, and from time to time they indulge in a "mutual display" ("parade mutuelle," after Sapin-Jaloustre and Bourlière, 1952) more or less intense ("loud and quiet mutual display," Sladen, 1958; "mutual epigamic display," Roberts, 1940). The birds face each other on or near the nest, bend toward the ground while thrusting their beaks at each other and then suddenly stand up and wave their heads held vertically. "Mutual displays" and "bowing" are shown most often alternately during nest reliefs of the incubating or brooding bird.

As for the Chinstrap penguin, Roberts (1940) provided three illustrations: a quite classic "ecstatic attitude," less developed than that of the Adélie penguin, and two phases of the "mutual epigamic display" identical to those of the "mutual display" of the Adélie penguin. Jouventin (1982) considered that the display of Chinstrap penguins was similar to that of Adélie penguins, although the head is not shaken in the same manner (Fig. 2.5). He thought that the "ecstatic displays" of these two species were very similar, as Sladen (1958) also pointed out. In the Chinstrap penguin, however, these displays are executed by both sexes, not only by males (the latter is usual in Adélie penguins).

In Gentoo penguins (Fig. 2.6), the "ecstatic" (Levick, 1915; Bagshawe, 1938), "bowing" (Roberts, 1940), and "mutual display" ("ecstatic display," Bagshawe, 1938) are less elaborate and lack such specialized variants as the "oblique stare bow" of Adélie penguins. A bachelor Gentoo penguin stands on its nest, flippers half-raised, lifts its beak vertically, and utters its song. van Zinderen Bakker (1971) recorded "low and high intensity" displays, as did Sladen (1958) for Adélie penguins. An identical mutual display occurs too after egg-laying, although the head is less shaken and both birds display instead of one. When meeting on the nest and particularly during reliefs, Gentoo penguins show much more elaborate bowing than the other two species ("bowing" Roberts, 1940). Curiously, mutual preening is absent in the genus *Pygoscelis*.

**FIGURE 2.5** Courtship displays of Chinstrap penguins (*Pygoscelis antarctica*): (A) ecstatic display of a male on its nest, (B) bowing of a partner returning to the nest, and (C) mutual display of a pair.

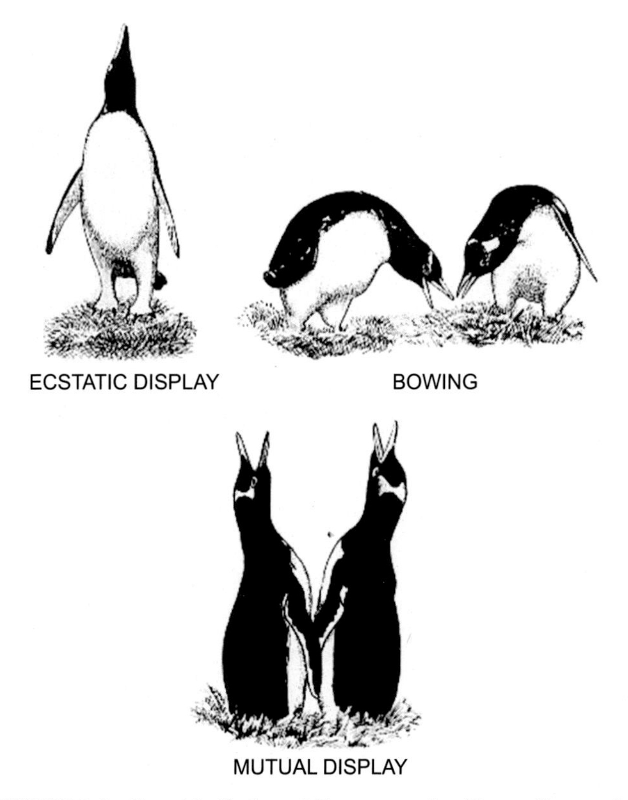


**FIGURE 2.6** Courtship displays of Gentoo penguins (*Pygoscelis papua*).

## Genus *Eudyptes*

Courtship displays of several species of crested penguins were described in detail by Warham (1971, 1972, 1974, 1975), superficially by Roberts (1940) in Rockhopper penguins, and briefly but precisely by Downes et al. (1959) in Macaroni penguins. From a comparative point of view, the displays of the different species (Northern Rockhopper, Southern Rockhopper, Schlegel, and Macaroni penguins that we observed) differ only in minor details. Their ritualized postures are fewer and less complex than those of three other genera (*Spheniscus*, *Pygoscelis*, and *Aptenodytes*), with the exception of Gentoo penguins. Only four types of postures having a dominant sexual context need to be distinguished for Macaroni penguins (Fig. 2.7). In the first, "ecstatic attitude" (Roberts, 1940), "vertical head−swinging" (Warham, 1975), or "ecstatic display" (Downes et al., 1959), after bowing the bird stretches up, flapping its flippers, and rapidly shakes its head from side to side while singing rhythmically. Secondly, "mutual epigamic displays" (Roberts, 1940) or "mutual forward and vertical trumpeting" (Warham, 1975) is performed by

**FIGURE 2.7** Courtship displays of Macaroni penguins (*Eudyptes chrysolophus*).

both partners during nest reliefs and when a parent comes to feed its chick(s). The display is longer, its intensity higher and the headshaking less pronounced than in the ecstatic display. Third is "pointing to the nest" (Downes et al., 1959); this term includes what Warham (1963, 1975) distinguished as "bowing" and the "shoulders-hunched posture," which were described under appeasement behavior. Pointing to the nest generally precedes "ecstatic" or "mutual epigamic" displays. Fourth is "mutual preening" ("mutual preening," Downes et al., 1959; "allo-preening," Warham, 1975). Headshaking and wing-flapping are much more rapid in the small Rockhopper penguin than in the larger Macaroni penguin species. According to Warham (1963), only the male Rockhopper penguin exhibits "vertical head–swinging" or "male display."

## Genus *Megadyptes*

Richdale (1951, 1957) gave detailed descriptions of 18 postures of the Yellow-eyed penguin and suggested their functions. But Jouventin (1982) observations suggested no more than a four-category classification of the different "love-habits" described by Richdale (Fig. 2.8). (1) In "full trumpet" the bird

**FIGURE 2.8** Courtship displays of Yellow-eyed penguins (*Megadyptes antipodes*).

thrusts its chest forward and sings with its beak pointing upward. (2) "Salute" is generally performed by males approaching a female. This curious short-step walk recalls the "slender walk" described in banded or crested penguins. The male points his beak down, flippers held forward, then suddenly stands erect and thrusts his beak upward, freezes for several seconds before assuming a more natural position while turning his head toward the female. The "gawky attitude" described by Richdale (1951, p. 263, Fig. 2.5) is used as a response to the "salute," which it resembles in its initiation. It is very close to what Richdale termed the "sheepish look," briefly described under appeasement postures. (3) Two patterns observed by Richdale ("half trumpet" and "ecstatic") are considered merely variants of the mutual display or "welcome ceremony." (4) "mutual preening" (according to Richdale "mutual preen" for the top of the head and "kiss preen" for the neck) is also observed in this species.

## Genus *Eudyptula*

Warham (1958) and Kinsky (1960) briefly described the courtship display of Little penguins. In the two subspecies (or species), which Jouventin (1982) observed, the postures seemed identical. It should be noted that the subspecies described by Kinsky and Falla (1976) differ in behavior only in minor details. Although careful observations are difficult in these nocturnal species, three main sexual postures were recognized (Fig. 2.9): (1) solitary singing of the erect bird (body 90 degrees from the horizontal), flippers spread, beak pointed forward; (2) mutual display (Warham, 1958; Kinsky, 1960), where the singer bows to 45 degrees from the horizontal at the height of excitement, its flippers are lifted and lowered synchronously with the song, beak following body; and (3) mutual preening (Warham, 1958; Kinsky, 1960).

## Genus *Aptenodytes*

The courtship display of Emperor penguin (Fig. 2.10) was described in part by Wilson (1907) and Stonehouse (1953), and in more detail by Prévost (1961), Isenman (1971), Jouventin (1971a,b), and quantitatively by Guillotin and Jouventin (1979). A bachelor Emperor penguin sings somewhere in the colony: it stops, lets its head fall on its chest ("cou en crosse," according to Prévost, 1961), takes a big breath, utters its song with lowered head, stands still for one or two seconds more, then continues its walk, repeating the behavior further on. The receptive female freezes, stretched as tall as possible, and is imitated

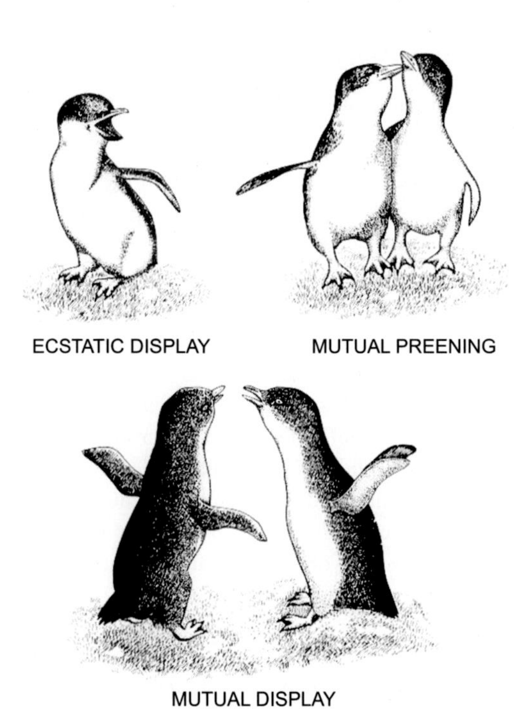


**FIGURE 2.9** Courtship displays of Little penguins (*Eudyptula minor*).

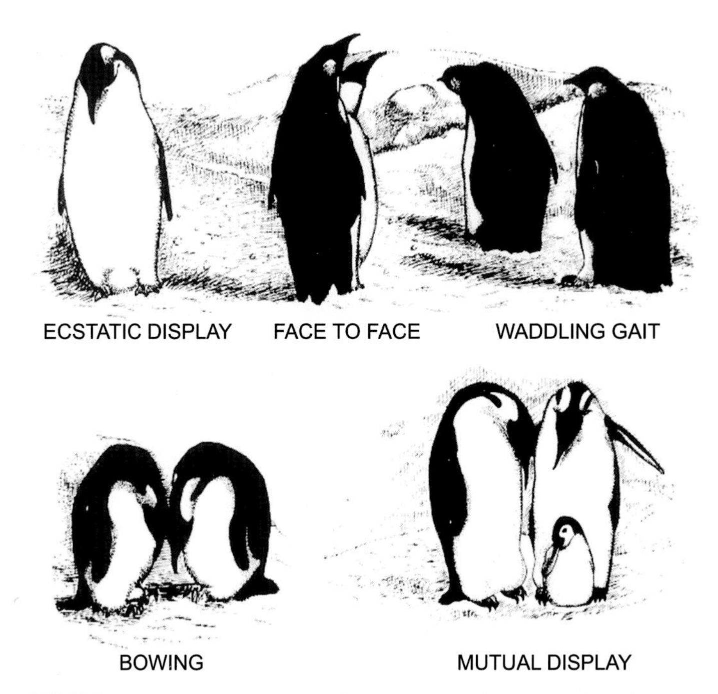


**FIGURE 2.10** Courtship displays of Emperor penguins (*Aptenodytes forsteri*).

by its partner, the birds more or less facing each other ("face to face,"—"face à face" according to Prévost, 1961). They slowly lift their heads while progressively contracting their neck muscles, freeze in this position for several minutes, then little by little relax and assume more normal attitudes. The pair then either separates, each bird searching for a mate on its own, or they stay together for hours and become partners, inseparable until the egg is laid. When moving about, the female follows in its partner's footsteps, both birds adopting what Prévost (1961) termed the "waddling gait" ("démarche balancée"). As days go by a new posture appears that Jouventin (1971a,b) called "bowing" ("la courbette") in this species: one of the birds bends down and is immediately imitated by its partner and so on (quantitative analysis of behavior before copulation in Fig. 2.11A). Finally, during one of the female bows, the male touches the back of its mate's neck with its beak. The female slides to the ground and the male mounts her back, steadying himself occasionally by pinching the female's neck with his beak and pressing his flippers against her sides. Staying on the back of the female is an awkward task for a male, and to copulate he must not fall off until during the act of intromission and ejaculation. Following copulation, they separate from congeners and sometimes stand chest to chest or lie flipper to flipper. Egg-laying occurs amid bows; once the egg is laid the first duets are heard, while the birds adopt a position we would

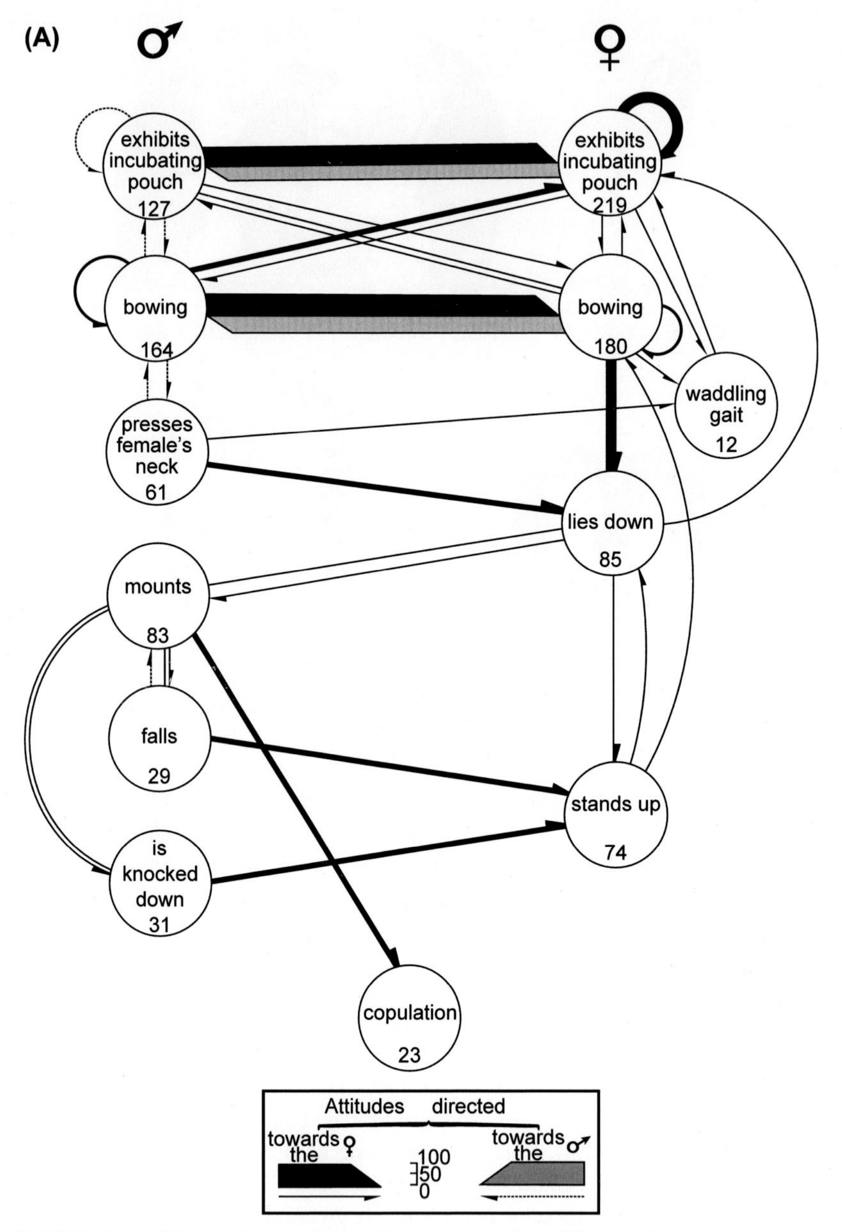


**FIGURE 2.11** Display frequencies in Emperor penguins: (A) preceding copulation and (B) resulting in relief of a partner during incubation. *From Guillotin, M., Jouventin, P., 1979. La parade nuptiale du manchot empereur Aptenodytes forsteri et sa signification biologique. Biologie du comportement 4, 249–267.*

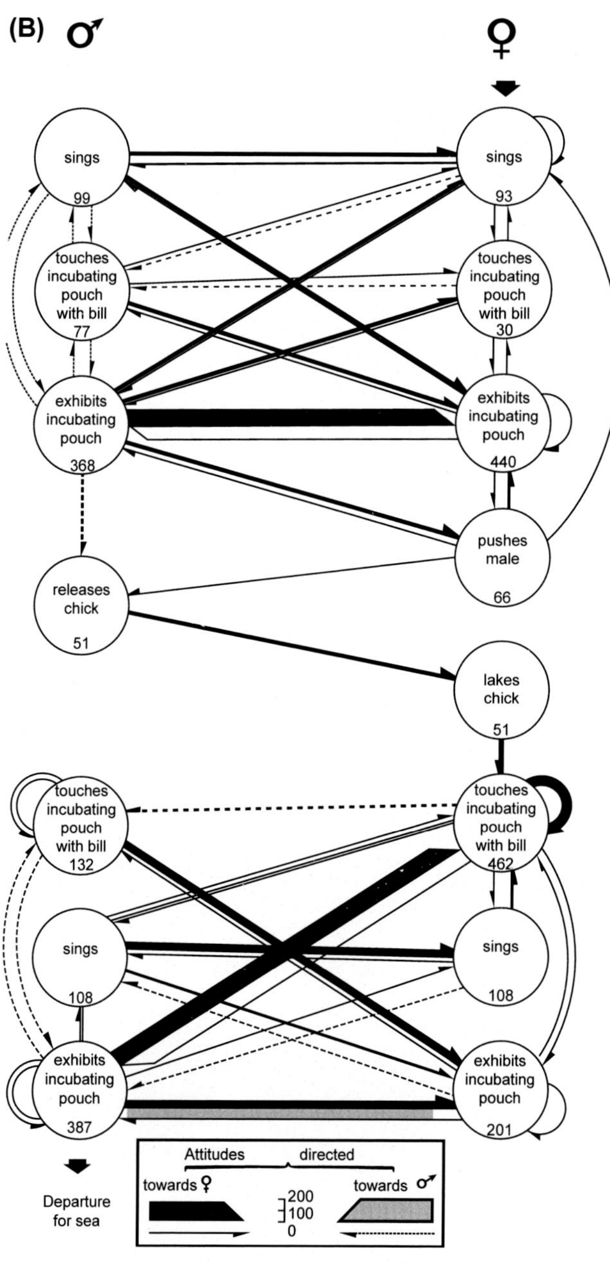


FIGURE 2.11 cont'd

call the "mutual display," the posture is more or less the same as when singing alone. Later on, during egg relief (quantitative analysis of behaviors during relief in Fig. 2.11B), the process is repeated, but omitting "face to face" and the "waddling gait." There remains courtship song, partner encounter, antiphonal duets, and bowing while exhibiting egg or chick (Fig. 2.12).

The same sequence exists in the King penguin (Fig. 2.13) but with two additional elements. Stonehouse (1960) described the sequence as the "advertisement posture." Although in the King penguin the head is lifted to sing, it is equivalent to the "courtship song" display in the Emperor. The "high pointing" is equivalent to "face to face," the "attraction walk" to "waddling gait," and "dabbling" to "bowing." "Beak-shaking" ("head-flagging" of Stonehouse) is present in both species but should be considered a displacement, not a ritualized display. The two new elements specific to the King penguin's display, absent in Emperor penguins, are "mutual preening" and "beak-clapping" during precopulation bows (Fig. 2.14).

**FIGURE 2.12** Lacking a nest location as a rendezvous site to help with meeting, each Emperor penguin must memorize the song of the partner and chick (and the chick must memorize the songs of both parents) to find each other in the colony, solely by the song and among thousands of birds.

**FIGURE 2.13** Courtship displays of King penguins (*Aptenodytes patagonicus*).

**FIGURE 2.14** On the beach, copulation occurs among the crowd in the King penguin. The male climbs on the back of the female with difficulty in the two species of large penguins.

## INTERPRETATION OF SEXUAL POSTURES

### Genus *Spheniscus*

In the banded penguins, the "ecstatic display" is mostly adopted by bachelors searching for a mate. Its function seems to be more similar to that of the "courtship song" in *Aptenodytes* than "ecstatic display" in *Pygoscelis*, for it is often performed outside the nest in banded penguins.

The motionless "turn-around and stare" recalls the appeasement gestures in crested penguins ("slender walk") and the Adélie penguin ("oblique stare bow"). They are observed only during pair formation, and potential partners often separate afterward. They contribute to mate acceptance and play a role similar to the "face to face" in the Emperor penguin. All these species use these displays when carefully choosing their sexual partner. They examine the partner's head pattern, indicating the species, sexual maturity, and body condition (as we will see in Chapter 3 on colored ornaments). The so-called "automaton" walk corresponds to the "waddling gait" in large penguins, and this too most likely serves to facilitate visual contact between newly associated partners. Sometimes during short pauses in the walk the birds stand "face to face," like Emperor penguins, but much more briefly. Walking and standing together probably strengthens the pair bond. Beak-rubbing ("knife-sharpening") occurs during or precedes copulation. These behavioral patterns in banded penguins likely promote sexual stimulation of the female, that is, mate synchronization at the beginning of the reproductive cycle. Bows with beak-clapping in the King penguin serve the same function, an interpretation confirmed by hormonal cycles of the sexes during nuptial displays (Jouventin and Mauget, 1996). The male is more active, while the female often becomes receptive only after several bouts of rubbing or clapping of the beak. Such signals are probably necessary because the birds copulate outside a territory in the large-sized and the banded penguins, and a prostrate female attracts other males that interfere with copulation. This stimulating behavior occurs frequently in Emperor and Jackass penguins but was never observed in the highly territorial Adélie penguins.

Reliefs of mates for banded penguins on the nest, during alternations of nest tending and foraging at sea, are preceded by repeated bows, which must have an appeasement function in this highly territorial genus. Mutual preening is frequent inside or outside of burrows. As in all birds in which it occurs, allo-preening serves to reduce partner aggressiveness and hence reinforces the pair bond. Finally, mutual display accompanied by song promotes individual recognition, as in other penguin species. Mutual display probably contributes to maintaining the pair bond, for it also occurs in other circumstances, such as when greeting an arriving partner.

### Genus *Eudyptes*

For many years, the only general interpretation of breeding behavior of crested penguins was that of Warham (1975, p. 263). According to him, the communication system includes the following: (1) postures apparently intended to convey the displaying bird's desire to avoid conflict and to lessen aggressiveness, for example the slender walk and mutual allo-preening; (2) loud songs, indicating an intention to resist aggression or trespass and to maintain the bird's territorial integrity via threat displays; (3) two-way recognition signals between parent and chick and between breeding partners via bowing, singing, and trumpeting; (4) the headshake, a sign that a communicatory phase is over; (5) bowing, mutual display, and mutual trumpeting, activities whose prime functions seem to contribute to pair bond maintenance and to emphasize the idiosyncrasies of one mated bird to its partner; and (6) nest site advertising via vertical head–swinging.

With regard to point (1) above, the "slender walk," with an agonistic tendency, is generally adopted by brooders of either sex, whereas "mutual allo-preening" has a sexual motivation and is executed by partners. This distinction is not absolute, since young females sometimes show "slender walk" when searching for a mate, and "mutual allo-preening" may occur between parents and chicks. At the rim of the nest "slender walk" imperceptibly grades into the "shoulders-hunched posture," which seems to be the last stage of the same basic behavior pattern. As in most other species, bowing and mutual preening tend to reduce aggressiveness, as they occur with "slender walk" outside of the territory and "shoulders-hunched posture" at the entrance of the territory. Mutual allo-preening occurs between partners or between parents and chicks near or on the nest.

Point (2) above was studied by Warham (1975), who interpreted it as an agonistic signal. For point (4) above, a headshake may not be a coda in a communication bout. The primary function of "beak-shaking" is getting rid of salt solution and this certainly occurs during "head shaking." It may also occur as a displacement activity in conflict situations, but there is no indication of a communicative function in the crested penguins. Point (6) agrees with Jouventin's (1982) interpretation of "vertical head–swinging" as being executed by bachelors on their nests. It likely indicates territoriality directed at other males, as it is contagious for other males. It seems equivalent to the ecstatic display of the Adélie penguin, which serves to repel males and attract females. It should be noted that in Rockhopper as well as Adélie penguins, vertical head–swinging is performed only by males and not, as in other species of these two genera, by both sexes. In Rockhopper penguins, therefore, the ecstatic display, or "vertical head–swinging," or "male display" behavior has a more specialized function than in other crested penguins. Interpretations of behaviors in points (3) and (5) are the most delicate. "Bowing" is the first phase of many mutual displays. We agree with Warham (1975, p. 250) that when it is performed by birds alone on their nest, it should be considered as an incomplete display and out of context. In the Emperor, King, and Gentoo penguins it comprises a series of almost identical bows. But bowing is more variable in the Adélie and crested penguins, depending on the context: (1)

during pair formation ("oblique stare bow" of Adélie penguins and "shoulders-hunched posture" of crested penguins); (2) before and after copulation; (3) during nest reliefs; and (4) after quarrels. Its function, as in the other species, is to reduce the partner's or chick's aggressiveness on the nest during important events. Bowing is observed (Jouventin, 1982) in 75% (n = 96) of the cases when a partner returns to the nest after disturbances and in 95% of the cases when the partner returns normally from the sea (n = 20). The presence of a bird on the nest (chick or partner) is crucial since the frequency of bows falls drastically, within 10 days if the newly emancipated chick has left the nest before the parent returns. "Mutual display," "mutual forward trumpeting," and "vertical trumpeting," were grouped by Warham in the category of mutual displays, and they seem to be merely variants of the same gesture. Bowing, the first part of mutual display, has an appeasement function, while the last phase, alternate headshaking and singing, reinforces the pair bond in such territorial species as the crested and Adélie penguins. Mutual displays also play a fundamental role in interpartner and parent–chick recognition.

## Genus *Megadyptes*

In Yellow-eyed penguins, Richdale's (1951) "full trumpet" is often adopted by solitary birds, mostly males outside of the nest and searching for a breeding partner. It is an ecstatic display having a sexual but no territorial function, just as in *Aptenodytes* and banded penguins. The "salute" is shown mostly by males approaching a female. It consists of two successive postures and resembles the "gawky attitude," to which the other bird responds. Characterizing the first interpartner contacts, it seems to be equivalent to the "turn-around stare" of banded penguins or the "shoulders-hunched posture" of crested penguins. Its appeasement function and head ornaments examination is evident. The "half trumpet," "ecstatic," and "welcome ceremony" must be variants of the mutual display, which is performed only by pairs during nest reliefs. This display is very similar to that of crested penguins, but the sequence is reversed: first the birds sing standing upright, then they bow down. As in crested penguins, it promotes individual recognition, reinforces the pair bond, and perhaps reduces aggressiveness. Mutual preening has the same appeasing role as in other penguin species.

## Genus *Eudyptula*

The so-called "solitary singing" of Little penguins is an ecstatic display performed at the burrow entrance and is followed by pair formation. As in Adélie penguins or song-birds, it repels same-sex individuals and attracts potential partners. Mutual displays generally occur near the nest to facilitate individual recognition and strengthen the pair bond. They also have a territorial function, for in playback experiments pairs will come out of their burrows and respond by mutual displays as often as the experimenter desires. Once again, mutual preening inhibits interpartner and parent–chick aggressiveness.

## Genus *Pygoscelis*

As shown by Penney (1968), Ainley (1975), and Spurr (1975) for the Adélie penguin, the "ecstatic display" signals to receptive females that a nest-owning male is searching for a mate. Sladen (1958) after observing birds of known sex confirmed this interpretation.

Bows ("bowing") appear when pairs are formed ("oblique stare bow," Penney, 1968), just before or after copulation, during partner encounters on the nest and following disturbances or quarrels. They become less frequent as the pair grows better acquainted, this gesture evidently playing a role in lessening partner aggressiveness. Bowing occurs much more rarely out of context than do ecstatic or mutual displays. "Handling" of nest-building materials (Roberts, 1940, p. 218) should rather be interpreted as displacement behavior. This may sound surprising, as this type of behavior might be thought of as "offerings to the mate" in penguin courtship displays. However, the behavior seems too inefficient to be nest-building and too infrequent to be a ritualized display. Bowing associated with "handling" is very rare and does not seem to have a specific function.

"Mutual display" is not present at the beginning of pair formation but generally occurs between well-acquainted partners or between parents and chicks. Its role as an appeasement ceremony (as proposed by Roberts, 1940, p. 214) is minor or even absent according to Sladen (1958, p. 74). Roberts (1940, p. 219) claimed that this display furthers sexual stimulation of partners, but then as such it cannot apply between parents and chicks. Moreover, it is not followed by copulation. On the other hand, its importance in individual vocal recognition, as shown by Sladen (1958), is undeniable. This was confirmed experimentally by Penney (1968, p. 121) by shifting incubating birds from one nest to another.

Another experiment that demonstrated that the mutual display is important to partner recognition was accomplished by preventing birds from singing by taping the beak closed, thus showing that identification was only vocal (Jouventin and Roux, 1979). Further, in the study, altering the physical appearance of a partner without apparent behavioral response of the second partner suggested that identification was not visual. These primary results reveal the importance of vocal signals for individual recognition. Our observations of many penguin species indicate that this conclusion can be extended to all penguins.

Sladen (1958) used the terms "confirmation" and "reinforcement and reassurance of individual recognition" to describe possible complementary behavioral functions of the mutual display besides simple recognition. The Adélie penguin's mutual display can be divided into three parts: first a bow, followed by some kind of frontal attack, and then greeting songs and movements. It is possible to assign an appeasement function to the first part while the second part, involving beak thrusting of one bird toward the other, has a dominant aggressive connotation, as shown by ruffled crests and off-center pupils. "Two-birds-in-the-same-nest" reflects a tense situation. This last stage serves to reinforce the pair bond, as well as mate recognition: it is very noisy and may be repeated several times. Thus, it seems that mutual display contributes first of all to recognition, as shown by singing, secondly to pair-bond reinforcement (hence repetition) and thirdly, instead of bowing, to appeasement (it has the same function as mutual preening in other species, but mutual preening is absent in Adélie penguins).

The Chinstrap penguin's displays are more similar to those of the Adélie than the Gentoo penguin. The Chinstrap penguin's display, however, is less varied. The "ecstatic display" is adopted by both sexes, indicating a less specialized function than it has in the Adélie penguin. In the Gentoo penguin, "ecstatic" and "mutual displays" are almost identical. They are extremely simple: the bird stands up, raises its beak, and utters a song. "Ecstatic displays" are performed only by males alone on their nests. Since the display occurs during the phase of attraction of mating partners, its role is similar to the "ecstatic" of the Adélie penguin. "Mutual display," on the other hand, is executed by pairs. It appears most commonly after egg-laying and continues between parents and chicks. The part it plays in individual recognition is clearly evident: a partner approaching the nest utters the "mutual display" song and is immediately recognized by its mate, who responds vocally even if still out of sight. Its functions seem identical to those of "mutual display" in the Adélie penguin.

"Bowing" is much more frequent in Gentoo penguins than in Adélie and Chinstrap penguins and is exhibited each time partners meet on the nest. At that moment, the arriving bird may sometimes drop a blade of grass. This is apparently a displacement activity, similar to when an Adélie penguin picks up pebbles. van Zinderen Bakker (1971) attributed the same functions as we did to the first two behaviors, but provided none for "bowing," although in this species its meaning seems at least as evident as in the other two. It is clearly an appeasement posture and its frequency, at up to 20 repetitions, shows that on the whole it takes the place of "mutual preening" in other species.

## Genus *Aptenodytes*

To carefully interpret the biological meaning of breeding behaviors, Guillotin and Jouventin (1979) observed and quantified the behaviors of Emperor penguins during their complete breeding season. During the 9-month period of reproduction, behavior patterns were recorded each day during 15 min focal samples, as shown in Fig. 2.15.

The "courtship song" of the Emperor penguin (and the equivalent "advertisement posture" in the King penguin) corresponds to the "ecstatic display" of the Adélie penguin. But in the two large penguins, the courtship song may be uttered by both sexes. It is frequent at the beginning of the cycle, disappears in the interval between pair formation and egg-laying, and then

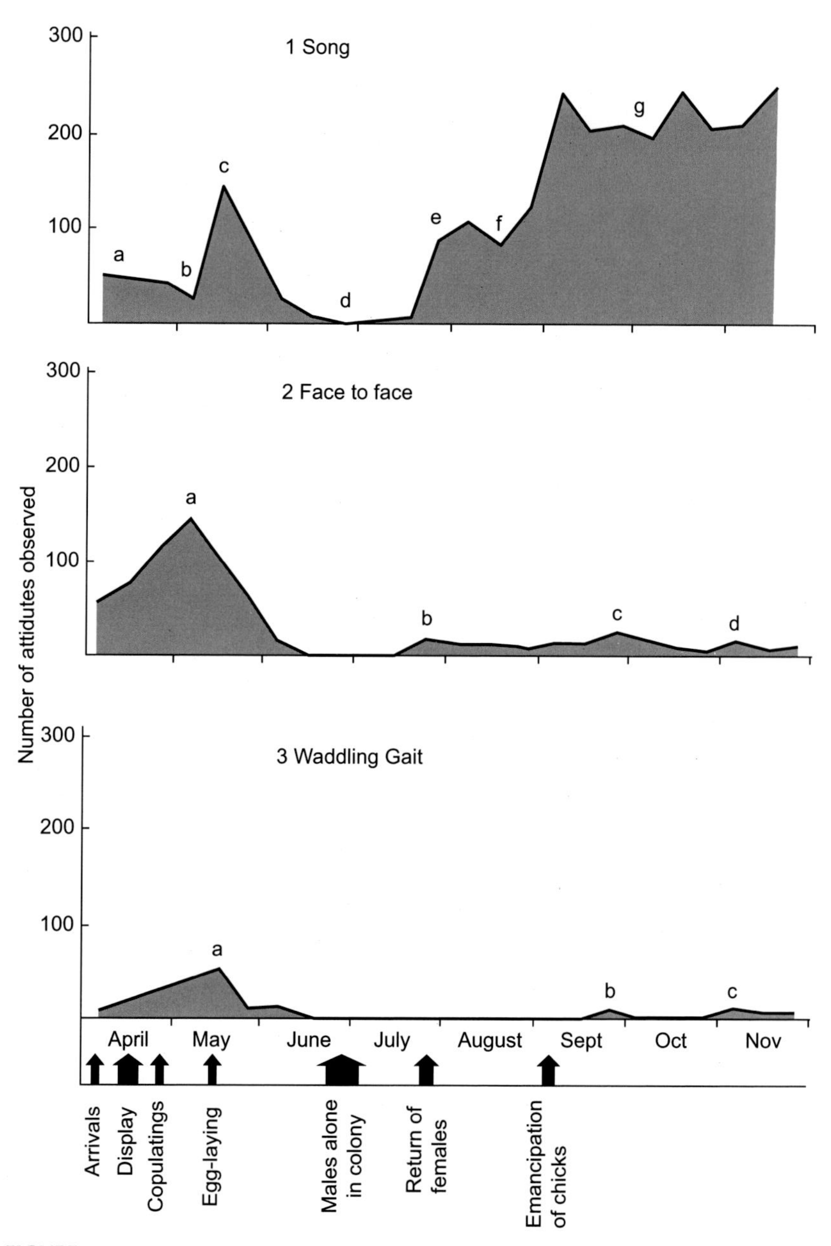


**FIGURE 2.15** Frequency of displays of Emperor penguins during the annual cycle of reproduction, summed over 15 min samples of focal observation: 1) song, 2) "face to face," 3) "waddling gait," 4) "bowing," 5) horizontal head—circling motion, and 6) a sum of all 12 behavior patterns ("attitudes"). Each point on the graphs represents a period of 10 days of observation, 15 min each day, for a sample of 40 penguins. *From Guillotin, M., Jouventin, P., 1979. La parade nuptiale du manchot empereur Aptenodytes forsteri et sa signification biologique. Biologie du comportement 4, 249–267.*

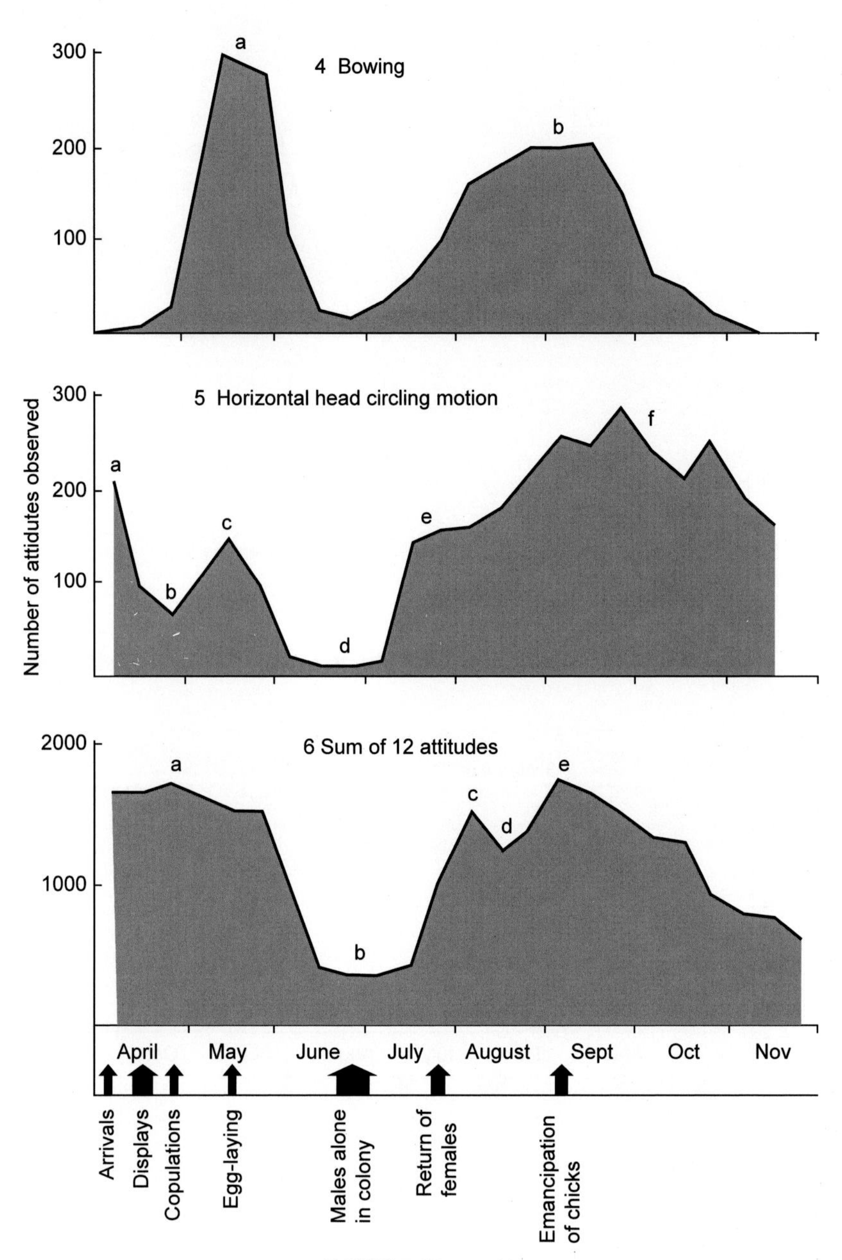


FIGURE 2.15 cont'd

reappears. The "courtship song," particularly in the nonterritorial Emperor, merely serves to attract potential mates and is not a territorial assertion, as in the Adélie penguin. "Face to face" in the Emperor and "high pointing" in the King penguin have no exact equivalents in the *Pygoscelis* species. After the "face to face" behavior pattern, the pair either unites or goes their separate ways. This display plays an important role in the choice of a mate as well as in reinforcing the pair bond, since it is present throughout the breeding cycle.

The "waddling gait" of the Emperor penguin is much more pronounced in the male, who most often leads the walk (this is called the "attraction walk" in the King penguin). The waddling gait usually follows the behaviors described above, and it occurs whenever the pair moves about as a couple. The birds are not inside a territory and hardly look at each other, and this special walk disappears after egg-laying. However, since pairs move about a lot and rarely sing just after their formation, it serves to distinguish partners from normal-walking neighbors and helps to keep them within sight of on another. If pairs are separated experimentally or naturally at the onset of a cycle, they almost always reunite if visual contact has been maintained. If visual contact is broken, couples can reunite if they have learned the song of the partner, as we will see in later chapters. Both Emperor and King penguins lack a territory during the mating period, so trios are frequent in both species when a supernumerary partner is attracted by the mating song (Isenmann and Jouventin, 1970; Jouventin, 1971b; Keddar et al., 2013). Trios are usually composed of two females and a male in Emperor penguins, and two males and a female in King penguins. Trios can last until a pair is formed and the supernumerary partner gives up the competition or is driven off.

The "face to face" posture of Emperor and King penguins reappears only from time to time after mating, but "bowing" ("dabbling" in the King penguin), which precedes copulation and egg or chick reliefs, gets more and more frequent. "Bowing" in large penguins cannot have the same function as in Pygoscelid species, since it appears only during egg-laying, and by this time the pair is firmly established. It does not primarily play an appeasement role; this could only be secondary in this genus because pair formation (especially in Emperor penguins) occurs on the edge of the breeding colony, and thus the "standing place" of the partners that are caring for an egg or young chick (the "territory" of *Aptenodytes* species). Both species of large penguins carry their egg or young chick on their feet, and thus are more mobile than nesting penguins. As well, the character of the territory is different in these mobile breeding species. In Emperor penguins, the territory is a personal space that moves frequently as the breeders stand with eggs or young chicks on their feet and jockey for sheltered space from the wind behind other individuals (Figs. 1.8 and 1.30). Territory is also a mobile personal space in King penguins, but if they are not disturbed by flooding or avian predators, they move about relatively little, usually only a few meters, during incubation and brooding (Lengagne et al., 1999a). Bowing reveals the brood patch, a pouch that is empty before copulation and contains egg or young chick later on. When a partner is relieved in the breeding colony, the waiting partner does not immediately respond with the egg transfer. The approaching partner bows, stops, and then resumes bowing. When the less active bird (usually the female during copulation and the brooder during later relief exchanges of egg or chick) finally bows back, the interval between bows gradually decreases (they can even follow uninterrupted) and copulation or relief takes place. Evidently, bowing plays an important role in female hormonal stimulation (Jouventin and Mauget 1996) and partner synchronization, necessary in birds where the egg or chick is in danger of rolling off the brooder's feet. When executed during duets, its synchronizing function complements song function in individual recognition.

The Emperor penguin's mutual display (or antiphonal duet) first occurs after egg-laying, and subsequently just before a departure from the breeding colony. Its contribution to appeasement seems minor or nonexistent as the performers are well acquainted. Nor does this display promote sexual stimulation, for duets occur only after egg-laying, and also occur between parents and chicks. On the other hand, its involvement in individual vocal recognition has been confirmed by observations and investigations by many authors, and this may be its only function. This effect is fundamental in the Emperor penguin, as it lacks a true territory and therefore any other clues for locating its partner. The problem will be reviewed in more detail in the chapter on vocal signals. Only the King penguin, territorial and hence more aggressive than the Emperor, shows mutual preening, however infrequently.

## SIGNIFICANCE OF SEXUAL POSTURES

If all sexual postures of the penguin species are to be classified, the system of Roberts (1940, p. 218) is the simplest. He distinguishes five categories: (1) offering nest-building materials ("handling of nest materials"); (2) "bowing"; (3) reliefs ("nest-relief-ceremony"); (4) mutual display ("mutual epigamic display"); and (5) ecstatic displays ("ecstatic attitude"). The first three categories may not be accurate. First, the communication value of handling nest materials, although the behavior is well known, has not been demonstrated. Second, bowing is absent in several penguin species. Finally, reliefs during incubation and brooding represent situations, not postures, though these events may give rise to mutual displays. The last two categories however are common to all penguin species. We shall then consider three groups (Table 2.2): ecstatic displays, mutual displays, and the rest. Table 2.3 gives a classification according to possible functions of the behavior patterns by sex and Table 2.4 give the names of all ritualized behaviors in all of the species (Jouventin, 1982).

**TABLE 2.2 Classification of Postures Described by Different Authors, Using the System of Roberts (1940)**

<table>
<tr><td>Aptenodytes patagonicus,<br>Pygoscelis papua,<br>Pygoscelis antarctica,<br>Spheniscus demersus,<br>Spheniscus magellanicus,<br>Eudyptes chrysocome<br>(Roberts, 1940)</td><td>Ecstatic attitude (5) (Ecstatic displays)</td><td>Mutual epigamic display (4) (Mutual displays)</td><td>• Handling of nest materials (1)<br>• Bowing (2)<br>• Nest-relief ceremony (3)</td></tr>
<tr><td>Aptenodytes forsteri<br>(Prevost, 1961)</td><td>Courtship song</td><td>Duo</td><td>• Face to face<br>• Waddling gait</td></tr>
<tr><td>A. patagonicus<br>(Stonehouse, 1960)</td><td colspan="2">Advertisement posture</td><td>• Dabbling<br>• High pointing<br>• Attraction walk</td></tr>
<tr><td>P. papua (van Zinderen Bakker, 1971)</td><td colspan="2">Ecstatic display</td><td>• Bowing</td></tr>
<tr><td>Pygoscelis adeliae<br>(Spurr, 1975)</td><td>Ecstatic display</td><td>Loud and quiet mutual display</td><td>• Sideways stare bow<br>• Bow display</td></tr>
<tr><td>S. demersus (Jouventin, 1982; Eggleton and Siegfried, 1979)</td><td>Ecstatic display</td><td>Mutual display</td><td>• Turn-around stare<br>• Automaton<br>• Face to face<br>• Bowing<br>• Mutual preening</td></tr>
<tr><td>E. chrysocome<br>(Warham, 1963)</td><td>Male display</td><td>Mutual display trumpeting</td><td>• Shoulders-hunched posture<br>• Bowing<br>• Mutual preening</td></tr>
<tr><td>Eudyptes chrysolophus<br>(Downes et al., 1959)</td><td colspan="2">Ecstatic display</td><td>• Pointing to the nest</td></tr>
<tr><td rowspan="3">Megadyptes antipodes<br>(Richdale, 1951)</td><td rowspan="3">Full trumpet</td><td>Welcome ceremony</td><td rowspan="3">• Salute<br>• Gawky attitude<br>• Mutual and kiss preen</td></tr>
<tr><td>Ecstatic</td></tr>
<tr><td>Half trumpet</td></tr>
<tr><td>Eudyptula minor<br>(Kinsky, 1960)</td><td></td><td>Mutual display</td><td>• Mutual preening</td></tr>
</table>

From Jouventin, P., 1982. Visual and vocal signals in penguins, their evolution and adaptive characters. Advances in Ethology 24, 1–149.

**TABLE 2.3** Contexts in Which Ritualized Postures Appear According to Sex

| | Display | Arrival At Empty Nest | Before Pair Formation | After 1st Meeting | Pair Moves Together | Preceding Copulation | Between Reliefs | Coming to Nest | Coming to Feed Chick |
|---|---|---|---|---|---|---|---|---|---|
| *Pygoscelis adélie* | Ecstatic | | M | | | | | | |
| | Stare-bow | MF | | MF | | MF | MF | | |
| | Epigamic | | | | | | | MF | MF |
| *Pygoscelis papua* | Ecstatic | | Mf | | | | | | |
| | Bowing | MF | | MF | | MF | MF | | |
| | Mutual | | | | | | | MF | MF |
| *Eudyptes chrysolophus* and *E.chrysocome* | Head swing | | Mf | | | | | | |
| | Pointing | MF | | MF | | MF | | | |
| | Preening | | | | | | MF | | |
| | Epigamic | | | | | | | MF | MF |
| *Megadyptes antipodes* | Full trumpet | | Mf | | | | | | |
| | Salute/ gawky | MF | | MF | | MF | | | |
| | Preening | | | | | | MF | | |
| | Mutual | | | | | | | MF | MF |
| *Eudyptula minor* | Ecstatic | | Mf? | | | | | | |
| | Preening | | | | | | MF | | |
| | Epigamic | | | | | | | MF | MF |

| | | | | | | | | | |
|---|---|---|---|---|---|---|---|---|---|
| *Spheniscus demersus* | Ecstatic | | Mf? | | | | | | |
| | Face-to-face | | | MF | MF (+auto) | | | | |
| | Bowing | MF | | | | | | | |
| | Preening | | | | | | MF | | |
| | Parade | | | | | | | MF | MF |
| *Aptenodytes fosteri* | Ecstatic | | MF | | | | | | |
| | Face-to-face | | | MF | | | | | |
| | Approach | | | | MF | | | | |
| | Bowing | | | | | MF | MF | | |
| | Parade | | | | | | | MF | MF |
| *Aptenodytes patagonicus* | Ecstatic | | MF | | | | | | |
| | High pointing | | | MF | | | | | |
| | Pair walk | | | | MF | | | | |
| | Dabbling | | | | | MF (+clack) | MF | | |
| | Parade | | | | | | | MF | MF |

**M**, postures of males alone; **MF**, shown by both sexes; **Mf**, mostly by males. All postures are types of displays described in the text by name, preening = mutual preening, **auto** = automaton walk, **clack** = beak clacking.
From Jouventin, P., 1982. Visual and vocal signals in penguins, their evolution and adaptive characters. Advances in Ethology, 24, 1–149.

**TABLE 2.4 Correspondence Between Posture Names Assigned by Different Authors and Our Own Classification**

| Species | First Name Given to Attitude and References | Category of Attitude |
|---|---|---|
| *Pycoscelis adeliae* | Ecstatic display (Wilson, 1907) | Ecstatic display |
| | Oblique stare bow (Penney, 1968) + Bowing (Roberts, 1940) | Bowing |
| | Mutual epigamic display (Roberts, 1940) | Mutual display |
| *Pygoscelis papua* | Ecstatic display (Levick, 1915) | Ecstatic display |
| | Bowing (Roberts, 1940) | Bowing |
| | Ecstatic display (Bagshawe, 1938) | Mutual display |
| *Eudyptes chrysolophus* and *Eudyptes chrysocome* | Head-swinging (Warham, 1971) | Ecstatic display |
| | Pointing to the nest (Downes et al., 1959); + Shoulders-hunched posture (Warham, 1963) | Bowing |
| | Mutual preening (Downes et al., 1959) | Mutual preening |
| | Mutual epigamic display (Roberts, 1940) | Mutual display |
| *Megadyptes antipodes* | Full trumpet (Richdale, 1951) | Ecstatic display |
| | Gawky attitude + Salute (Richdale, 1951) | Bowing |
| | Mutual preen + Kiss preen (Richdale, 1951) | Mutual preening |
| | Welcome ceremony + Ecstatic + Half trumpet (Richdale, 1951) | Mutual display |
| *Eudyptula minor* | | Ecstatic display |
| | Mutual preening (Kinsky, 1960) | Mutual preening |
| | Mutual display (Kinsky, 1960) | Mutual display |
| *Spheniscus demersus* | Ecstastic display (Roberts, 1940); Ecstatic (Eggleton and Siegfried, 1979) | Ecstatic display |
| | Ronde + Face à face + Automate (Jouventin, 1982) | Gait |
| | Vibratory head shake (Eggleton and Siegfried, 1979) | Bowing |
| | Allo-preening (Eggleton and Siegfried, 1979) | Mutual preening |
| | Mutual ecstatic (Eggleton and Siegfried, 1979) | Mutual display |

**TABLE 2.4 Correspondence Between Posture Names Assigned by Different Authors and Our Own Classification—cont'd**

| Species | First Name Given to Attitude and References | Category of Attitude |
|---|---|---|
| *Apenodytes forsteri* | Chant de cour (Prevost, 1961) | Ecstatic display |
| | Face à face (Prevost, 1961) | Face to face |
| | Démarche balancée (Prevost, 1961) | Gait |
| | Courbette (Jouventin, 1971a,b) | Bowing |
| | Parade mutuelle (Jouventin, 1971a,b) | Mutual display |
| *Apenodytes patagonicus* | Advertisement posture (Stonehouse, 1960) | Ecstatic display |
| | High pointing (Stonehouse, 1960) | Face to face |
| | Attraction walk (Stonehouse, 1960) | Gait |
| | Dabbling (Stonehouse, 1960) | Bowing |
| | Parade mutuelle (Jouventin, 1982) | Mutual display |

From Jouventin, P., 1982. Visual and vocal signals in penguins, their evolution and adaptive characters. Advances in Ethology, 24, 1–149.

The prime function of the "ecstatic display" is sex recognition. It is ensured mainly by the song component, as will be shown under "vocal signals," but even posture alone generally indicates a male in several species. In at least four genera of penguins, the ecstatic display also plays a role in territorial assertion. Ecstatic displays form a homogeneous group, as they are very similar in all penguin species except the Emperor penguin: the bird is erect, beak vertical, and flippers half-opened. In territorial species, the bird chooses an open site or the nest to perform the "ecstatic" display, and remains in the position long enough to be located. The nonnesting Emperor penguin also moves to an open site to perform its "ecstatic" display.

The main functions of "mutual displays" are individual recognition of mate or chick and pair-bond reinforcement. The posture is nearly identical with the "ecstatic displays" in the King, Emperor, and Gentoo penguins but different from that of the Adélie penguin. In all penguin species (except obviously Emperor and King penguins) it is performed on or near the nest. It also serves territorial assertion. The "mutual display" category is also highly homogeneous.

The other postures are difficult to classify, for although similar in form they may have different functions (or similar functions may relate to different behavior patterns), and they may not have exact equivalents in all species. All beak movements (e.g., "beak-clapping" in the King, "beak-sharpening" in the Jackass penguin, "beak-rubbing" in most species) precede or accompany copulation and physiologically stimulating the mating partner (Jouventin and Mauget, 1996). All "mutual preening" seems to reduce aggressiveness and to

## TABLE 2.5 Biological Function of Different Sexual Attitudes in Penguins

| Species or Genus | Category of Attitude | Search for Partner | | | Pair Formation | | | Procreation and Chick Breeding | | | |
|---|---|---|---|---|---|---|---|---|---|---|---|
| | | Territorial Assertion | Identification of Sex | Localization of Singers | Reproductive Isolating Mechanism | Body Condition | Visual Contact | Partner Synchroni-zation | Reinforce-ment | Identific-ation | Territorial Assertion |
| *Pygoscelis adélie* | Ecstatic | ■ | ■ | ■ | ■ | | | | | | |
| | Bowing | | | | ■ | ■ | | | ■ | | |
| | Mutual | | | | | | | | ■ | ■ | ■ |
| *Pygoscelis papua* | Ecstatic | ■ | ■ | ■ | ■ | | | | | | |
| | Bowing | | | | ■ | ■ | | | ■ | | |
| | Mutual | | | | | | | | ■ | ■ | ■ |
| *Eudyptes* | Ecstatic | ■ | ■ | ■ | ■ | | | | | | |
| | Bowing | | | | ■ | ■ | | | | | |
| | Preening | | | | | | | | ■ | | |
| | Mutual | | | | | | | | ■ | ■ | ■ |
| *Megadyptes* | Ecstatic | | ■ | ■ | ■ | | | | | | |
| | Bowing | | | | ■ | ■ | | | | | |
| | Preening | | | | | | | | ■ | | |
| | Mutual | | | | | | | | ■ | ■ | ■ |
| *Eudyptula* | Ecstatic | ■ | ■ | ■ | ■ | | | | | | |
| | Preening | | | | | ■ | | | ■ | | |
| | Mutual | | | | | | | | ■ | ■ | ■ |

| | | | | | | | | | | | |
|---|---|---|---|---|---|---|---|---|---|---|---|
| *Spheniscus* | Ecstatic | | | | | | | | | | |
| | Gait | | | | | | | | | | |
| | Bowing | | | | | | | | | | |
| | Preening | | | | | | | | | | |
| | Mutual | | | | | | | | | | |
| *Aptenodytes forsteri* | Ecstatic | | | | | | | | | | |
| | Face to face | | | | | | | | | | |
| | Gait | | | | | | | | | | |
| | Bowing | | | | | | | | | | |
| | Mutual | | | | | | | | | | |
| *Aptenodytes patagonicus* | Ecstatic | | | | | | | | | | |
| | Face to face | | | | | | | | | | |
| | Gait | | | | | | | | | | |
| | Bowing | | | | | | | | | | |
| | Mutual | | | | | | | | | | |

*Shaded boxes = behavior present, open boxes = behavior absent. Ecstatic = Ecstatic display, Preening = Mutual Preening, and Mutual = Mutual display.
From Jouventin, P., 1982. Visual and vocal signals in penguins, their evolution and adaptive characters. Advances in Ethology, 24, 1–149.

favor intermate or parent–chick contacts. They are important in Little (*Eudyptula*), crested (*Eudyptes*), and banded penguins (*Spheniscus*). "Beak-rubbing" and "mutual preening," although apparently unambiguous, are absent in some species. "Mutual preening" displays, which are frequent and original, form a behavioral class of their own.

"Bowing" displays have an appeasement function in the *Pygoscelis* species, crested, and banded penguins. They include the "oblique stare bow" in the Adélie penguin, and the "shoulders-hunched attitude" exhibited in crested penguins during pair formation. They reappear after copulation, during the reliefs of partners at the nest during incubation and brooding, and even after disturbances (though generally in a less ritualized form). In the Yellow-eyed penguin, "bowing" is part of the "salute" and the "ecstatic" display, where its appeasement role is preserved. However, in the large penguins, it serves mainly to synchronize mates, along with beak-clapping in King penguins, and ends in copulation or relief between shifts in the colony. It is infrequent and nonritualized in Little penguins.

There has been some confusion about some behavior patterns in some of the pioneering research works. Richdale (1951) constantly mistook the "ecstatic" for the "mutual display." He considered the "full trumpet" of the Yellow-eyed penguin a displacement activity (p. 212). He failed to distinguish "trumpeting" of the Emperor from the "ecstatic display" of the King penguin (p. 24) and the "ecstatic display" of the Emperor penguin from "bowing." In fact, this confusion shows that the function of a behavior pattern comes before its form. For example, although in the Emperor penguin the "ecstatic" display is almost identical to "bowing," the displays occur at different times, the "ecstatic" being associated with a vocal signal. The "automaton" display of banded penguins and the "gawky attitude" of the Yellow-eyed penguin are very similar behavior patterns to the "shoulders-hunched posture" of crested penguins and probably share the same appeasement function. We would put these behavior patterns in the 'bowing' category. The "waddling gait" of the large penguins and "automaton" display of banded penguins resemble each other in form and function as do the "face to face" and "turn-around stare" behaviors, although to a lesser degree. They should be placed in separate categories because of their originality. The waddling gait of the two large penguins is a surprising adaptation for keeping mates in visual contact when the pairs move about. During this time, the pair cannot sing without attracting new competitors for matings, and these competitors are difficulty to keep away because the large penguins lack the territorial landscape that goes with having a nest.

Thus, it is now possible to bridge the gap between different descriptions (Table 2.4), and to propose a functional classification of all penguin species with respect to their sexual postures (Table 2.5).

Chapter 3

# Experiments on Visual Signals

## WHY PENGUINS HAVE ORNAMENTS

The species of penguins exhibit a diversity of markings and color patterns. In this chapter, we turn to the biological meaning of these markings, especially those that are the most colorful. Such markings help human observers distinguish the different species of penguins, but for the birds themselves they have several special biological meanings. Colorful or even black-and-white markings can be thought of as "ornaments." The term ornament may well be anthropomorphic, as it suggests an aesthetic function. But such characteristics have in fact an adaptive significance for animals because they reflect information about the individuals that carry them. For example, ornaments of birds often indicate species and sexual identity, age, or breeding experience (i.e., juvenile vs. adult) and body condition to a prospective mate (Andersson, 1994). Alternatively, when associated with a stereotyped display, behavioral dominance or aggressiveness may be communicated. From a Darwinian evolutionary viewpoint, this might assist mating, aid in defense of territory, or contribute to the protection of offspring, and thus promote individual fitness. If body ornaments are used in communication or identification among individuals, then the ornaments are signals from one bird to another. But why should the birds signal one another, and what information do their ornaments convey?

To gain answers to questions like these, it is necessary to study the birds in their natural habitats, where the context of the information signaled includes the surrounding social environment of birds that receive the signals and the ecological environment in which communication occurs. Rather than simply speculating about the possible use of ornaments from a human viewpoint, we need to do experiments to understand how and why natural selection (or a form of natural selection, sexual selection) has produced such traits in the field and in natural populations. For example, flowers are beautiful to the human eye, but natural selection has produced flowers as signals to pollinators to come to the flower, thus contributing to the reproduction of the plant. The bright colors of birds may attract our notice, but sexual selection has favored ornamental colors because of their roles in mate choice and competition for matings. Thus, the ornaments that we thrill to see are adaptations that have been favored by

**Why Penguins Communicate. http://dx.doi.org/10.1016/B978-0-12-811178-9.00003-2**

selection, and the natural environment is the key to how such traits produce advantages in survival and reproduction. Studies in the laboratory or in zoos may help us also understand the basic properties of ornaments, such as their chemical composition and development during growth and maturity, how they change over different seasons, the influence of diet and nutrition, and other factors. But in their environmental context, we can ask whether these ornaments seem to be useful to the individuals that carry them. We might also ask how they are used to communicate, a descriptive study of information transfer between the signaling and receiving individuals. If information is shared between or among individuals, the social environment is immediately involved. And we might also investigate the evolution of the ornament, that is, whether it is an adaptation, and the use or function that it serves at the present time. When we try to understand why an ornament has evolved, the context of the social and ecological environment is especially important.

## NATURAL AND SEXUAL SELECTION

As originally envisioned by Charles Darwin (1859), natural selection is a process that modifies the frequency of traits, or characteristics of organisms, over time in a population (Endler, 1986). So when ornaments such as the color patches or black-and-white markings of penguins are observed, we can look to the elements of natural selection to investigate the adaptive use of the ornaments. Within a population, the ornament must be variable among individuals, and not just in terms of the differences that might occur due to variations in nutrition, while individuals are growing up, or accidents that occur along the way. According to the evolutionary theoretical background, differences among individuals must have a genetic basis. This genetic basis is then passed down from parents to offspring so that aspects of ornaments, such as a bright yellow color of the crest feathers on the head, might be associated with greater survival or increased production of offspring. We know from the scientific literature, and particularly from studies of birds, that brighter colors are usually associated with good physical condition. A bird arriving at the breeding grounds will often increase its capacity to rear chicks if it finds a partner in good condition (and that has, say, a brighter crest), and thus the fitness of the bird is increased if it produces more or higher quality offspring. If so, in the natural population those individuals in the best condition will slowly become brighter yellow in the head crest, as offspring replace their parents. This process of replacement reflects the fitness difference that favors the brighter yellow crest color. If we want to know why the brighter ornament is favored, then we can look to the environment to see what factors seem to facilitate the greater success (or fitness) of the brighter ornament. For example, the brighter yellow crest might be associated with fighting ability or general health, and thus facilitate behavioral dominance of the bird in protecting its nest. Finally, the evolutionary advantage of the brighter color might simply be that it is attractive to members of the opposite sex. All of these scenarios contain associations of the trait with behavioral or environmental factors that we can test with experiments. In this way, the penguins provide facts that enable evaluation of our questions about the evolution of traits, and we can sift our correct conjectures from our poor ones.

For penguins, males and females have very similar appearance of their ornaments (a phenomenon called "monomorphism"); so much so that it is difficult or impossible for human observers to tell the males and females apart (Fig. 3.1). This sexual monomorphism differs from the patterning of ornamental traits for most species of animals and particularly birds, as in the songbirds, where males usually exhibit bright ornaments and females have very reduced ornaments or lack ornaments entirely. This latter situation is the most widely studied and known in birds; and it contrasts with the similar ornaments of monomorphic male and female seabirds, about 300 species of

**FIGURE 3.1** (A) A pair of Northern rockhopper penguins, showing the similarity in appearance of the male and female ornaments. (B) A group of King penguins, showing that it is not possible to visually distinguish the sexes, the ornaments of all adults having a similar appearance.

them, poorly studied and thus poorly understood (but see the review by Kraaijeveld et al., 2007). In the better studied cases in birds, where males and females exhibit different appearances (i.e., where males are often brightly colored and females are drab), it is hard to see how the sexes could be similar in their ecological niche. To explain differences in ornaments of males and females in general and across many groups of species, Charles Darwin suggested a special kind of natural selection that reflects the differences in social environment for the sexes. He called this sexual selection (Darwin, 1871), and it arises from competition for reproductive advantages. One sex, usually males in songbirds, has fewer chances to produce young than the other sex. This can be because of an unbalanced adult sex ratio or because a few males mate with many females, leaving many other males without mates.

When mating opportunities for males and females are unbalanced, there are two ways in which males can improve their chances of mating with as many females as possible. One way is to aggressively defend mating opportunities by fighting with other males and excluding them from the competition for matings. This process may be widespread in mammals, where males are often larger and have more developed aggressive armaments than females, like antlers or horns. The alternative is to attract the notice of females by showing them brightly colored ornaments that are physiologically difficult to produce and impossible to develop if an individual is in bad health or inefficient at finding food. The classic example of this is the peacock's tail, brightly colored with iridescent blues, greens and shades of gold, long and lustrous; and much different than the relatively short dull tail of the peahen. Our human idea may be that the tail of the peacock is a beautiful showy ornament, but in fact it is a signal that reveals to females the good body condition or perhaps the genetic quality of the male. The most ornamented peacock attracts most of the peahens and does most of the mating, and he thus produces more descendent offspring than other males (Petrie and Halliday, 1994). Offspring of the peacock and peahen inherit the genetic basis of the father's glorious tail, and the genetic basis for the mother's preference for the showy tail (Fisher, 1934). The ornament of the showy tail evolves via sexual selection for both the ornament itself (in the sender of the signal) and the preference for it (in the receiver of the signal). These traits in males and females are genetically associated with each other and get passed down through the generations together. The male trait and the female preference (also a trait) are genetically linked. Because the two traits are more associated than random, we say that the traits exhibit linkage disequilibrium.

The story of sexual selection in songbirds and peacocks leaves a basic question unanswered. In species like penguins, males and females look very similar, so similar that we cannot tell the sexes apart by visual inspection. Does this mean that their ornaments have some other use besides mate attraction? Not necessarily. Over 100 years ago, Julian Huxley (1914) pointed out that in many species of waterbirds, both males and females provide parental care. In fact, offspring will not survive unless both parents provide for them. In such cases, finding a good mate that is a cooperative and successful partner, or even a superior partner, might be important to both sexes. Thus, sexual selection in some species may become a mutual process of the sexes. An alternative is that both sexes benefit from an ornament in an environmental context that is not about competition for reproductive advantages in mating with the opposite sex. A less likely alternative is that different or alternate environmental contexts are favoring the same elaboration of the trait, thus producing sexual monomorphism. Of course, regardless of the aspect of the environment that is favoring selection for the trait, we must first test for the use of a potential ornament by showing that it has signal content that is acted upon by other individuals. Some apparent ornaments of birds appear to be little used as signals, and the best evidence of signal reception in the wild is a change in behavior of the receiver individual. Thus, by examining the body condition or other advantages of mates and the context in which ornaments convey information in nature, it should be possible to test hypotheses of mutual sexual selection (that is, the ornament is used during mate choice by both sexes) and nonsexual natural selection (the ornament signals information during other times than when mates are chosen, and similarly in both sexes or not).

## SEXUAL SELECTION IN PENGUINS

The process of finding a mate is complex and has several parts. Initially, it is important to ensure that a member of the same species is selected. Because penguins come in different sizes and appearances, this may not seem difficult. But we have to keep in mind that an oceanic bird like a penguin can easily travel far from its feeding ground, thus increasing the danger of interbreeding with another, perhaps similar species of penguin. The ecological niches of penguins are differentiated by water mass, different ocean regions differing greatly in physical characteristics such as temperature and salinity. Nonetheless, southern oceans lack strong boundaries, and thus a penguin will occasionally arrive at a relatively new island to its species, as we recorded in several places. Such individuals may not be the best mates for penguin species that already breed on the island. Thus, after speciation has occurred, any hybrid offspring may not be well adapted to local ecological niches. Over evolutionary time, the importance of finding a mating partner that can produce viable or even superior offspring may have produced the differences in sexual ornaments among species that now seem so obvious.

Next, only members of the opposite sex are suitable as successful mating partners. Penguins are notorious among zoological gardens for pairs of males that have raised young (Zuk, 2006), and if human observers have trouble differentiating the sexes, perhaps it might not be easy for the penguins either. We need to understand how they are able to differentiate their congeners from other species, as well as how they differentiate sexes and why they sometimes behaviorally pair with same sex, though without producing offspring (Pincemy et al., 2010), and how they choose the best mate, using their ornaments.

Members of the correct species, and of the opposite sex, must also have suitable age and experience. Colorful ornaments often indicate breeding status in birds (Hawkins et al., 2012).

Penguins, like many birds, have plumage development such that ornaments do not mature until individuals are ready to breed (Nicolaus et al., 2007). In all species of penguins, as in the majority of birds, juveniles have either no colored ornaments or a different pattern of black-and-white feathers from the adults (Fig. 3.2). This allows breeding birds to exclude juveniles from nuptial displays, as we observed in the field. In addition, among individuals with breeding plumage, first-time breeders are often unsuccessful and appear to be best avoided if possible. These birds often have weakly developed breeding plumage. Finally, potential mating partners may differ in quality; either by being more experienced at breeding or by having superior body condition, and thus may be preferred as mates. Body condition may reflect stored resources, fishing and diving capacities, greater immunocompetence, or other beneficial aspects of individuals at prime breeding age. These characteristics are often associated with strong development of colored ornaments, so the degree of ornament elaboration may be a signal of condition to a potential mate.

Further aspects of signaling aid our understanding of information exchange. The first is that attraction occurs over a variety of distances. On display grounds, penguins are colonial birds that live in impressive rookeries

**FIGURE 3.2** An immature King penguin, showing the absence of coloration of the beak, auricular feather patch, and breast feather patch. Adults in the background provide comparison.

that contain several thousands of breeding birds. Two species, Emperor and King penguins, lack a nest and stand with their single egg on their feet. Thus, these parents can move about the breeding colony over time, short distances for King penguins (Lengagne et al., 1999a) and much longer distances for Emperor penguins (Jouventin et al., 1979). Finding and keeping track of a good mate of the same species, of the opposite sex, at an appropriate physiological state for breeding, and in good body condition may be extremely difficult in a crowd of thousands. In territorial birds, such as almost all seabirds and most of the penguins, the unpaired female can easily evaluate the breeding ability of males and likely nesting success by searching throughout the colony for a good nest site (reviewed by Rowley, 1983). If a displaying male has a nest, and moreover is in the center of the colony, thus protected from egg-predators as Skuas (*Stercorarius maccormicki* on the Antarctic continent or *S. lonnbergi* on sub-Antarctic islands), and if the nest (stones or piled seaweed) is big, it may indicate that this male is a "good match." Helped by this physical measure of the male's capacity to reproduce, females should consequently initiate mutual displays, then pair, and finally mate and breed with the male. But in nonnesting penguins, there is no nest to give an indication of which male might be the best possible and most cooperative mating partner, or even just an acceptable partner. While many suitable males may be present, overload of information and a crowded environment may make discrimination difficult when the topographical cues given by the nest are lacking. After arriving from the sea, an individual must first approach an area where several birds are displaying, whether an open beach, sea ice, or rocky cliff. The display ground must be identified from a distance. Once a bird is close to other potential mates, identification of the attributes of birds of the same species, opposite sex, and high quality becomes important.

Another aspect of signaling is that ornaments can be time dependent. Feather ornaments often best reflect the condition of the bird at or just before the time of molt, especially if they result from pigments that were deposited in developing feathers. Other potential ornaments, such as skin, beak, and feet colors can reflect the more immediate condition of birds, particularly if these tissues are vascularized. Each species of penguin not only has a variety of possible ornaments, with colored feather patches and yellow head plumes perhaps the most likely, but also including black-and-white markings such as breast bands. Of these, colored patches seem the strongest candidates as ornaments because their colors depend on dietary pigments that may be difficult to obtain, and should be good indicators of fishing abilities and thus the capacity to provide food resources to chicks.

## THE ROLE OF ORNAMENTS IN MATING

Penguins are easy to lure and capture as part of a research program. First, penguins have relatively small brains compared to mammals. Thus, many of their behaviors are innate (i.e., genetically programmed) rather than learned.

Penguins are numerous in the field and it is possible to acquire large samples for measurements and tests. This is important for statistical analyses. Penguins are easy to catch and to manipulate because they are unable to escape via flight. Consequently, they can be followed in their natural habitats. Of course, penguins are strongly constrained by their original mode of life as diving predators at sea and colony nesters on land. And finally, most penguins are relatively unafraid of researchers, since they live on islands that are free of mammalian predators. Thus, they behave naturally if observed from a short distance of several meters. This means that penguins are about as easy to observe in their natural habitats as they are in captivity, but without the biases from captivity that can mask the influences of natural selection.

For the most part, penguins are territorial nesters, and thus good models for most birds and all other seabirds that share the nesting mode of reproduction. What's more, for one genus with two species, they are good models of non-nesting and nonterritorial birds, although these modes of life are unique among bird species. Nevertheless, the resulting variation in behavioral adaptations to alternative nesting habits provides extremely interesting cases for investigation of an extreme way of life. We can use this difference as a "natural experiment" to investigate the use of the nest by comparison to the species that don't have one, something unique to penguins, since nearly all birds use nests for breeding. To study animal behavior objectively, of course, we have to apply relevant experiments that are designed to pose specific questions, and by inference understand and interpret the behavioral responses that we observe.

Preliminary experiments on optical signals in penguins were first conducted by Stonehouse (1960) when he modified the head pattern of two unmated King penguins, painting over and thus masking their auricular orange patches (i.e., orange ear patches). The sample of tested birds was too small to be other than anecdotal, but it suggested that one bird was unable to find a partner, while the other mated after the paint fell off. In Adélieland, Jouventin (1982) tried the same with Adélie penguins. The results, however, were more ambiguous. Partners apparently recognized their last year's mate from the vocal signature of their song (see Chapter 5). So the visual mechanism of mating seems to be most important during the first partner encounter and is then supplanted by identification of a vocal signature.

At the Crozet Archipelago, Jouventin (1982, p. 96) increased the sample of manipulated individuals and of species by capturing 200 couples of three penguin species (two species that nest, Rockhopper and Macaroni penguins, and one species that carries its egg on its feet, the King penguin). The crested penguins had their yellow crest feathers experimentally removed. This experiment was performed just before the mating season and the previous mate was held in captivity to prevented repairing from the vocal signature. A week later, the released birds with the head feathers modified or not (controls) were examined for pair formation. By-species samples were small and results were not statistically significant for Rockhopper and Macaroni penguins. When the species were combined, however, penguins without yellow crests (64% of 39 shaved) formed significantly fewer pairs than the control group (91% of 22 controls; Fisher exact test, one-tailed $P = .02$). The study colony of King penguins contained several tens of thousands of pairs, providing a much larger sample of penguins for pair formation, and thus a broader choice of potential mates for pair formation. In this case, birds had the tips of their orange auricular spots shaved off, rendering them white. After a week passed, significantly fewer of the shaved birds had paired (29% of 45 shaved) than the control group (93% of 30 controls; Fisher exact test, one-tailed $P < .0001$). These experiments, while difficult to conduct in the field, strongly suggested the importance of colored auricular feathers for pair formation in crested and King penguins. In all three species, the original partners were released at the end of the experiment, and they tried to reunite with their partner by singing, even if their partner was modified from having colored feathers removed. The release set up a conflict situation in which the original pair displayed song recognition with their original partner. Clearly, the head pattern influenced pair formation, but individual recognition of the song was also evident. Just as in the earlier experiments with Adélie penguins, the visual mechanism appeared to be used only for initial pair formation.

These pioneering experiments produced suggestive preliminary results and indicated that further experimentation might prove illuminating. Cutting colored feathers clearly retarded pair formation, but the biological significance of these ornaments was unclear. In any unnatural manipulation like these, it is possible that species recognition is altered (especially for the similar crested penguins; each species has its own particular pattern of head plumage, and the head plumage is typical of all individuals in the same species). Further, altered feather patches might change the apparent breeding status of the individuals from adult with bright colors that is ready to reproduce, to an absence of colored feathers (leaving the white down feathers showing) that might indicate that the bird is a juvenile or subadult without previous breeding experience. Altered feather patches might also alter indications of the sex of individuals or make sex more difficult to identify. And finally, the apparent condition of an individual, signaling whether it should be preferred as a mate, might be altered. This last possibility is something that one would like to test, but a clear result of decreased preference for pairing might still indicate the other alternative interpretations. So cutting all the colored feathers or painting plumage shows their primary role in the initial pairing of individuals but disturbs the experimented bird in an unnatural way. Nevertheless, such experiments were useful because we repeated them several times to be sure of the outcomes and in better observed circumstances where we could subsequently be sure of the responses to possible alterations as possible species, developmental stage, and sexual identity. We also used natural variations in colors within populations to search for such natural patterns as "assortative mating" (see below).

## COLOR MEASUREMENT AND UV DISCOVERY

Over the last 25 years, colors and ornaments have been particularly well studied in birds (e.g., Hill and McGraw, 2006). Most of these studies have focused on songbirds. Studies of ornamental traits in other groups of birds that are socially monogamous and monomorphic, such as penguins and other seabirds, are much less common. Massaro et al. (2003) demonstrated in Yellow-eyed penguins that the intensity of plumage and eye coloration reveals the parental quality of individuals. McGraw et al. (2009) showed that unusual fluorescent pigments color the yellow feathers of crested penguins (specifically, Snares penguins) varied according to the year (reflecting variation in marine resources), the sex of the bird, and body condition.

These results on penguins followed the same patterns of relationship to environmental food resources as in the extensive experimental evidence from passerines birds (Hill and McGraw, 2006). The two species of large penguins that do not have a nest (Emperor and King penguins) were consequently a focus of our research due to the number and complexity of their ornaments and their associated ecological and social constraints. The most highly developed ornaments of the penguins are the feather and beak patches of these two species of the *Aptenodytes* genus. These penguins have more variations in ornaments on different parts of the body and also different sources of colors than any of the species of nesting penguins. Both Emperor and King penguins have two pigments in their feathers and hard beak covering of the lower mandible, as well as both pigmented and structural coloring of the beak. The King penguin is the most ornamented, with yellow–orange auricular patches on both sizes of the head, and a bright breast patch of feathers that grades from black and brown at the throat to bright yellow on the upper breast, and finally fading to bright white on the mid-to-upper breast. Carotenoid pigments, the most common yellow pigments in the feathers of songbirds are not present in the yellow feathers of the King penguin (McGraw et al., 2004, 2007, 2009). The yellow feather colors are primarily spheniscin pigments, similar in their properties to pterin compounds (Thomas et al., 2013). Pterins are antioxidant pigments, though perhaps of more limited antioxidant reactivity than carotene pigments (Martinez and Barbosa, 2010).

The colors of the beaks of King penguins initially seemed a curiosity. The upper mandible is black, but the lower part of the beak is black at the tip, with a colored spot that starts about a quarter of the way back from the tip and continues to the gape of the mouth (Fig. 3.3). This beak spot varies from yellow to orange, to pink, and at times even violet. In addition, it is unusual in being molted as a tissue plate, around the time that the feathers are molted. Feather molt lasts about 3–4 weeks, and then molted birds return to the sea to forage (Weimerskirch et al., 1992). The beak spots are molted just after feather molt (Schull et al., 2016). The black part of the beak is not molted, only the colored part of the lower mandible (i.e., the "beak plate"). The yellow color of the beak appears to come from deposited carotenoid pigment compounds in the lower beak tissues (Dresp et al., 2005; McGraw et al., 2007). Carotenoids are antioxidants and in many bird species reflect the strength of the immune system (McGraw, 2005; Hill and McGraw, 2006). In songbirds such as house

**FIGURE 3.3** An adult King penguin, showing the yellow−orange beak spot on the lower mandible (that is also ultraviolet in color), as well as the colored auricular and breast patches.

finches, the carotenoid colors of males (red in house finches; Hill, 2002) are strongly attractive to females during mate choice.

When the beak spots of King penguins are molted, they are shed and fall onto the beach or into the surf at the breeding colony, though they might also be shed at sea, as the newly molted adults forage for 3−4 weeks before mating. The casts of shed beak spots have the appearance of a thin sheet of plastic-like keratin material, and they are somewhat iridescent when held up to the light. We first examined the colors of the beak spots with a Colortron color-sensitive measuring device attached to a computer for viewing the reflectance of colors at different wavelengths (Nolan et al., 2006). The Colortron measures colors that are visible to humans, from about 400 (violet light) to 700 nm (red light) in wavelength. We discovered something quite surprising; that there was a high peak in the light emitted from the beak spot on live King penguins at about 400 nm in wavelength. The Colortron did not measure reflected light much below 400 nm, perhaps to 390 nm, but it was clear to us that there was a strong reflectance of light from the beak that was in the ultraviolet (UV) range. This turned out to be true for every adult King penguin that we measured (young birds have beaks that are completely black) and for similar beak spots of adult Emperor penguins as well, but not for the beaks of any of the other species of penguins, which vary from black to orange (Fig. 3.4; Jouventin et al., 2005).

UV color presents a complication for the study of ornaments and visual signaling in birds. First, we cannot see them because we do not have eye cone

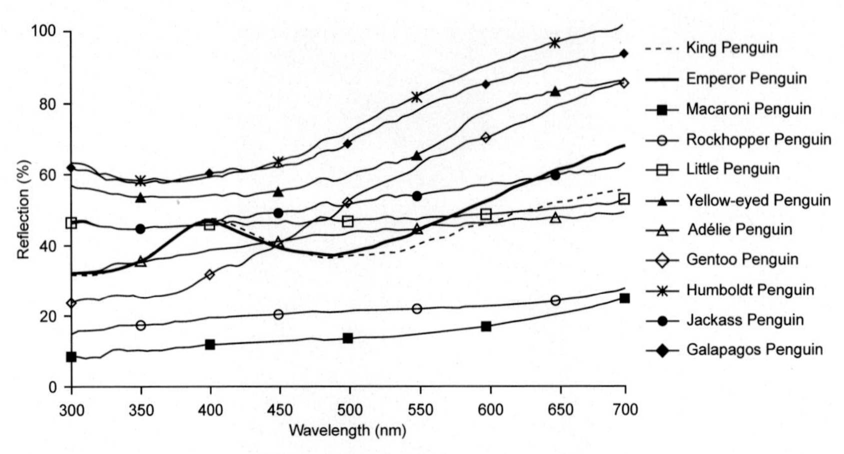


**FIGURE 3.4** Spectral reflectance measurements (y-axis) of color wavelengths (x-axis) for beaks (lower mandibles) of 11 penguin species, showing Ultraviolet reflectance peaks are present for only the large Emperor and King penguins (specimens examined at the National Natural History Museum in Paris). *From Jouventin, P., Nolan, P.M., Ornborg, J., Dobson, F.S., 2005. Ultraviolet beak spots in King and Emperor penguins. Condor 107, 144–150.*

cells that are within the peak sensitivity range of light at 300–400 nm. Nonetheless, many if not most bird species have UV-sensitive cones that are in or very close to this range of light frequencies (Hart, 2001). There are two types of UV-sensitive color cones in the eyes of birds, with peak sensitivities at about 370 nm (UV-sensitive) and at about 405 nm (violet-sensitive) (Bowmaker and Martin, 1985; Capuska et al., 2011). Penguins have violet-sensitive cones that perceive wavelengths into the UV range (Vorobyev et al., 1998; Osorio and Vorobyev, 2008). For comparison, human blue-sensitive eye cones peak in sensitivity at 420–440 nm and we cannot perceive light below about 390 nm. The mean peak wavelength of short-wave light reflected by the King penguin beak spot is about 386 nm.

Second, the structural origin of UV vision is different from the usual colored feathers or skin. UV color is produced not by pigment, but by the microstructure of the tissues of feathers or integument (Shawkey et al., 2003; for King penguins, Dresp et al., 2005 and Dresp and Langley, 2006), and is thus termed "structural color." Unlike humans that have three color cone types in their eyes, birds commonly have four types of color cones and one extends vision into the UV range of light wavelengths (Cuthill et al., 1999). Thus, many bird species with blue colored feathers also have UV coloring that human observers cannot see. This explains why for many years, UV reflecting ornaments were ignored by ornithologists, including those studying penguins. Also, UV reflectance cannot be measured with equipment that is designed for the human visual system, so more sensitive measuring devices are necessary. We acquired a spectrophotometer (from Ocean Optics, Inc.) that measures

**FIGURE 3.5** Illustration of the Ultraviolet beak and pigmented feather colors of the head and breast of adults of the large *Aptenodytes* penguins. *From Jouventin, P., Nolan, P.M., Ornborg, J., Dobson, F.S., 2005. Ultraviolet beak spots in King and Emperor penguins. Condor 107, 144–150.*

wavelengths of light from 300 to 700 nm in wavelength, and used this more sensitive device in further studies. We quickly found UV reflectance of the beak spots of the two large *Aptenodytes* penguin species (Fig. 3.5). A second problem created by the presence of UV in the beak spot ornament is that other colors are reflected, specifically the yellow–orange appearance that is easily visible to human observers. The color of the beak spot that King penguins perceive is some combination of the light frequencies that they can see, and pigmented and structural colors that interact to heighten one another (Shawkey and Hill, 2006). An advantage of the Ocean Optics spectrophotometer is that both UV and yellow–orange colors of the beak can be measured (Fig. 3.6).

How are UV (i.e., structural) colors produced in animals, and why are structural colors different from the colors that arise from pigments? Pigments are chemical materials that are stored within cells, and they reflect the same color when viewed from any angle. A pigment, like chlorophyll, absorbs wavelengths of light, except for wavelengths for the green color that we see. So for chlorophyll, it reflects green, but absorbs, for example, blue and red.

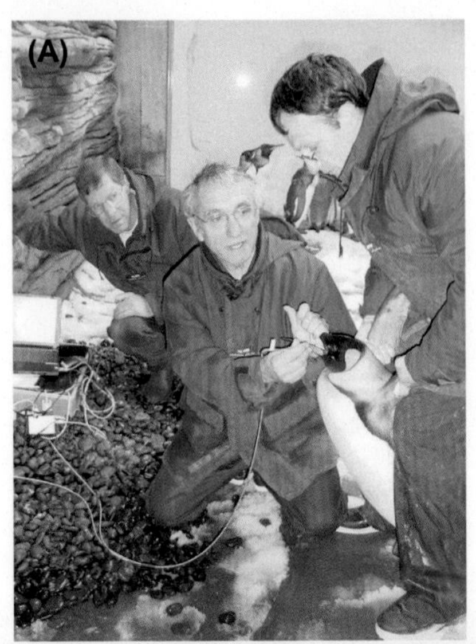


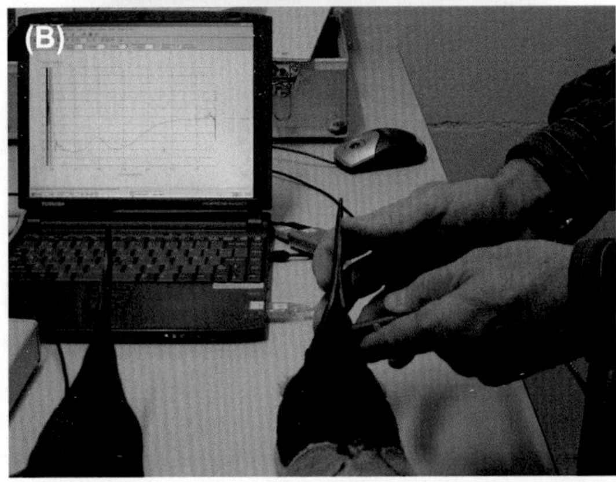


**FIGURE 3.6** (A) Measurement of King penguin beak at a marine park. The fiber-optic sensor is placed directly on the lower mandible of the penguin and colors are recorded on laptop computer. (B) Measuring the beak of a King penguin specimen at the Museum of Natural History in Paris, with the spectrophotometer readout on the computer screen. Notice the Ultraviolet peak of the color graph on the left-hand side.

Structural color comes not from pigments but from molecular-level photonic structures, microstructures that produce interference from light reflecting from upper and lower surfaces of thin layers of microscopic materials. As the viewing angle changes so does the degree of interference and thus the reflected color. UV is reflected by stacked lamellae in the keratin material of the outer layer of the beak tissue of large penguins (Fig. 3.7). Most wavelengths are subject to wave interference in the physical structure of the keratin lamellae, but some light wavelengths are reflected back, thus producing visible color, in

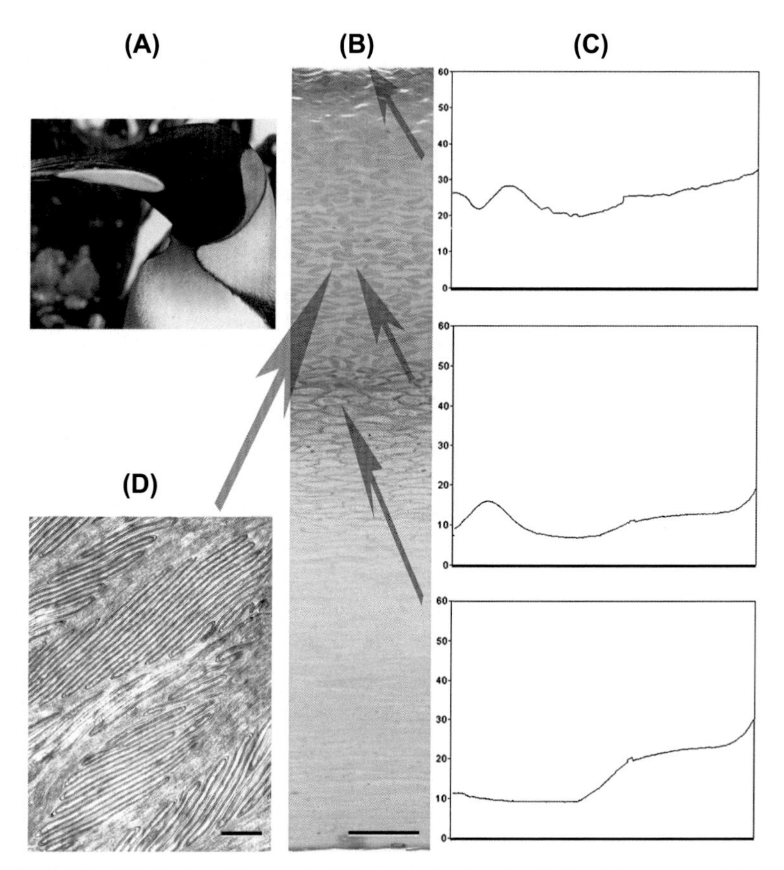


**FIGURE 3.7** (A) Head and breast of a King penguin, showing the hard, horny material of the beak spot. (B) Histological transverse section of beak horn, but parallel to the short axis, with the outer surface at the top. Scale bar, 25 μm. (C) Reflectance spectra obtained on intact and scraped beak horn. Top shows outer surface with Ultraviolet (UV) peak (left) and yellow–orange plateau (right). Middle shows midlayer with main UV peak. Bottom shows lower layer with old cells lacking UV color but with a yellow–orange plateau from carotene pigment. (D) Transmission electron micrograph of the upper region of the beak horn, showing packed microstructures of folded stacks of multilayer folded membrane. Scale bar, 1.0 μm. *Modified from Dresp, B., Jouventin, P., Langley, K., 2005. Ultraviolet reflecting photonic microstructures in the King penguin beak. Biol. Lett. 1, 310–313.*

this case UV. This color is not the same if the view is from the side of the microstructures as at right angles to the material (the usual view for a King penguin). Of course, in some tissues pigments and photonic microstructures occur together, thus producing a variety of colors. The beak of the King penguin contains both yellow–orange carotene pigment and UV color from the layers of keratin lamellae. Unfortunately, we currently do not know what color the penguins actually “see,” since we can perceive only a small part of the range of light wavelengths. In addition, integration of information from the color cones occurs in the brain of the penguin, providing another avenue for investigation of the mystery of what animals see.

The measurement of color is not simple, whether based on the human visual system or that of birds. The most common way to think about colors is to break them down into some underlying characteristics. We used a modification of the system developed by Albert Munsell about 100 years ago (Montgomerie, 2006), which defined three aspects of color: hue, saturation, and brightness. Hue is a property of color that we generally "see" as, for example, violet, blue, green, yellow, orange, or red. These and other hues can be measured from the intensity of light that is reflected over different wavelengths. While there are several ways to typify hue and other characteristics of color (Montgomerie, 2006), we broke the colors of the beak spot into two parts. We measured hue as the frequency at which the reflectance of light is half way between its maximum and minimum values for yellow to red colors (after Pryke et al., 2001; 500–700 nm, the right-hand side of Fig. 3.8), and as the peak frequency of reflectance for the UV to violet colors (300–500 nm, on the left of Fig. 3.8). This allowed us to test the influence of the structural and pigmented colors separately. For the hue values of the feather patch ornaments of the ear and breast, we used the method for the yellow to red colors over the entire range of reflected wavelengths, since these ornaments lacked a UV component.

Another aspect of color is its brightness, also called reflectance or the intensity of a color. Brightness is the total reflectance across the visual spectrum, but we estimated it instead from the average height of the line that shows wavelength-specific reflectance in the graphic output from the

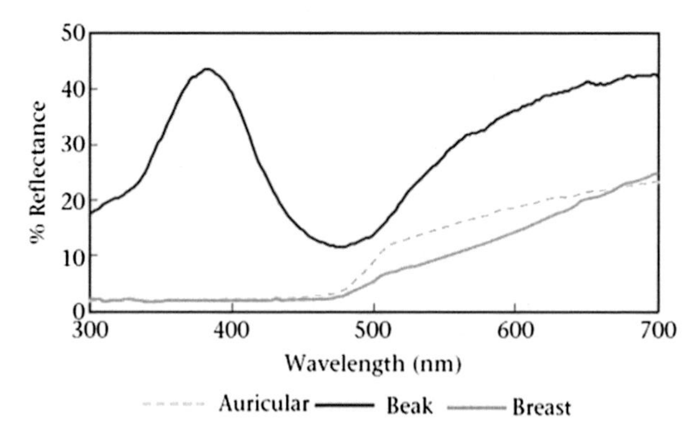


**FIGURE 3.8** Spectral representations of color in the ear (auricular) patch (*dashed line*), the beak patch (*solid line*), and the breast patch (*gray line*) of King penguins. The three ornaments appear yellow to orange to human observers, but the beak also shows a bimodal distribution with a maximum reflectance peak in the ultraviolet spectrum on the left. *From Pincemy, G., Dobson, F.S., Jouventin, P., 2009. Experiments on colour ornaments and mate choice in King penguins. Anim. Behav. 78, 1247–1253.*

spectrophotometer (the same number of measurements was used to construct such lines for every individual penguin, yielding unique values). White, of course, is much brighter than black, but the same could be said of light blue and dark blue or lighter and darker shades of any color. Saturation, also called chroma, is a measure of the purity of a color. When most of the light reflectance comes from a narrow band of wavelengths, we see a strikingly pure color; and when this is not so we perceive a bleached or washed-out color. We measured color saturation as the difference between the maximum and minimum values of reflectance, relative to the brightness of the color (Montgomerie, 2006). Obviously, hue, saturation, and brightness can be strongly associated with each other. As saturation declines, for example, brightness may tend to increase. Importantly, UV and yellow–orange brightness of the beak spot were highly associated (that is, a bright beak is bright in both colors), moderately associated for saturation, and weakly but significantly for hue (more UV in color was slightly more yellow than orange).

The human visual system is based on colors that roughly coincide with the peak sensitivity of the cone cells of our eyes: blue, green, and red. An alternative to the human system is to focus on the sensitivities of cones in the eyes of birds: UV or violet, blue, green, and red. The peak sensitivities of birds and human observers are somewhat different, but the primary difference is that birds have a UV/violet cone type that extends their vision into a range that we cannot see. This raises the fascinating question of what they do see. For example, much of the purple that we see is not violet but rather a mixture of red and blue. King penguins likely see a color of the beak spot that is a mixture of UV and yellow–orange, and it is curious to think about what that color might be. Nonetheless, we can examine the impact of the UV and yellow–orange colors by teasing them apart experimentally. The integration, or perception, of color occurs in the brain. So we may not be able to imagine what specific colors penguins see, especially when UV is involved, but we can measure its aspects by separating UV and yellow–orange parts of the beak color.

A system that integrates visual information uses the bird color vision system of four cones to give us a tetrahedral "bird model" of color vision (Goldsmith, 1990; Endler and Mielke, 2005; Stoddard and Prum, 2008). This model produces a 3-dimensional color space, the center of which is an achromatic origin or gray point. The gray point represents equal stimulation of the four color cones at once. One direction away from this point represents saturation of the color of the measured object (usually called chroma in the bird vision model, and measured as a proportion of the maximum value). There are two other directions away from this point (we could think of these three directions as x-, y-, and z-axes in a 3-dimensional Cartesian space); one representing stimulation of the UV–violet cone (called phi, positive values reflect greater stimulation) and one representing the stimulation of the other three cones combined (called theta; for King penguins, negative values indicate reds, values close to zero indicate yellow, and positive values indicated greens and blues). Taken together, phi and theta can be thought of as a hue vector. The third color axis is the distance of the color from the achromatic origin in 3-dimensional space; and it is termed chroma, a measure of color purity. When studying the beak spot of King penguins, we used this model to test the similarity of male and female breeding pairs by examining the distance between individuals in all three measures of color at once (that is, how close they were in 3-dimensional space) compared to how close randomly generated pairs from the total sample of birds of opposite sexes (Keddar et al., 2015c).

## WHERE DO COLORS COME FROM?

Before describing our experiments on the role of color ornaments in the mating patterns of King penguins, we should ask whether the ornaments have potential for conveying information about individual quality. If so, then models of sexual selection that invoke "indicator mechanisms" of quality, such as superior ability to survive, are appropriate (Andersson, 1986; Grafen, 1990; Kokko, 1997, 1998). In other words, if an ornament was not linked in some way to the potential benefit of mating with an individual, the ornament would not qualify as "honest advertising" of the value of mating with the holder of the ornament (Kodric-Brown and Brown, 1984). To evaluate the honesty of an ornament as an accurate signal of quality, the mechanisms of ornament production might be used. Honest signals should indicate whether they are costly in terms of energy and other resources, so that they reveal the quality of the carrier. The ornament can be used to indicate that the bearer is in such good condition that it can produce a costly signal (e.g., Zahavi, 1975; Grafen, 1990). In this, it is an advantage that colored ornaments are produced by either pigments or accurately constructed microstructures (the latter for UV coloration). In birds in general, yellow to red ornaments are often produced by carotenoid pigments that the animals cannot synthesized themselves but must gain through the foods that they consume. Pterin-like compounds such as spheniscins, however, are yellow pigments that can be synthesized in non-integument cells of the body, especially cells associated with the immune system (McGraw, 2005).

For structural colors (UV−violet), hue appears to depend on the regularity of the stacking of keratin lamellae in the hard tissue of the beak, especially the distance between the lamellae that determines the purity of the UV−violet color (Dresp and Langley, 2006). Since these structures are replaced when the beak plates are molted, materials and energy are involved, so the birds in best condition are likely most capable of producing the strongest UV−violet color (Cuthill et al., 1999). Exacting control of the distance between lamellae likely requires extra energy and indicates whether growth was regular or variable, the latter likely due to periods of stress of poor food resources. Both pigments and energy for structural color development are most likely influenced by environmental resources. Yellow and red pigments in many species of birds come from food resources (Hill and McGraw, 2006), such as the crustacean foods that give flamingoes their red colored feathers (Fox et al., 1967).

To study the dietary origin of pigment colorations in penguins, we conducted tests of color acquisition in Gentoo penguins in a marine park (Jouventin et al., 2007b). In Gentoo penguins, beak color is orange and highly variable among individuals in the same colony. It was unclear whether the variation in beak orange was due to sex, body condition, migrants among populations, or the food resources of Gentoo penguins. Red, orange, and yellow carotenoid-based coloration abound in birds, with over half of all avian orders known to display it in some form (Hill and McGraw, 2006). Penguins, however, are an order of birds for which the proximal causation of ornaments is unclear. The yellow to orange colors of feathers, skin, and beak surfaces are not well studied as to their chemical composition and properties. While pterin-like spheniscin compounds appear to occur in yellow—orange feathers (a new pigment that occurs only in penguins; Thomas et al., 2013), the beak of Gentoo penguins and beak spots of King and Macaroni penguins contain carotenoid pigments (McGraw et al., 2007). We experimentally supplemented the diet of captive Gentoo Penguins for 2 months with extracts of krill, a common carotenoid-rich food source for these animals in the wild, to determine whether orange coloration in the beak and feet was influenced by carotenoid content of the diet. Using UV-visible reflectance spectrophotometry, we found that dietary carotenoid enrichment elevated the brightness of the yellow—orange color of the beak but not of the foot (Fig. 3.9A and B). This suggests that the crustacean part of the diet is at least partly responsible for orange beak coloration (but curiously, not the similar color of their feet) in Gentoo Penguins. Like other carotenoid signals, beak color has the potential to reveal important aspects of mate quality (e.g., nutrition, health, or at least foraging ability). This was the first experimental evidence that carotenoids confer bright colors in seabirds.

Environmental resources appeared to have dramatic effects on the color ornaments of King penguins. Our studies of King penguins took place over several years, from our preliminary experiments on feather ornaments in the mid- to late-seventies, to the last color tests of the beak spots. At Cape Ratmanoff on Kerguelen Island in 2008 and 2010, we had the chance to compare ornaments in a normal breeding year (2008—09) with a year in which breeding was exceptionally late and survival of chicks was extremely low (2010—11, chick production virtually failed; Keddar et al., 2015a). Late breeding and low chick survival likely reflected extremely poor foraging conditions for adults. Adults that came to the breeding colony were only slightly lower in body mass, and not significantly; but this was not surprising since only birds in the best body condition were likely to initiate breeding. The source of the differences in chick survival might well be shifts in the abundance or spatial location of food resources of King penguins, and it is also likely that the differences among years reflected associated changes in global climate (Le Bohec et al., 2008).

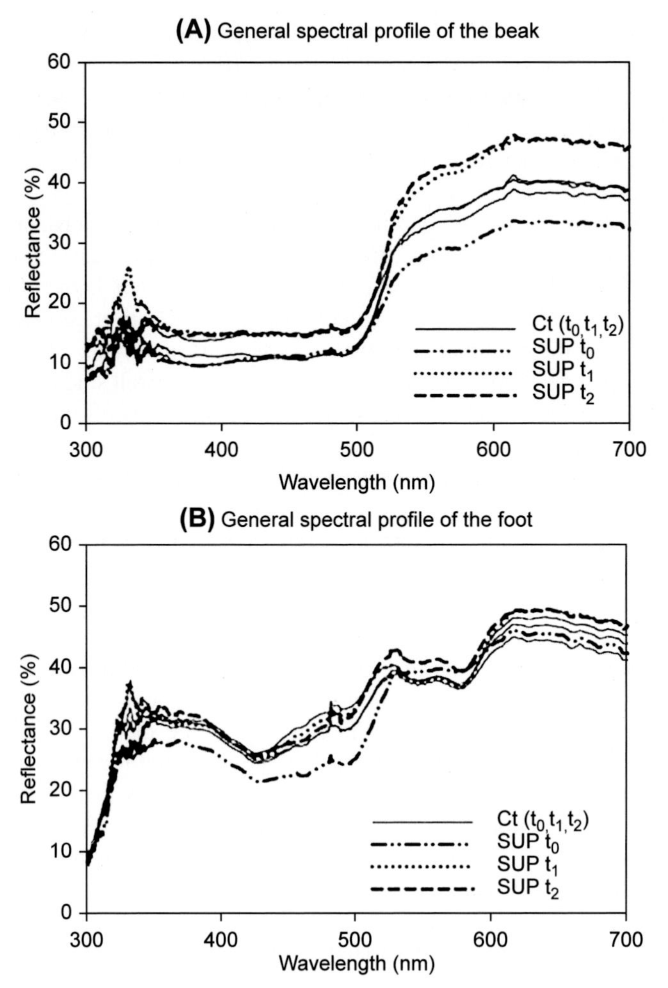


**FIGURE 3.9** Carotenoid supplementation experiment on Gentoo penguins. Dietary carotenoid enrichment elevated beak but not foot brightness. Top: (A) Spectral representations of color of the beak (top) and (B) foot, under control (Ct), beginning of carotenoid supplementation (SUP $t_0$), after 1 month of supplementation (SUP $t_1$) and after 2 months of supplementation (SUP $t_2$). Bottom: Temporal changes in beak brightness and % reflectance. ***, $P < .001$ from $t_0$; #, $P < .05$ from control. *From Jouventin, P., McGraw, K.J., Morel, M., Celerier, A., 2007b. Dietary carotenoid supplementation affects orange beak but not foot coloration in Gentoo Penguins* Pygoscelis papua. *Waterbirds 30, 573–578.*

Color differences in the ornaments of adult King penguins between years were dramatic. The poorer year had much lower saturation of the brown and yellow parts of the breast patch of feathers (by about 14% and 5%, respectively), much lower saturation of the yellow–orange color of auricular feather patches

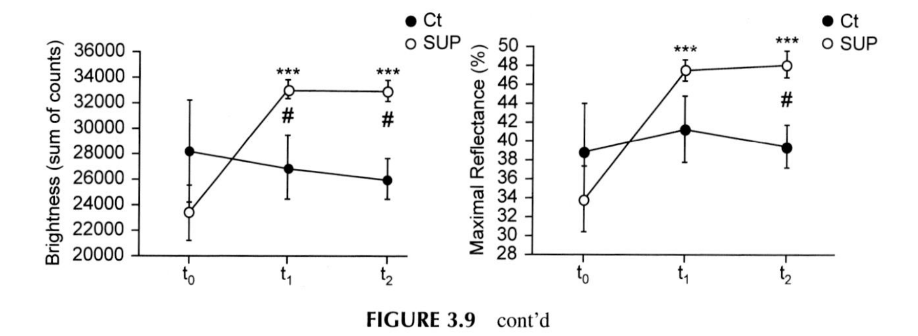


**FIGURE 3.9** cont'd

(by about 15%; Fig. 3.10A), the auricular patches were much smaller (by about 13%), and there was a reduced brightness of the UV and yellow–orange colors of the beak spot (by about 6% and 9%, respectively; Fig. 3.10B). In addition, the hue of the breast and auricular patches was more light yellow than deep yellow or orange, and the same was true of the beak spot (Fig. 3.10C). This last result is especially interesting because the feather patches are due to the pterin-like spheniscin pigment, and the yellow–orange color of the beak patch is from deposition of carotene pigment. Thus, it was clear that in 2010 the colors of ornaments had changed, even in birds that had recently molted, likely due to poor resource conditions associated with changes in the marine environment. These results support the idea that the colored ornaments of King penguins reflected honest information about the quality of the birds and their diets, useful information for making the choice of a good mate.

## PATCHES OF COLORED FEATHERS

Three genera of penguins are primarily black-and-white (with some blue/gray-feathered populations in Little penguins), and three genera are colored with yellow feathers or crests. Colored breast patches are unusual in penguins, occurring only in the largest species, the King and Emperor penguins of the genus *Aptenodytes*. The color patches on the breast of Emperor penguins are greatly reduced and pale yellow compared to the more extensive and colorful breast patches of King penguins. The breast patch of the latter species changes in color, from brown at the throat to bright yellow, and then to fading to white on the upper breast. This makes documentation and measurement of the breast patch of the King penguin especially difficult. These patches are not predominately visible in nuptial displays, as the head is extended far above the breast in the face to face display (Fig. 3.11). The breast may, however, be attractive at a distance, drawing displaying birds together. Later in the crowded colony, when breeding adults stand with their egg on their feet, the breast may be a more visible ornament during aggressive displays. In the 2010–11 reproductive season at Cape Ratmanoff (Kerguelen Island), we found circumstantial evidence of this in the breeding colony. When standing with the

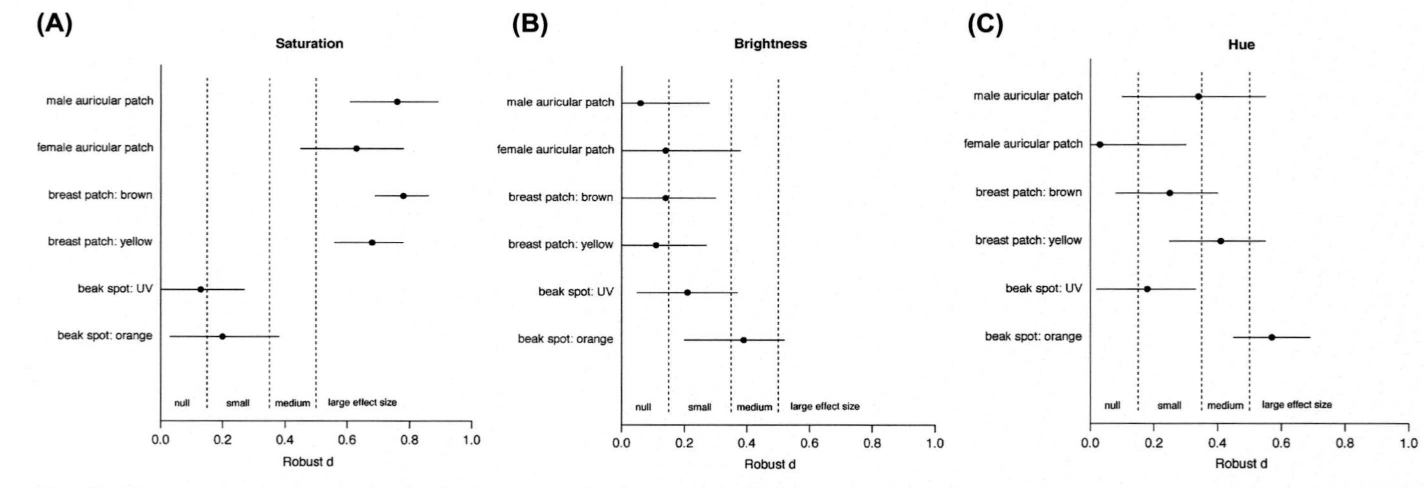


**FIGURE 3.10** For colored ornaments of the King penguin, the effect of a good foraging year compared to a poor one, showing which aspects of color were stronger in the good year. Robust d varies between 0 and 1, showing the importance of the difference in (A) saturation, (B) brightness, and (C) hue between years, and positive values show the biggest changes. *From Keddar, I., Couchoux, C., Jouventin, P., Dobson, F.S., 2015a. Variation of mutual colour ornaments of King penguins in response to winter resource availability. Behaviour 152, 1684–1705.*

**FIGURE 3.11** When a female King penguin on the beach meets a bachelor that sings the short song (ecstatic display), the first mutual posture is, here, the "face to face," used to know the body condition of the mate by looking at "honest signals": both heads are raised slowly with neck progressively inflated to show auricular patch size and ultraviolet colors of the beak. If their mate choice is positive, they stay and move together, usually mute to prevent attracting a new male (and thus forming a trio that can lead to breeding delay or failure).

egg on their feet, both male and female King penguins that were further from the edge of the breeding colony, and thus less susceptible to predation, had upper breasts that were deeper brown and less reflective than birds near the edge of the colony (Fig. 3.12). In addition, for King penguins breeding early in the austral spring, when pairs are most successful, lower breast hue is more strongly yellow and the yellow color is more saturated, compared to late breeding pairs that most often fail at reproduction (Weimerskirch et al., 1992; van Heezik et al., 1994; Jouventin and Lagarde, 1995; Olsson, 1996; Brodin et al., 1998). Males have a deeper and more pure yellow compared to females at any time in the breeding season.

Early in the 2001–02 breeding season at Possession Island in the Crozet Archipelago, we injected an irritant into the foot webs of male King penguins (we used a harmless plant lectin that causes a minor inflammation, much like getting a splinter; Nolan et al., 2006). Then we looked to see whether the degree of inflammation, as indicated by the degree of swelling caused by the injection, was associated with aspects of the color ornaments. The response to the inflammation showed a single association with ornamental colors: the greatest response to the inflammation was by males with the yellowest (as opposed to more orange) breasts. Thus, the breast may carry a signal of quality

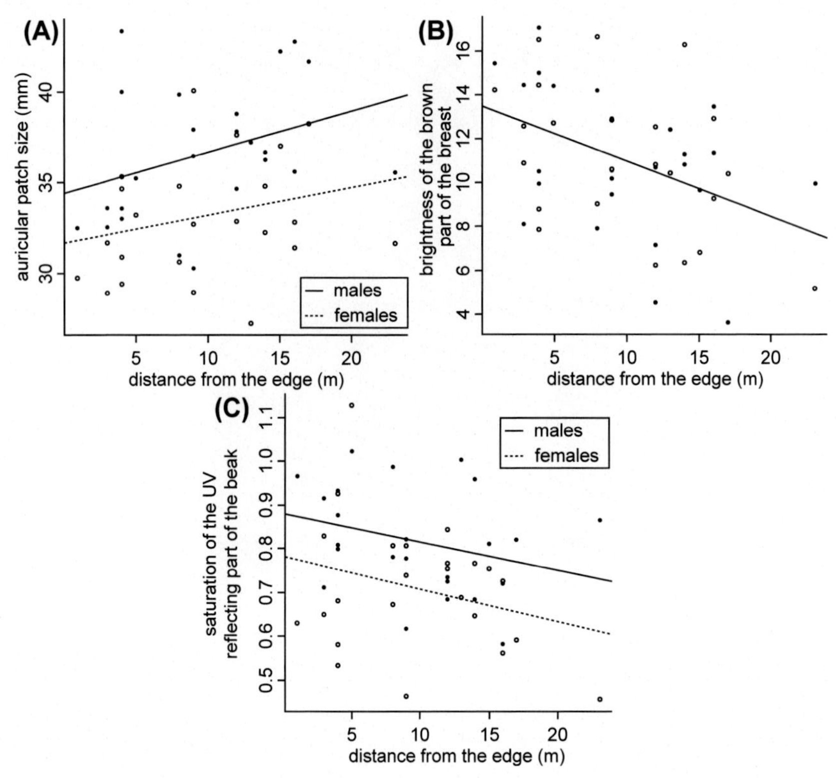


**FIGURE 3.12** For King penguins, relationship between the distance at which males (*black dots*) and females (*open circles*) established their territory from the edge of the colony (and thus nearer the center) and (A) the size of their auricular patches, (B) brightness of the brown part of the breast feathers (males and females pooled due to a lack of difference), and (C) saturation of the ultraviolet reflectance of the beak spot. *From Keddar, I., Jouventin, P., Dobson, F.S., 2015b. Color ornaments and territory position in King penguins. Behav. Process. 119, 32–37.*

of the immune system, but there was no indication of influence on mate choice for variations in the breast ornament. Though we did not demonstrate an influence on mate choice, the breast may well reflect aspects of body condition of individual King penguins.

In one way, however, the color of the breast may be important to mate choice. In many birds, "nuptial plumage" of feathers, usually not only their colors but also various fancy plumes, do not fully develop until an individual is both old enough and in good enough body condition to successfully breed (reviewed by Hawkins et al., 2012). We did an experiment where we shaved the tips of breast feathers, cutting off only the colored tips of the feathers (Pincemy et al., 2009). This leaves the underlying parts of the feathers exposed, giving a white appearance. Younger, preadult King penguins (2–4 years of age; Weimerskirch et al., 1992) and those that are in exceptionally poor condition lack strong breast color but visit the breeding colony and walk about. In addition, the size of the breast patch of younger birds is often reduced, though the breast patches of younger birds do not differ in color form those of older birds (Nicolaus et al., 2007). It is easy for older adults in better condition to avoid these individuals as mating partners. Our experiment took adult birds, ready to breed, and gave them the appearance of birds that were young, sickly, or otherwise in poor condition by giving them white breasts. The manipulated birds were then returned to the breeding colony, and we measured the number of days that it took them to find an initial mating partner. When a bird finds a suitable partner for breeding, it learns the song of the partner (see Chapter 5). So we could test for pairing by walking between the members of a pair, so that visual contact was broken. To reunite, the pair members had to sing; and if they reunited we knew that they had at least initially paired.

Surprisingly, our results in the "whitening" experiment were different for males and females (Fig. 3.13). For males, shaving the colored tips of the breast feathers produced significant delays in finding and initiating pairing with a female when compared to controls that were not shaved, and almost half of the breast-clipped males could not successfully pair at all. We concluded that we had made the clipped males look young or "sick," and that females thus

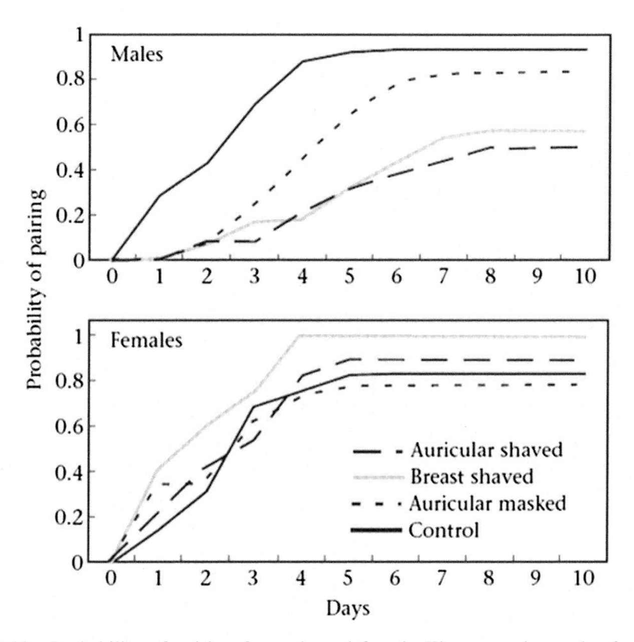


**FIGURE 3.13** Probability of pairing for male and female King penguins under four different treatments: ear (auricular) patch shaved, breast shaved, ear (auricular) patch masked with black dye, and control. *From Pincemy, G., Dobson, F.S., Jouventin, P., 2009. Experiments on colour ornaments and mate choice in King penguins. Anim. Behav. 78, 1247–1253.*

avoided them as mating partners. For females, however, clipping the breast feathers made little or no difference to the rate of initial pairing with males, and all shaved females at least initially found partners and paired (only about 80% of control females paired within 10 days). In other words, the males did not appear to react to females that appeared younger or sickly via the breast patch of feathers. For example, birds with whiter breasts are often smaller than birds with full color, and our shaved birds were about the same body size, on average, as control birds that were captured and measured, but not treated with feather clipping. This was our first hint that perhaps males were competing most strongly in this species for the mating favors of females. As in other bird species, females appeared to be the choosier sex, in spite of the monomorphism of colored ornaments.

The yellow auricular patches of adult King penguins are duller yellow than the bright yellow of the breast, but they are still strikingly colored. Auricular patches of colored feathers occur only in the King and Emperor penguins, and they are most highly developed in the King penguin. For this latter species, they are clearly visible from a distance as well as during nuptial displays ("face to face" described in Chapter 2; Fig. 3.11), when two birds extend and elevate their heads far above the body and sway about together with their eyes squinting or closed, taking occasional wide-eyed "peaks" at each other. Mates can easily see the auricular feather patches of the potential partner in this display. These yellow—orange auricular feather patches do not differ significantly between the sexes in color (Dobson et al., 2008; Pincemy et al., 2009). The auricular patch, however, is clearly important to mate choice, particularly the choice of the male partner by females. This was suggested by the two historical experiments that we saw in the beginning of this chapter that were previously conducted by early researchers on King penguins. Stonehouse (1960) and Jouventin (1982) shaved the auricular patches of King penguins to turn them white and rendered individuals unable to attract mates. Young and poor quality birds often have lightly colored auricular patches, and such individuals can only rarely attract a partner for pairing.

We repeated the treatment of shaving the ends of feathers and applied it to the auricular patches of King penguins. We found very similar results to those of shaving the breast. Males with shaved auricular patches of took longer to pair and were less likely to pair compared to control males with unmanipulated auricular patches, but there was not significant difference in time to pair or likelihood of pairing for females (Fig. 3.13). In a sample of 18 control birds, those with less saturated and more yellow (as opposed to nearly orange) colors of the auricular patch paired significantly more slowly. It appears that a yellow-orange auricular patch is something of a prerequisite for mating, but only for males in the King penguin.

We conducted two sets of experimental reductions to test the importance of the size of auricular patches of King penguins in two different colonies and locations, one in a colony on Possession Island (Crozet Archipelago) and one on Kerguelen Island (Jouventin et al., 2008; Pincemy et al., 2009; Nolan et al., 2010). The sizes of the auricular patches were reduced by about 50% by coloring over the yellow feathers with a permanent black marker. An equal area of black head feathers was colored over on control birds, to control for the presence of the permanent black ink. The results of these replicated experiments were virtually identical: males with reduced auricular patches took longer to pair when compared to control males, but females with reduced auricular patches did not take significantly longer to pair (Fig. 3.13). In general, control males took 20%–30% longer to pair than females among control birds, and our statistical analyses took this fact into account. Nonetheless, the longer time to pair for males suggested that males were competing more strongly for mates than females. In conclusion, the results of these replicated experiments indicated that females preferred to mate with males that had larger auricular patches, but that males were less choosey about the appearance of auricular patches of females with whom they chose to pair.

If female King penguins consistently prefer males with larger auricular patches as mating partners, then there should be sexual selection for larger auricular patches in males. That is, we can predict that the auricular patches of the males will be larger than those of the females, even if differences in male and female body size are taken into account. Males are 5%–10% larger than females (a combined measure of flipper and beak dimensions of 73 males and 73 females yielded a 6.2% difference), at least for displaying birds in the breeding colony and among paired birds (Pincemy et al., 2009; Dobson et al., 2011). If sexual selection favors larger male auricular patches (but not larger female auricular patches), then males should exhibit auricular patches that are significantly larger after body size is taken into account. When we looked at this, we found that the auricular patches of males were 7.7% larger than those of females, a highly significant difference (Fig. 3.14). But if the auricular patches of males are larger than those of females, then the distance between them over the top of the head should be shorter. They are, with the distance between auricular patches at the top of the head being 7.5% shorter in males (adjusted for body size) than for females. These differences in size and placement of the auricular patches are best explained by the preference of females for males with larger auricular patches, supporting the idea that female choice selects for larger auricular patches of males; and thus, that sexual selection may be contributing to the evolutionary maintenance of large auricular patches in males in this species.

The conclusion about sexual selection favoring larger patches of colored auricular feathers in males than in females seemed to contradict mutual mate choice as an explanation of the auricular patch ornaments of male and female King penguins. Before turning to how the auricular patches might be maintained by natural selection in both sexes, it is worth considering whether King penguins show some similar patterns in competition for mates to other species of birds, especially songbirds where males often exhibit ornaments such as

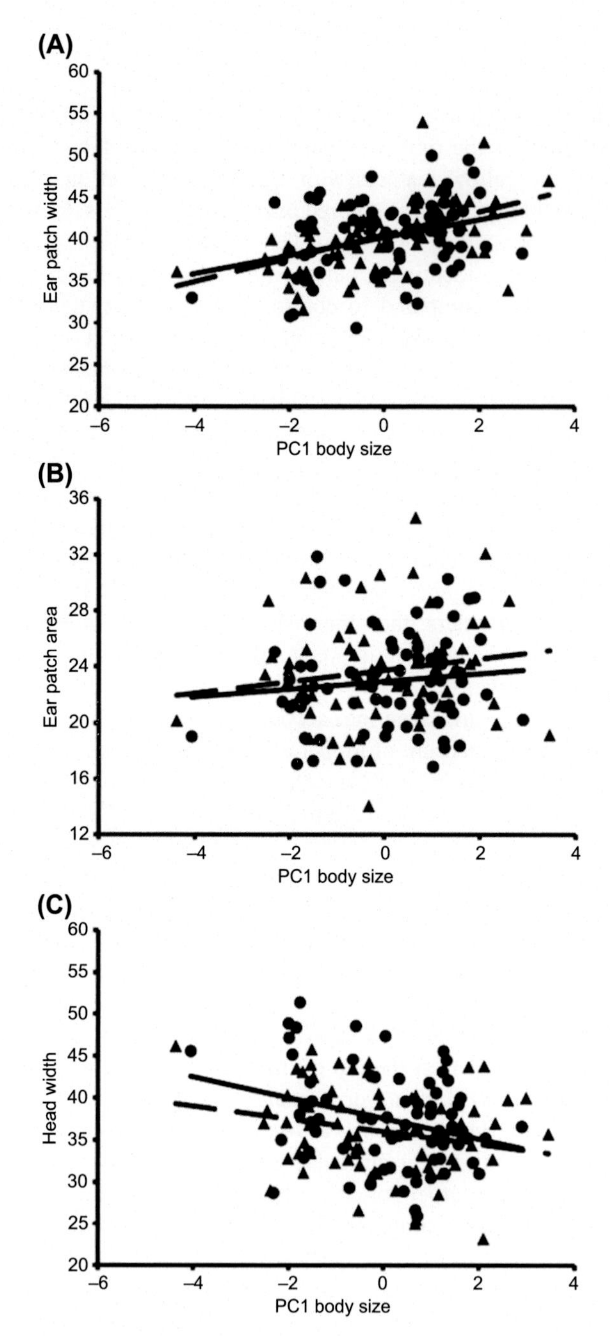


**FIGURE 3.14** The positive relationship between auricular patch size and body size in King penguins. Body size is estimated by a principle components analysis (PC1) of several beak measurements, flipper length, and body mass. (A) Width across the auricular patch (in mm). (B) Area of the yellow−orange feathers of the auricular patch (in $cm^2$). (C) Distance across the black feathers over the top of the head, between the yellow auricular patches (in mm). Males are *triangles* and females are *circles*. Best-fit lines are *dashed* for males and *solid* for females. *From Dobson, F.S., Couchoux, C., Jouventin, P., 2011. Sexual selection on a coloured ornament in King penguins. Ethology 117, 872−879.*

colored feathers and females do not. What sorts of conditions lead to greater competition among males for mates than among females? One obvious factor that would increase the competition for mates among males is if there were not enough females to go around. That is, if there was a sex ratio bias among the displaying birds, that is in the initial breeding population, towards too many males. Another obvious way that competition for mates could be greater for males would be if the species were not monogamous. But in an obligate monogamous bird like the King penguin, the necessary high degree of cooperation between male and female partners for successful breeding ensures that both sexes need to have a good and faithful partner. For this to occur, males must have a high degree of confidence in their paternity, so that paternal care is not redirected onto nonoffspring. Natural selection should thus favor behavioral traits, such as close attendance of the female mate, while she is reproductively receptive, which serve to increase confidence in paternity.

The question of the sex ratio of displaying birds at courtship is difficult to answer because the two sexes of King penguins look so similar. Sexing in the field can be accomplished by finding paired birds where one is much larger than the other, and assuming that the larger bird is the male. Alternatively, birds can be sexed by differences in the voice and behavior (Jouventin, 1982). In both of these cases, however, there is strong overlap of individuals in diagnostic traits that can be seen or heard by observers so that mistakes are likely. Nonetheless, a study that applied a combination of behavioral criteria found male-biased sex ratios among displaying birds of 1.53 males per female (Bried et al., 1999), thus suggesting that relatively rare females should be choosier than more numerous males. An alternative to behavioral methods is to use DNA markers for molecular sexing (Fridolfsson and Ellegren, 1999). Studies that applied DNA sexing estimated sex ratio of displaying King penguins at 1.67–1.96 males per female (Pincemy et al., 2009, 2010). Thus, there appears to be an excess of males among the displaying birds, and this suggests that males compete more strongly for mating and breeding partners than females in this species. This conclusion is fortified by the surprising fact that about a quarter of displaying pairs are males displaying to other males, a homosexual imbalance that usually gets sorted out before pair members learn each other's vocal signature (Pincemy et al., 2010).

A further test of the strength of competition for mates between the sexes involves a remarkable behavior of the penguins that was once poorly understood. One often sees trios and rarely quadruplets of penguins marching about the colony in single file. Some of these trios walk through the breeding colony among the birds standing with eggs, suffering pecks from the birds that are protecting their eggs. Trios form when a third bird approaches a male–female pair. Commonly, one of the pair pecks the other, and the parade is on, with the pecked bird leading. These trios are one of the first things that a casual visitor notices when arriving at a King penguin breeding colony, since the leading and following birds appear so coordinated. We reasoned that if males were competing more strongly for females, we might be able to explain why these trios of marching penguins occur (Jouventin, 1982; Keddar et al., 2013). The trios might reflect competition by two males for the attentions of a female. If so, we would expect the trios to be composed of two males and a female, and that females invariably lead the march.

We found that 19 of 20 trios were indeed two males and a female, the lone exception being an all-male trio. In the 19 cases where a female was involved, she always led the march, with two fighting males close behind (Fig. 3.15). The moving female may be easier for her initial male partner to defend, leading to these slow "chases" around the colony, with the two lagging males fighting over access to the female. So it seems that the parading trio phenomenon reflects stronger competition for mates among males than among females. We will return to these surprising trios or quartets of displaying penguins in Chapter 6. Indeed, they are also found in Emperor penguins, but with the opposite sex imbalance (i.e., usually two females following a male) in a species where females outnumber males among penguins that are displaying to find a breeding partner (Isenmann and Jouventin, 1970; Isenmann, 1971). So we see several males for a female in the King penguin, but several females for a male in the Emperor penguin. We will see that the ecology of the breeding cycle of each species of large penguin is different and uniquely so, and that the difference in sexual imbalance can explain the opposite results in the composition of trios.

In conclusion, these experiments on the biological significance of feather ornaments in King penguins suggest that the yellow auricular patches of males are evolutionarily maintained by female choice and sexual selection on the males. Since the auricular patches, like the breast, can be seen at a distance, it might also provide an attractant to females that are coming to display to males on the beach, adjacent to the breeding colony. But none of the evidence explains why females have such similar (though not identical) yellow auricular patches or brown-to-yellow breasts as the males. What maintains these ornaments of females? Of course, it is possible that genetic divergence of ornaments between the sexes has not had time to develop (Lande, 1980), but this seems unlikely given the long evolutionary history of penguins and the diversity of ornaments among the species.

Sexual selection is not the only evolutionary process that maintains colored ornaments, such as the ornamental colored feathers of female King penguins. Viera et al. (2008) found that both males and females with larger auricular feather patches were found standing with their egg in the center of the breeding colony, and that these birds were more aggressive than birds on the edge of the breeding colony. Incubating birds in the center of the breeding colony are perhaps less likely to lose their egg or young chick to predacious Skuas or Giant petrels. We found that pairs with larger auricular feather patches settled closer to the center of the breeding colony after pairing and mating (Fig. 3.12A; Keddar et al., 2015b), perhaps facilitated by their greater

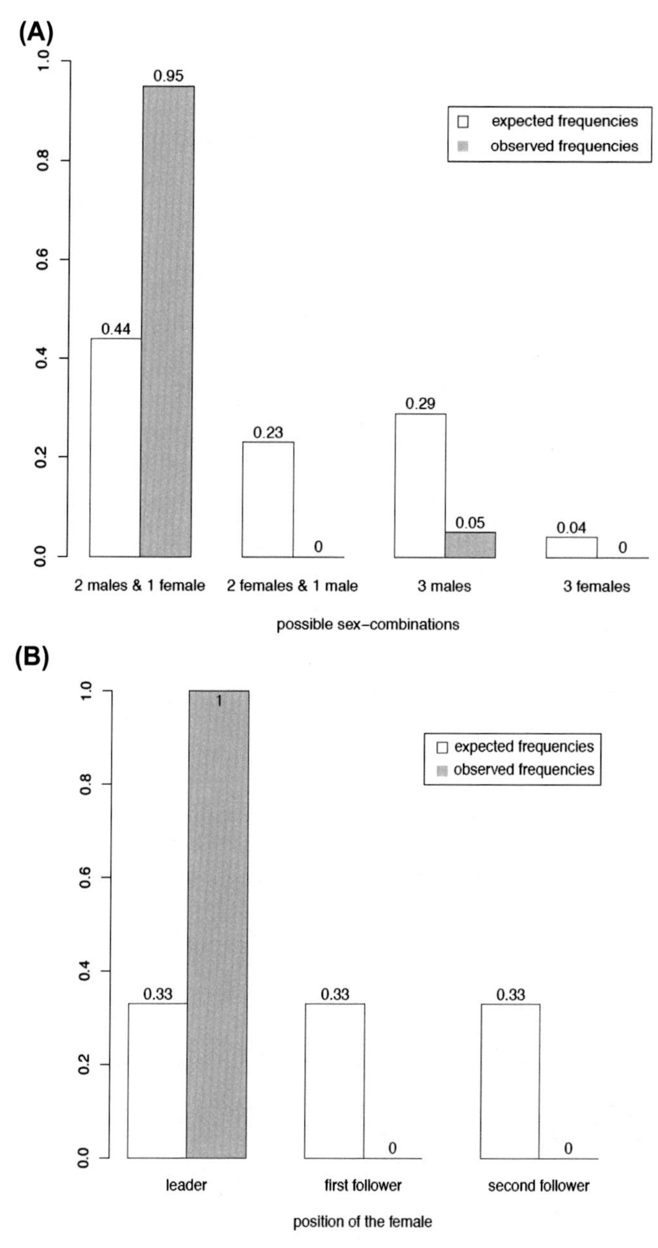


**FIGURE 3.15** Composition of trios of King penguins. (A) The sex makeup of trios is compared to a random expectation for trios made up of two males and a female, two females and a male, three males, or three females. Nineteen of 20 trios were made up of two males and a female. (B) Position of the female in 19 trios when they are on parade, compared to a random expectation. In all 19 cases, the female led the trio parade. *From Keddar, I., Andris, M., Bonadonna, F., Dobson, F.S., 2013. Male-biased mate competition in King penguin trio parades. Ethology 119, 389–396.*

behavioral dominance and more aggressive behavior. If the size of the auricular or breast patch is a useful signal at other times than during mate choice, then it is possible that the female ornaments are maintained by social selection outside of the mating season (reviewed by Kraaijeveld et al., 2007). Of course, social selection could be influencing the size of the auricular patch ornament in males too. Sexual and social selection can both occur during the reproductive period and complement each other.

## BEAK SPOTS AND MUTUAL SEXUAL SELECTION

UV and yellow–orange beak spots occur only on the lower mandible of large penguins and are most highly developed in King penguins, the lower mandible being larger than the upper mandible. UV reflectance from the beak spots of King penguins was discovered at the Marin Bay colony, Possession Island (Crozet Archipelago), in late 2001 (Jouventin et al., 2005). Subsequent measurements of museum study skins revealed that only the beak spots of Emperor and King penguins reflect UV, although some other penguin species have orange beaks. Both the Emperor and King penguins shed "plates" of plastic-like translucent material from these beak spots right after feather molt (Jouventin, 1982; Schull et al., 2016). Juvenile Emperor and King penguins have black beaks that do not reflect UV wavelengths, and both the UV and yellow–orange colored parts of the beak develop with age (Jouventin et al., 2005; Nicolaus et al., 2007). Much of the UV color, but not the yellow–orange color, is lost during molt and then returns in the new beak plate and increases in saturation over time (Fig. 3.16). When comparing birds during nuptial displays and paired birds that were successful at finding a suitable mate, the latter paired birds had significantly stronger UV reflectance than the average for displaying birds (63% reflectance for paired birds; 54% reflectance for displaying birds; Jouventin et al., 2005). This was our first hint that UV reflectance was important to mate choice. In addition, UV reflectance appeared high for both males and females (65% reflectance for males and 61% reflectance for females).

But the UV colors are not the only thing that the birds see, since there is yellow–orange color as well. Fortunately, the UV and yellow–orange colors tend to change together, giving somewhat consistent differences among individuals. In a sample of 125 birds, brightness of the UV and the yellow–orange parts of the beak spot are correlated by over 90% ($r = 0.922$), the saturation of the colors by over 50% ($r = 0.559$) and the hue by over 25% ($r = 0.283$) (Dobson et al., 2008). In general, a change in one of the color parameters accompanies a similar change in the other part of the beak color. However, there is quite a bit of variation in changes in UV and yellow–orange colored parts of the beak, and in different years the colors may change in slightly different ways (Keddar et al., 2015a). Birds in better overall condition are brighter in the colors of their beak spots (Dobson et al., 2008). The

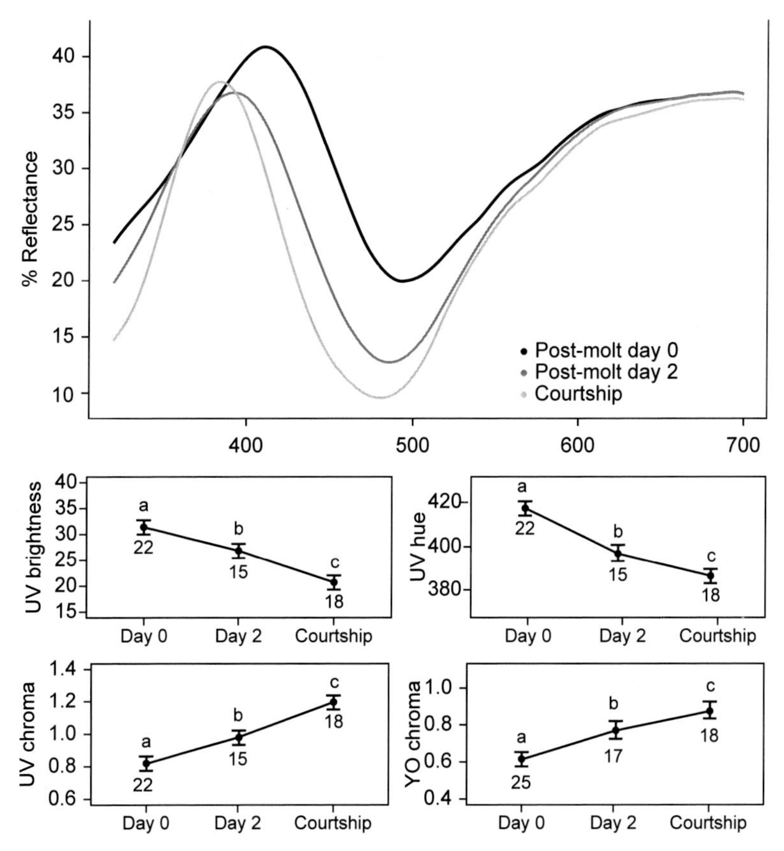


**FIGURE 3.16** Changes in the beak coloration of adult King penguins following the molt. Average changes in raw spectral data over all birds are presented for illustrative purposes. Top, beak color was measured on the day the old beak spot was shed (day 0), 2 days later (day 2), and after birds returned from their postmolt foraging trip for breeding (courtship). Bottom, changes in brightness, hue, and saturation were assessed using linear mixed models, with bird identity specified as a random variable. Least-Square means $\pm$ SE are presented. Values not sharing a common letter are significantly different at $P < .05$. Sample sizes are given below the means. *From Schull, Q., Dobson, F.S., Stier, A., Robin, J.-P., Bize, P., Viblanc, V.A., 2016. Beak color dynamically signals changes in fasting status and parasite loads in King penguins. Behav. Ecol. 27, 1684–1693.*

yellow–orange color of the beak spot reflects concentration of carotene pigment, an antioxidant compound (McGraw, 2005; McGraw et al., 2007), and the microstructures that produce UV color in birds are known to require good body condition (Cuthill et al., 1999).

If male and female King penguins were both selecting the most colorful members of the opposite sex as mating partners, we might expect to see a phenomenon called assortative mating. Assortative mating produces a correlation between males and females in trait value. This is a pattern in which we expect to find the most ornamented males mated with the most ornamented females, leaving relatively poorly ornamented males and females to also pair with one another. While this pattern is not guaranteed under mutual mate choice, it might indicate that mutual mate choice is likely. We found significant assortative mating of males and females for body mass ($r = 0.465$, $N = 19$ pairs), body condition ($r = 0.451$), and UV saturation and yellow–orange saturation of the beak spot ($r = 0.526$ and $0.774$, respectively) at Possession Island early in the 2003–04 breeding season (Fig. 3.17). We found a similar assortative mating pattern in the 2006–07 breeding season at Cape

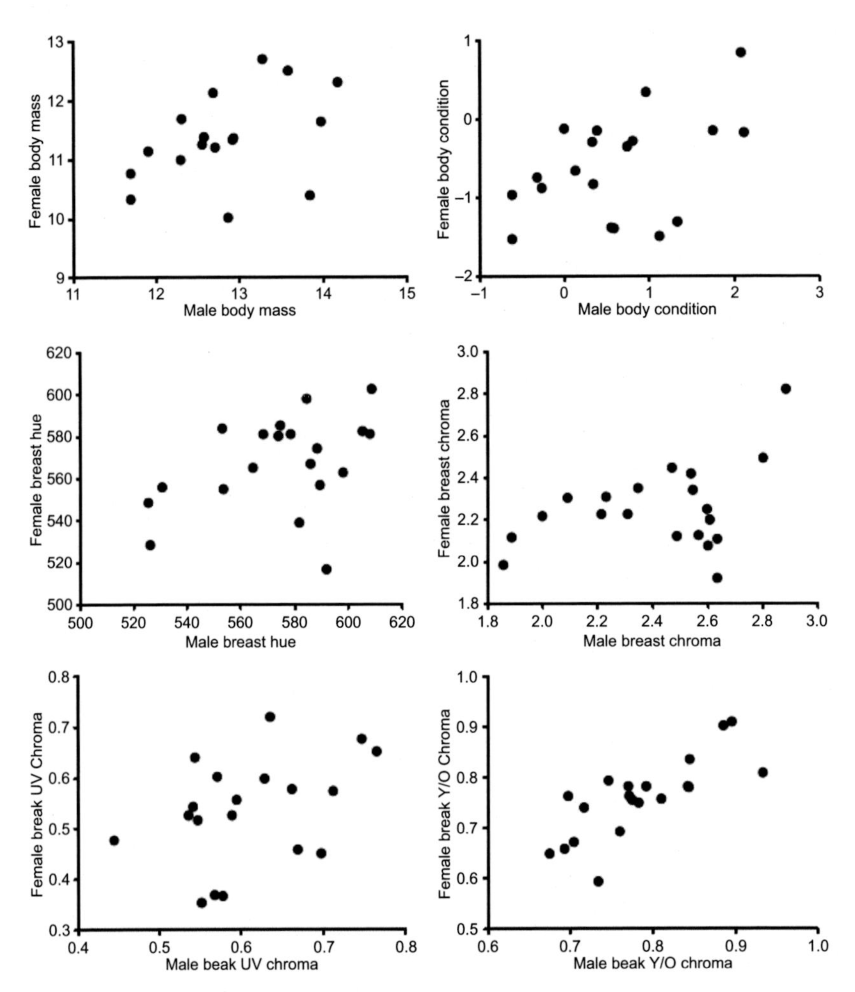


**FIGURE 3.17** Associations between male and female King penguins for body size and color traits. ultraviolet (UV) indicates the UV part of the color spectrum and Y/O the yellow–orange part for the beak spot. *From Dobson, F.S., Nolan, P.M., Nicolaus, M., Bajzak, C., Coquel, A.-S., Jouventin, P., 2008. Comparison of color and body condition between early and late breeding King penguins. Ethology 114, 925–933.*

Ratmanoff, Kerguelen Island, for body mass (r = 0.255, 75 pairs), and UV saturation and yellow–orange saturation of the beak spot (0.575 and 0.370, respectively; G. Pincemy, F.S. Dobson, and P. Jouventin, *unpublished results*). Finally, we examined the penguin colony at Cape Ratmanoff again in 2008–09, and found using the avian vision model based on the eye cone types of birds, that pairs of males and females that produced eggs were significantly more similar in color than expected at random (N = 37 pairs; Keddar et al., 2015c). In addition, pairs that successfully produced eggs were larger and had more colorful beak spots (hue more strongly in the UV range and more yellow than orange) than pairs that failed at reproduction (N = 37 and 36 pairs, respectively).

Comparisons of males and females among displaying and paired birds suggested that the colored beak spots are an important signal for males and females during nuptial displays. An experiment from the Marin Bay King penguin colony in the 2001–02 and 2003–04 breeding seasons supported this conclusion from more natural comparisons. We painted the beak spots of male and female King penguins with an emulsion of powdered chalk and varnish that blocked about 30% of the UV reflectance, though yellow–orange color appeared to be unchanged (Nolan et al., 2010). This was possible because the UV color is caused by the cellular and structure of the beak material, rather than by a pigment. The chalk/varnish coated the beak spot with a layer of material that absorbed much of the UV-reflected light wavelengths. This was done with birds that were displaying and searching for partners; and we measured the length of time that it took these birds to attract mates after the experimental treatment and compared them to control individuals. From the delay in finding a partner, and differences in the likelihood of pairing, we were able to estimate the probability of pairing for treated and control birds, both males and females. We found that treated birds were significantly less likely to pair and took longer to pair, compared to controls (average time to pair was 18.6 and 11.1 days, respectively; Fig. 3.18). This pattern was different for males and females (average time to pair was increased by 77% for males and 23% for females; Keddar et al., 2015c). Thus, the combination of comparative and experimental results suggests that, if the biological use of the yellow color of feathers was not obvious in females, UV reflectance was an important mate-choice signal for both male and female King penguins, though female choice appears stronger than male choice. These results support Huxley's (1914) mutual mate choice hypothesis for similarity of ornaments, in this case the color of beak spots, of males and females; but with the caveat that mate choice may operate in both sexes, but be stronger in one sex than the other (Kokko and Johnstone, 2002).

Information about individual condition of King penguins may be reflected by the UV part of the beak color. Within the Marin Bay breeding colony in 2011–12, an experimental parasite removal using application of an anti-parasite solution showed that for birds of both sexes, UV brightness was greater, UV hue greater, and UV saturation less than in control birds

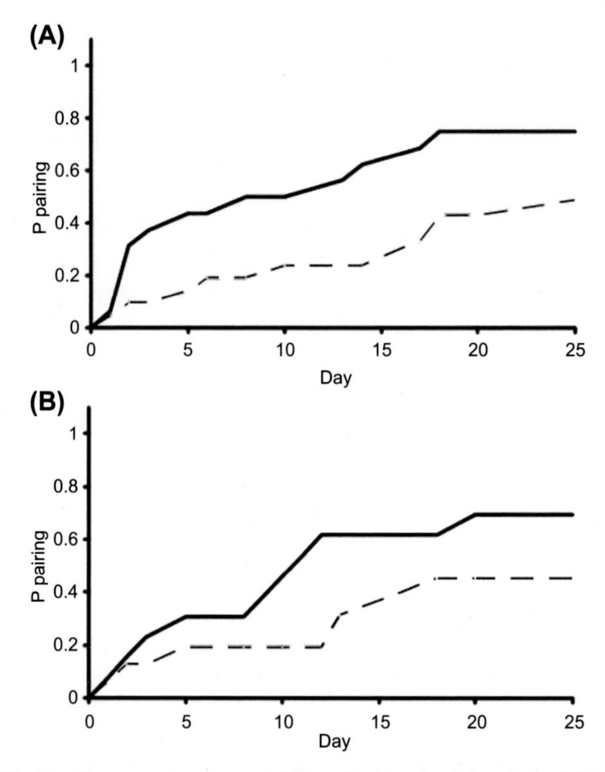


**FIGURE 3.18** For King penguins, the probability of pairing for (A) males and (B) females after 30% experimental reduction of ultraviolet (UV) reflectance from the beak spot. *Solid lines* are control individuals, and *dashed lines* are individuals with reduced UV reflectance. *From Nolan, P.M., Dobson, F.S., Nicolaus, M., Karels, T.J., McGraw, K.J., Jouventin, P., 2010. Mutual mate choice for colorful traits in King penguins. Ethology 116, 635–644.*

(Fig. 3.19). For the same birds, there was no significant difference between the treatment and control in the yellow–orange coloration of the beak. Beak color changes as birds prepare for nuptial display. From the molt onward, beak UV and yellow–orange colors decrease in brightness and hue, and increase in saturation, patterns that are stronger for UV than for yellow-orange colors. In addition, UV brightness may reflect changes in body condition over time. For 36 males that were standing with an egg on their feet in the first incubation shift, a significant decrease in UV brightness occurred over a 10-day period (Fig. 3.20).

Signals used for mating choices are thought to reveal the quality of individuals, either their body condition or aspects of good parenting. This should be especially important in King penguins because successful breeding in this species is a particularly long and complex process that requires both parents to invest for more than a year. This breeding cycle is especially arduous, as

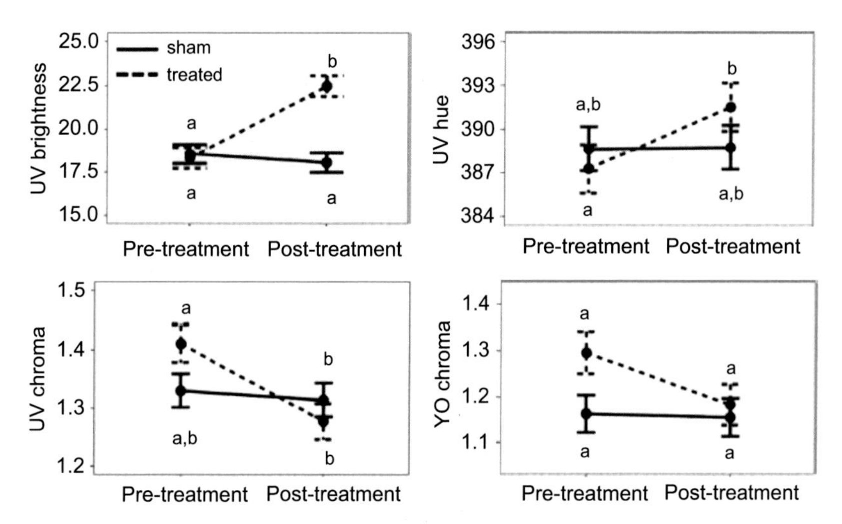


**FIGURE 3.19** Interactive effects of an experimental antiparasitic treatment on the beak coloration of adult King penguins. Changes in brightness, hue, and saturation were assessed. Treatment and control (sham) and time period are shown. Values not sharing a common letter were significantly different at the $P < .05$ level. *From Schull, Q., Dobson, F.S., Stier, A., Robin, J.-P., Bize, P., Viblanc, V.A., 2016. Beak color dynamically signals changes in fasting status and parasite loads in King penguins. Behav. Ecol. 27, 1684–1693.*

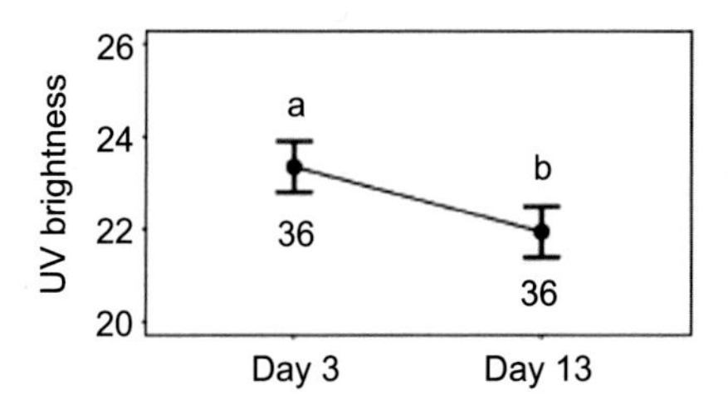


**FIGURE 3.20** Changes in mean beak ultraviolet brightness for male King penguins that were measured on day 3 of their first incubation shift and on day 13, after having experienced a 10-day fast while standing with an egg in the colony. Means ± SE are presented. Values not sharing a common letter are significantly different for $P < .05$. Sample sizes are given below the means. *From Schull, Q., Dobson, F.S., Stier, A., Robin, J.-P., Bize, P., Viblanc, V.A., 2016. Beak color dynamically signals changes in fasting status and parasite loads in King penguins. Behav. Ecol. 27, 1684–1693.*

parents have to care for the completely dependent egg and young chick, then feed the semiindependent chick, while it waits on the beach for sustenance, and finally feed the chick through the energy-expensive process of feather molt. Only after this extended work by both parents is accomplished can the single chick fledge into the sea. Thus, finding a breeding partner in good condition should be extremely important. Links between body condition or body mass and colored ornaments were difficult to document, however, because body mass is both highly changeable in penguins, and difficult to measure in the field without disturbance to the birds. While we devised ways to measure body mass in the field using digital scales, documenting patterns of changing mass were more problematic due to feeding at sea and fasting on shore. After molting their feathers near the breeding colony, both males and females forage at sea before returning to the colony to display and commence the breeding cycle. Males invariably take the first shift of standing with the single egg, and thus they fast for a long time on land (during the whole mating period of about 2 weeks and for another two to 3 weeks standing for the first shift with the egg, before being relieved by the female partner). Thus, males in particular come to the breeding grounds to find a mate with stomachs full of fish and squid, increasing the variation in body mass. Nonetheless, males that arrived early in the breeding season, when chances of breeding success are greatest (Weimerskirch et al., 1992), were both heavier than females and later displaying males, and also much brighter in UV reflectance (Fig. 3.21).

## SIGNIFICANCE OF VISUAL SIGNALS

Our starting problem was to try to understand why male and female penguins look so similar in their colored or black-and-white ornaments. This requires identification of markings that are used as ornaments, and then tests of the meaning or use of the ornaments as signals. We chose to focus on King penguins because, similar to Emperor penguins, they have the most highly developed ornaments. They are very colorful and have three colored patches of either feathers or hard integument (the beak spots). An additional advantage of studying King penguins was that they breed during the austral spring and summer, as opposed to Emperor penguins that breed in winter on sea ice, a harsh environment for experimental field studies and impossible to reach during the season of bad weather. In birds, males often have colored ornaments that attract the mating choices of females. So it was natural to turn to King penguins and focus initially on the beginning of the breeding season, when mating choices of both males and females are being made. This led to the exciting discovery of the UV reflectance of the beak spot, which added an element of mystery to our research, since this color is not visible to humans. But there are other hypotheses to explain the monomorphic color patterns of King penguins.

The most likely of these is social selection for behavioral dominance, since dominance likely contributes to the location of the small breeding territory within the colony where it is safer than around the edges, where skuas and giant petrels can get at the penguins or their eggs (Viera et al., 2008; Keddar et al., 2015b). But we can consider that for ornaments that indicate body condition, a bright coloration is not a signal of social dominance for all of the other birds in a colony, but a signal only for local competitors. In this, the stronger birds occupy the best positions, safe from predators, in the center of

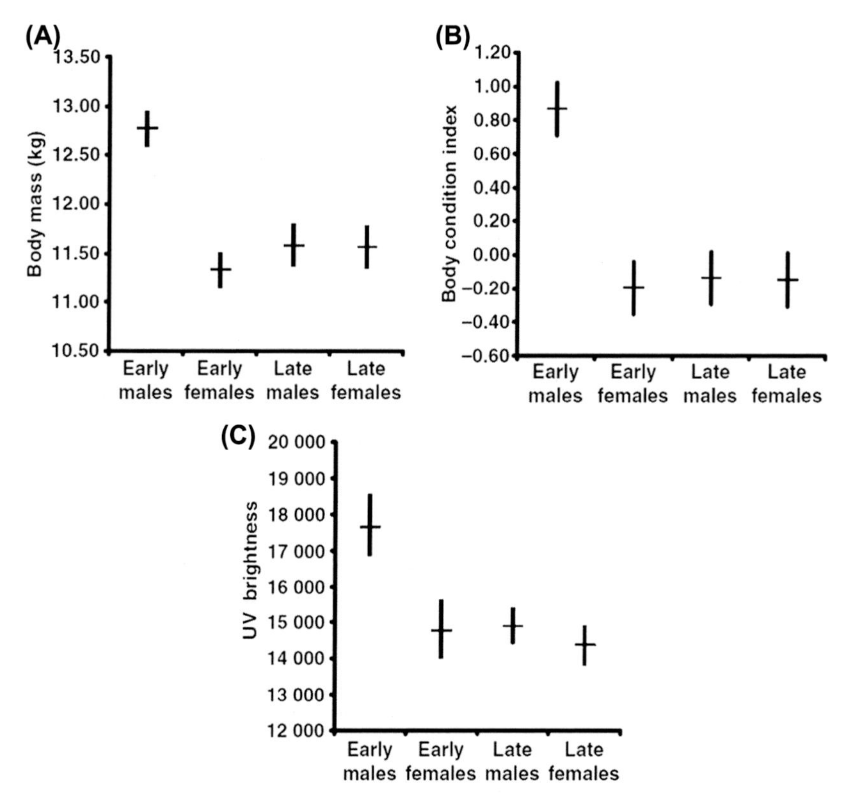


**FIGURE 3.21** Mean body mass (A), condition (B), and ultraviolet (UV) brightness of beak spot (C), of King penguins (±SE) by sex and season. *From Dobson, F.S., Nolan, P.M., Nicolaus, M., Bajzak, C., Coquel, A.-S., Jouventin, P., 2008. Comparison of color and body condition between early and late breeding King penguins. Ethology 114, 925–933.*

the colony. This position must be maintained by force, as evidenced by the high rates of aggression among neighbors (Viera et al., 2008). Thus, any ornament used as a social signal can only partly defray the need to maintain spatial position by aggressive conflict. Of course, another possibility is that individuals in the center of a colony are more aggressive because they are more "crowded," so correlations of position and rate of aggression may be a consequence of settlement, rather than aggression causing the topographical position (Keddar et al., 2015b). In either case, penguins in the center of the colony must be of sufficient condition or quality to withstand high rates of aggression.

King penguins have three colored ornaments, the brown-to-yellow feather patch of the breast, the yellow–orange auricular feather patch, and the UV and yellow–orange beak spot. It is natural to ask why King penguins have so much color, when many other penguin species have few or no colored ornaments.

Two conceptual ideas may help us understand this wealth of coloration. The first is that different ornaments may be more visible at a distance versus up close, and when relaxed versus in nuptial display. The second is that ornaments may reflect individual condition at different times. The breast patch is visible from a distance, and is clearly visible as birds move about the display areas adjacent to the breeding colony. The breast may attract potential partners to a bird that is adult and ready to mate, perhaps a male that shows a good immune system. If the breast is visible from the front and from a distance giving the breeding status and roughly the body condition, the nuptial display ensures that auricular patches and the beak spot are visible during the pairing process and yield much more accurate information about the possible mate.

The auricular feathers and beak spot, especially the structural color of the beak spot, likely convey different information. Feathers are replaced at the time of molt, three to 4 weeks before the nuptial displays. While feathers may change over time, the pigments deposited in feathers, and thus the colors of the breast and auricular patches, likely reflect individual condition at the time of molt (Table 3.1). The tissues of the beak, however, are vascularized and because of their blood supply the carotenoids of the beak can probably be mobilized, at least to some degree. The structural UV color of the beak has also been shown to change over time. Thus, feathers may reflect longer-term indications of individual quality or condition, and the beak spot more short-term and dynamic changes in quality or condition. The difference between the female's strong choice of a large auricular patch in a male mate and the mutual preference for UV and yellow–orange beak color might relate to a basic difference of males and females. As noted before, males invariably take the first incubation shift of standing with the egg. Thus, King penguin males spend about 11–12 days on land before mating and about 16 days during the first shift of incubation (Weimerskirch et al., 1992). Thus, the male must fast on land for 4–5 weeks before being relieved by his female breeding partner (Olsson, 1997). A breeding female needs to find a male who can suffer this long fast without abandoning the egg. At the same time, both sexes need to find partners that can successfully cooperate during the 14 months that it takes to raise a chick to be independent in the sea.

It is notable that Emperor and King penguins have multiple color ornaments, while other species of penguins have only yellow head plumes (the crested penguins), or black and white markings such as chin stripes. Interestingly, the division of multiple colors versus single ornament/marking follows the difference between penguins without a nest (Emperor and King penguins, the only birds with no nest) and other penguins that have at least a nest of pebbles. The nest gives the location of a breeding partner and the young that are being raised. Why do nesting penguins have only pigment in their ornaments when the nonnesting species have an additional UV signal? Is it an original feature inherited by the species from a common ancestor, or is the concentration of pigments in feathers an adaptation to a more complicated ecological problem? Similar to other seabirds, territorial penguins begin the breeding cycle by going to a breeding colony to find a nest site, the male arriving before the female at the start of the breeding season. The position of the nest, peripheral or central, can provide an indication of the quality of the male to the female. Only the best males are able to acquire a nest, and a good one that is well placed near the center of the colony.

In nonnesting penguins of the genus *Aptenodytes*, pairs probably have the most difficult breeding problem to solve. Moreover, they have to choose a very good mate that is able to maintain the egg or young chick during very long feeding trips by their partner, through incubation and brooding. But in the case of Emperor and King penguins, there is no signal of male quality that can be found in an evaluation of the nest. Furthermore, when a nest is not present, additional time and effort must be expended to locate and identify the correct partner or dependent young (Dobson and Jouventin, 2003a). This is even a problem for nesting penguins, thus the need for visual ornaments in those species. We suggest that the UV of the beak gives a complementary signal to pigment-based ornaments. The beak spot is more efficient than a pigment deposited in feathers because structural colors indicate the continual growth of cells, and not the past body condition as reflected by pigments that have been deposited in feathers. Thus, for the largest penguins that have no nest, it is no surprise that advanced forms of communication have evolved in the form of original visual signals. During the "face to face" display, when for several minutes mates inflate their neck and raise their head, looking carefully at the colored head (auricular patches) and beak spots of the partner, each bird can make a close examination of the head ornaments and associated body condition of the opposite sexed partner. Clear and abundant evidence, both from natural measurements and experimental reductions of the size of auricular patches, indicates strong female preference for the size of auricular feather patches of males in King penguins.

UV and yellow–orange colors of the beak spot appear to be important influences on the mating choices of both males and female King penguins. This is not to say that the strength of mate choice is equal. One sex may have a stronger preference for the sexual ornaments than the other, yet both sexes exert a preference. Thus, beak spots may reflect the influence of mutual mate choice and sexual selection acting on both sexes, and a stronger preference by females. Females clearly exercise stronger discrimination among males than vice versa in this species. This is likely due to the strongly male-biased sex ratio among adult King penguins that are displaying to attract a mate. This bias was revealed by our results from trio parades, which showed that males were competing most strongly for females on the display ground. And the stronger preference of females for males is shown in the results on auricular patch size, which reflects ongoing selection on males via female choice. At the same time, evidence from the beak spot supports the possibility of some level of male preference for females. Results showing that some females are more successful than others at moving quickly into the breeding colony and producing an egg for their male suggest that males should prefer them to the less successful females, perhaps contributing to pairs that break up, or "divorce" (Olsson, 1998; Olsson et al., 2001; Keddar et al., 2015c). There are assortative mating patterns for UV and yellow–orange saturation of color of the beak spot (from 3 years of results and two island populations), and assortative mating patterns are less likely to be caused by sexually one-sided ornament preferences. Finally, an experimental reduction of the UV reflection of the beak spot appeared to affect both sexes, and in the expected direction. Nonetheless, males with a UV-block on the beak spot took longer to pair than females (Nolan et al., 2010). Links of UV characteristics of beak spots to body mass or condition occurred for both sexes (Dobson et al., 2008; Keddar et al., 2015c; Viblanc et al., 2016), suggesting that honest information about individual size or quality may be reflected by these ornaments.

The stronger mate choice by females that we discovered in King penguins occurs only in this single species on which we conducted most of our visual experiments. The other nonnesting species, the Emperor penguin exhibits exactly the opposite pattern. Trios in the Emperor penguin are composed not of two males following a female but of two females and one male. This indicates that the role of mate choice between the sexes in the Antarctic species is reversed. We will come back to this topic in Chapter 6, but it appears that differential mortality on the sexes in Emperor penguins causes a difference in adult sex ratio, leading to the difference in which sex competes most strongly for a partner of the opposite sex. To conclude briefly on the sexual selection in these strictly monogamous birds: penguins, as for nearly all all of the more than 300 seabirds, are monomorphic, both sexes being ornamented and thus perhaps suggesting mutual mate choice in all species. Nevertheless, in spite of the strong influence of required sharing of parental duties, dividing feeding at sea and breeding on land, one sex can be choosier, particularly in the two large penguins where the female cannot be helped by information about the location and quality of the nest when selecting a male partner.

To summarize our interpretation of our numerous and complex experiments, feather pigments of penguins can be synthesized by the birds themselves. Once deposited in feathers, however, they cannot be retrieved and used in the immune system. The brightness of feather colors (and the size of auricular patches in King penguins) are linked to the foraging abilities of the particular individual, during the whole of the past year and until feather molt, that influences mating choices by giving information to potential partners. In the King penguin, foraging ability and body condition are also linked to the color of the beak patch, a dynamic ornament (see Fig. 3.22 for the carotenoid color of the Gentoo penguin beak). The results of our experiments showed that the beak patch in King penguins is an "honest" signal that visually indicates to the partner: sexual maturity, present body condition, and probably past

**FIGURE 3.22** Gentoo penguins are colored but it is highly variable and several hypotheses were formulated to explain its determinism and significance, including sexual and species identity, body condition, and general health. We demonstrated experimentally with birds from a marine park that this color comes from carotenoids found in their foods (Jouventin et al., 2007b).

condition or breeding success. All of these messages are indicators to potential partners about whether the bird will be a good parent. Why does structural UV occur only in large penguins, and particularly the King penguin? Differently from pigments, UV color allows a short-term and dynamic evaluation of individual quality, complementary and more accurate than feather color due to the continuous growth and wearing away of the outer tissue of the beak.

Before giving a nuptial display, nesting penguins have more sources of information about individual quality than nonnesting penguins. In territorial species, males come to the breeding colony, and then find and defend a nest. Central locations in the colony are safer from avian predators (Bried and Jouventin, 2001; Massaro et al., 2003; Keddar et al., 2015b) as is true of seabirds in general (Coulson, 1968). Thus, the location, size, or perhaps components of the nest itself may show the quality of the male. Females arrive and thus have both the nest and the ornaments of the male to examine. This is not true of the nonnesting penguins, since they display away from the incubation site and move into the breeding colony only after they have paired. Thus, nonnesting penguins require more information from ornaments. The Emperor penguin has to carefully choose a mate, often a new one each year (86% of breeding partners each year are usually new; Isenmann and Jouventin, 1970; Isenmann, 1971). King penguins have similar difficulty but increased by a long breeding season and successful reproduction only once every 2 years (about 80% of breeding partners are new; Weimerskirch et al., 1992; Olsson, 1998). We suggest that the large nonnesting penguins of the genus *Aptenodytes* have multiple color ornaments, including UV color of the beak, to facilitate mate choice via evaluation of individual condition by partners that are not assisted in their choices by signals given by a nest. Table 3.1 summarizes experiments and conclusions on color ornaments in King penguins.

**TABLE 3.1** Large Nonnesting Penguins Have the Most Sophisticated Ornaments Among Penguins; With Both Pigments and Structural Colors. During Nuptial Displays, Ornaments Indicate Both Short- and Long-Term Body Condition

| | Body Part | Physiological Basis | Evidence/Experiment | Hypothesized Biological Significance |
|---|---|---|---|---|
| Long-term effects (months and year) | Ear patch | Spheniscin color produced once a year during molt | Observations and delay in pair formation, size or brightness of the color reduced, age or maturity, location in colony | Main signal of yearly body condition among birds of the same sex (reveals sexual maturity) and between sexes (males sexually selected), closely inspected during the "face to face" display |
| | Chest | Spheniscin color, also during molt | Experiments on age/sexual maturity: males more sexually selected than females and immunocompetence | Another yearly signal of body condition among birds of the same sex (reveals sexual maturity and body condition), likely viewed at a distance prior to nuptial display |
| Short-term effects (weeks and months) | Beak color | Carotenoids (deposited from food at beak molt | Unpublished observations | Special color signal for large penguins; shows individual quality/condition by carotenoids and faster to give new information to the partner |
| | Beak UV | Ultrastructural color (renewed during beak molt; continuously maintained) | Information on body condition and dynamic short-term changes | External conditions (as food or stress) modify the alignment of cells and change dynamic UV signal in days or weeks, specifies present body condition in the absence of a nest that provides information on individual quality |

Chapter 4

# Description of Vocal Signals

## MODELS FOR ACOUSTIC SIGNATURES

When the study of penguins began more than half a century ago, researchers confirmed experimentally that penguins use vision to find the colony and the nest. But surprisingly, during our comparative experiments on vocal signatures, it became clear that all penguins use only audition to identify their mate and chick (Jouventin, 1982). This identification occurs from learning the "vocal signature" of the mate and chick. Surprisingly, penguins are unable to identify their partner or chick visually. It is a challenge to hear and correctly identify these vocal signatures in the noisy environment of a huge colony. Penguin colonies are so noisy that it seems impossible to clearly hear and correctly identify the song of another bird, especially if the distance is more than a few meters. In this chapter, we will demonstrate this difficulty by describing experiments that were conducted in nature. Penguins are extraordinary models for understanding individual vocal recognition. Over a period of about 40 years, we conducted experiments to understand the complexities of vocal communication in this exceptionally diverse family in terms of sonic characteristics within a great diversity of ecological and social contexts for communication.

In the penguin family, there are species that nest in easily located burrows under bushes (Little penguins), under rocks (Rockhopper penguins), above rocks (Macaroni penguins), on a nest of collected stones (Adélie penguins), on grassy knolls (Gentoo penguins), and for the large penguins even on their feet, either on sandy or pebble beaches without landmarks (King penguins) or on the sea-ice (Emperor penguins). The last species, the Emperor penguin, walks about extensively with its egg on feet, so that there is little topographical information that can be used to find the breeding partner, other than the general location of the breeding colony. In the penguin family, we consequently observe a gradient of difficulties, according to the species, in the ability to find and identify the partner and chick. The penguins exhibit a range of natural constraints that provide ideal comparisons of individual recognition by vocal signatures, a situation almost experimental, but without human interference.

**Why Penguins Communicate. http://dx.doi.org/10.1016/B978-0-12-811178-9.00004-4**

The first aspect of vocal signals that we can ponder is that they likely indicate species identity to potential mates of the species. A unique song prevents confusion between species, since all of the species have different songs, and thus protects the singer from inbreeding (see Chapter 6). In turn, these species-unique songs are specially adapted to a penguin's way of life. Vocal information confirms the visual information provided by the unique head pattern of colored feathers in each species, as we saw in Chapter 1 (Fig. 1.4). A second and more surprising aspect is that the song can give information about sex and social status (e.g., whether unpaired, a paired breeder, or the owner of a nest) of the singer, thus facilitating efficient pair formation when the breeding season commences.

A third aspect of songs, perhaps the most critical to successful reproduction, is for identification of the breeding partner and between parents and chicks. Penguins are the bird family most strongly adapted for life in the water. But since they are birds, they must breed on land, a scarce resource in the southern hemisphere. Thus, all penguins are colonial breeders. Penguin colonies are often composed of thousands of pairs. But breeding in such a crowded place generates a cascade of problems. First, individuals have to find a place or location in which to conduct their breeding activities. Second, several times during a breeding cycle they have to locate and navigate to their nest among thousands of territorial birds. This last aspect might be especially difficult when both parents leave the nest to forage in the sea, and then return to feed the semiindependent chick. In some species, the chick wait for the returning parents in a crèche with many other similar chicks. Penguins are much noisier on land than at sea, where they rarely sing while swimming. But even at sea they stay in communication using a contact song. This grouping song assists collective swimming and discovery of the best fishing areas. When returning to the site of the nest each year, breeders have to defend their territory by aggressive postures and agonistic songs, a pattern common to all seabirds.

## SMALL NESTING PENGUINS

There are three types of vocalizations in penguins: (1) contact songs that enable isolated birds to locate their group at sea, (2) agonistic songs, mostly uttered in territory defense, and (3) the display song of single birds (ecstatic display song) or pairs (mutual display song) that are used for individual identification. Within a song, the smallest units will be termed "syllables," which may be combined into "phrases" (Fig. 4.1). Several phrases and their intervals form a "sequence." Sonograms give a picture of sound in three dimensions, frequency in kilohertz on a y-axis, intensity represented by blackness, and time passing along an x-axis. Sometimes, a graph that represents the amplitude of the sound, taken from an oscilloscope, appears below the sonogram, with the same timing as the sonogram.

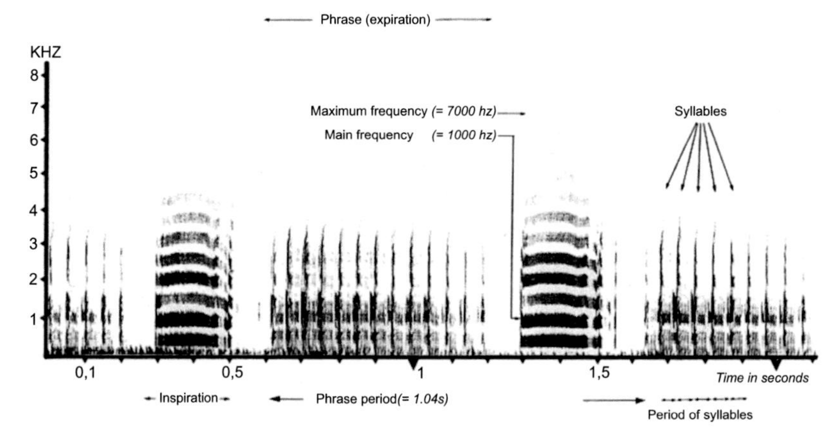


**FIGURE 4.1** Sonogram of the Gentoo penguin from South Georgia Island showing the structure of a song.

## Pygoscelids Penguins (Genus *Pygoscelis*)

The contact song of the Adélie penguin (Fig. 4.2A) is often heard in groups of penguins at sea, but rarely on land. It is a relatively quiet grunt, rising to about 2000 Hz and lasting 0.3 s, and is less frequent than contact songs of other species. The agonistic song (Fig. 4.2B), used when threatening other birds, is quite variable (Fig. 4.2C) and may transition into the display song. The agonistic posture ("beak-to-axilla") is accompanied by a repeated dull grunt, while the bird whirls its beak at the base of its flippers (Fig. 4.2D).

The Adélie penguin is the only penguin with two completely different display songs, one expressed before mating when males are alone, and the other one given after pair formation between two partners. Bachelor males on the nest in ecstatic display begin with a series of clapping sounds (Fig. 4.2E). Later, after pair formation, ecstatic displays end with the mutual display song with their partners (Fig. 4.2F, H and J). The two sexes may sing a duet together (Fig. 4.2I). To record vocal behavior within the colony, especially duets, a microphone was tied to a pole and positioned as close as possible to an individual penguin. Sometimes the partner's beak was taped closed to prevent sound mixing because the main problem recording penguins is to isolate a song from the background noise. Recordings of ecstatic and mutual display songs were made with a bare microphone held several cm from the bird's beak. These two signals (Fig. 4.2G and H) clearly showed the same structure, that is, a sequence of syllables. The complete song (Fig. 4.3, first two lines) may last 1–10 s, depending on the animal's motivation, and consists of a series of very similar phrases, six in Fig. 4.3. Redundancy facilitates recognition in the surrounding noise of the colony. In

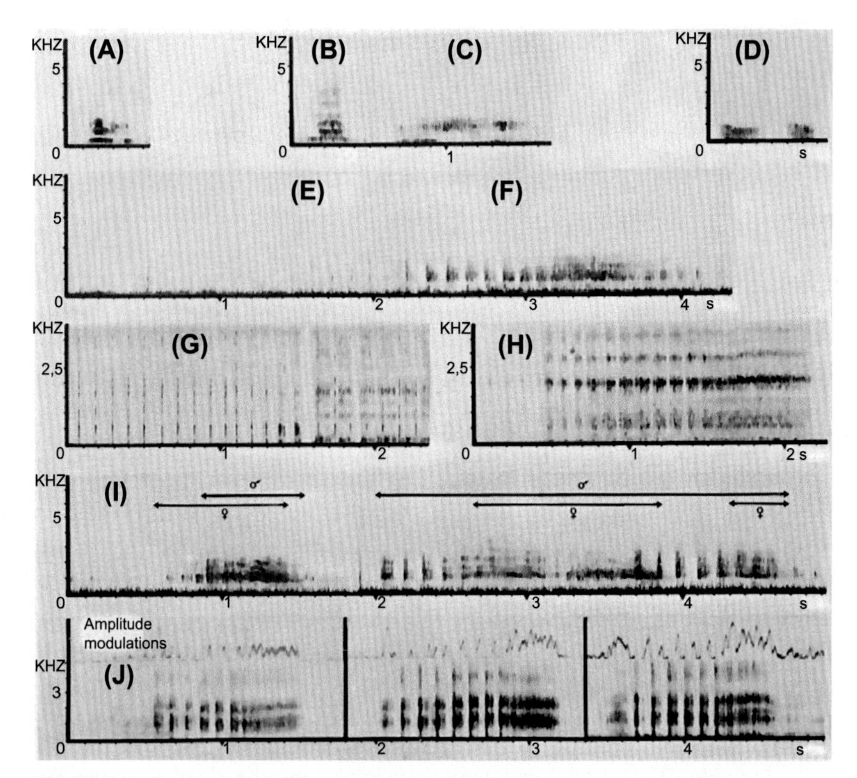


**FIGURE 4.2** Songs of the Adélie penguin. (A) Typical contact song of the Adélie penguin. (B) Typical agonistic song. (C) Less typical agonistic song. (D) Song uttered in the "beak-to-axilla" agonistic behavior. (E) First part of the ecstatic display song (series of clappings) given by males on the nest to attract females. (F) Last part of the same song (identical to mutual display song). (G) The ecstatic display song is enlarged showing the clappings. (H) The ecstatic display song is enlarged showing the mutual display song. (I) Duo during mutual display. (J) Three mutual display songs uttered by one individual (an amplitude modulations graph is shown above the song analysis). *From Jouventin, P., 1982. Visual and vocal signals in penguins, their evolution and adaptive characters. Advances in Ethology,* ***24****, 1–149.*

many species of penguins, chick songs are uttered in series of threes for the same reason.

Vocal sexual dimorphism was not evident in the Adélie penguin, either by ear or from examination of sonograms. The singer's sex was determined whenever possible, however, from other evidence. The ecstatic display, for example, is specific to males, and thus may suffice to differentiate the sexes, for human observers as well as for penguin females. On the other hand, sonogram and amplitude modulation analyses (Fig. 4.2J) show that although the number of syllables varies from phrase to phrase, interindividual variation in the mutual display song (Fig. 4.3, bottom left) is greater than intraindividual

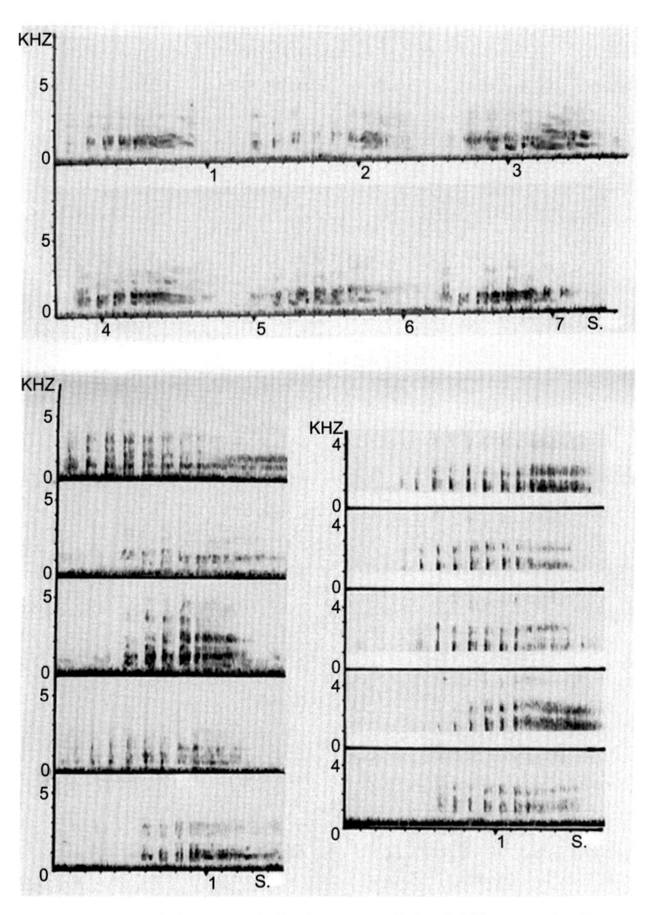


**FIGURE 4.3** Sonograms of the mutual display song of the Adélie penguin. Top, same-phrase repetition, evident in this analysis of six repetitions of the whole song. Bottom-left, songs of five individuals. Bottom-right, five phrases of different songs of one individual. *From Jouventin, P., Roux, P., 1979. Le chant du manchot adelie (*Pygoscelis adeliae*). Role dans la reconnaissance individuelle et comparaison avec le manchot empereur non territorial. Oiseau et La Revue Francaise d'Ornithologie 49, 31–37.*

variation (Fig. 4.3, bottom right). The frequency distribution of mutual display songs is also quite stable within individuals. The statistical analysis of sonogram measures (Table 4.1) yields the following results. The mean coefficient of variation (a mean-standardized measure of variation) for parameters relative to sequence (length of syllables) is 9.91% for songs within individuals (15 songs, two individuals) and 19.38% for songs of different individuals (n = 15), giving a population to individual ratio of 1.95. The mean of coefficient variation for parameters relative to maximum and main frequencies is 9.28% for

**TABLE 4.1 For Adélie Penguins, Comparisons of General Song Characteristics, With Song Measurements of 15 Males, and Then Multiple Songs of Two Individual Males**

| | | Main Frequency in Hz | Maximum Frequency in Hz | Length of Phrase in 1/100 s |
|---|---|---|---|---|
| Songs of 15 different individuals | Mean (n = 15) | 1432 | 4326 | 113 |
| | Extreme values | 1000–1200 | 3000–5200 | 84–190 |
| | Coefficient of variation | 15.57% | 14.77% | 20.70% |
| 9 songs of one individual | Mean (n = 9) | 1345 | 3980 | 96 |
| | Extreme values | 800–1800 | 3500–4500 | 84–115 |
| | Coefficient of variation | 9.59% | 9.53% | 9.81% |
| 6 songs of another individual | Mean (n = 6) | 1756 | 4666 | 113 |
| | Extreme values | 1000–2500 | 4000–5000 | 97–132 |
| | Coefficient of variation | 9.28% | 8.74% | 11.94% |

From Jouventin, P., Roux, P., 1979. Le chant du manchot adelie (*Pygoscelis adeliae*). Role dans la reconnaissance individuelle et comparaison avec le manchot empereur non territorial. Oiseau et La Revue Francaise d'Ornithologie 49, 31–37.

individual songs and 15.17% for songs of different individuals, giving a lower population to individual ratio of 1.63. However, it should be noted that the mutual display song in the Adélie, a fiercely territorial species, serves only to identify the bird beside the nest as the partner (further discussed in Chapter 6).

Jouventin and Roux (1979) conducted several field experiments to examine the mechanism of individual recognition. Obstruction of olfactory ducts and painting the plumage of Adélie penguins did not interfere at all with partner recognition. But once a bird's beak was taped closed, it was pecked at by its partner when returning to the nest, especially if its partner was the more aggressive male. At the same time, an incubating bird on the nest that is silenced with tape around the beak will leave the nest to the singing incoming partner. The incoming bird's song is clearly indispensable for an incubating or brooding bird to accept relief from its returning partner at the nest. But this also depends on the incoming bird's behavior: if it hesitates, as an intruder would normally, the bird on the nest retaliates more violently than if the incoming bird adopts the insistent posture of an owner, whether or not it is more determined (males) or submissive (frequent in females). The nest plays an important role too. When Adélie penguin pairs were placed together in an enclosure outside the colony, partners behaved as strangers and pecked at each other.

Field observations of Adélie penguins (Sladen, 1958; Penney, 1968) indicated that the mutual display song provides partner recognition. This was confirmed by the following observations and experiments (Fig. 4.4, from

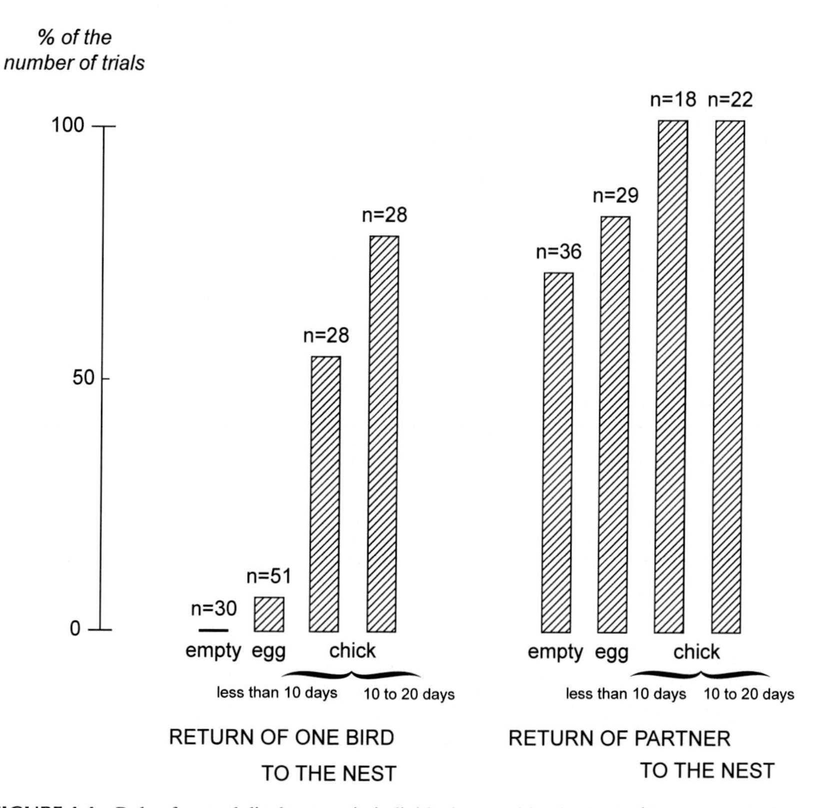


**FIGURE 4.4** Role of mutual display song in individual recognition in the Adélie penguin. Left, a breeding bird, returning to its nest, sings the mutual display song (*hatched rectangles*) only when the chick is hatched and more often when the chick becomes older. Right, the return of a partner almost always involves mutual display singing with the nesting partner. *From Jouventin, P., Roux, P., 1979. Le chant du manchot adelie (*Pygoscelis adeliae*). Role dans la reconnaissance individuelle et comparaison avec le manchot empereur non territorial. Oiseau et La Revue Francaise d'Ornithologie 49, 31–37.*

Jouventin and Roux, 1979). Adélie penguins returning to their nest (on the left in Fig. 4.4) do not sing if it is empty (n = 30), sing in 3 out of 51 cases if it contains an egg, in 15 out of 28 if the chick is still young, and in 22 out of 28 when the chick is older ($\chi^2_3 = 66.9$, n = 137, $P < .0001$). If the partner is present upon its return (right side in Fig. 4.4), the frequency of display songs greatly increases but continues to rise nevertheless from an otherwise empty nest (25 out of 36) to a nest occupied by a large chick (22 out of 22; Fisher exact test, n = 105, $P < .002$).

Furthermore, the mutual display song (Fig. 4.2I) is used in parent–chick recognition as described before by Sladen (1958), Penney (1968), and Thompson and Emlen (1968). When a parent returns from foraging at sea to an abandoned nest (Fig. 4.5E), it utters the display song (Fig. 4.5F). Nearby chicks are attracted and harass the returning parent and a chase may be initiated (Fig. 4.5G). If the chick responds to the parent's song, foreign chicks are pecked away and the parent feeds its own chick (Fig. 4.5H). Parent and chick then return to the nest where the two engage in duets. The final decision concerning aggression and feeding lies with the parent. The apparent disorder involved in selecting the chicks for aggression or feeding is in fact highly efficient. Encounters occur inside the territory where other potential partners are excluded, and the choice among the starving chicks is at first grossly accomplished by chasing nonoffspring chicks away. But the process depends on the chick's response to the parent's song and parental identification of the chick from the chick's song (both memorized before separation).

According to Penney's experiments (1968, p. 124), Adélie penguin chicks return to the nesting site when they heard a parent's song. Penney believed that vocal recognition was enhanced by visual recognition but provided no evidence to support this idea. Visual recognition seems unlikely because chicks that have their appearance changed with paint are readily accepted (Jouventin, 1982). If both a legitimate and a foreign chick are beak-taped so that they cannot sing, the adult is easily confused about which chick to accept, though four of six adults eventually recovered their own chick. The chick's behavior greatly influenced this result because the legitimate chick was generally much more insistent. On the other hand, a parent recently separated from its chick easily adopts a foreign one if it is silent, and may even accept it, although less readily, if it sings. Nevertheless, parents may be motivated to feed alternative and unrelated chicks, probably due to parental hormones, when their own chicks are not present, as was demonstrated in King and Emperor penguins (Mauget et al., 1995; Jouventin and Mauget, 1996; Garcia et al., 1996; Jouventin et al., 1995; Lormée et al., 1999; Angelier et al., 2006).

When do parents learn to recognize their young? Fig. 4.6 shows sonograms of four Adélie penguin chick songs at different ages. At 3–10 days, chick songs are highly variable in the same individual, not very complex, and differ little between individuals (Jouventin, 1982 gives measurements that confirmed this). At 22 days after the chick has left the nest, a chick's song is highly

**FIGURE 4.5** Parent–chick individual recognition. (A) Emperor penguins memorize their partner's song to identify their partner after a feeding period at sea. (B) Parent and chick memorize their respective songs that will allow them to identify each other later on. (C) An Emperor penguin returning from the sea sings in front of a group of chicks, to find and feed its own chick. (D) Adélie penguins also forage at sea. (E) They then return to their colony where the chicks await them near the nest. (F) On the site of the nest, the arriving bird sings. (G) Various chicks run toward the singer to be fed. (H) The parent feeds the legitimate chick, once it recognized the chick's song and chases the others away. *From Jouventin, P., 1982. Visual and vocal signals in penguins, their evolution and adaptive characters. Advances in Ethology,* ***24****, 1–149.*

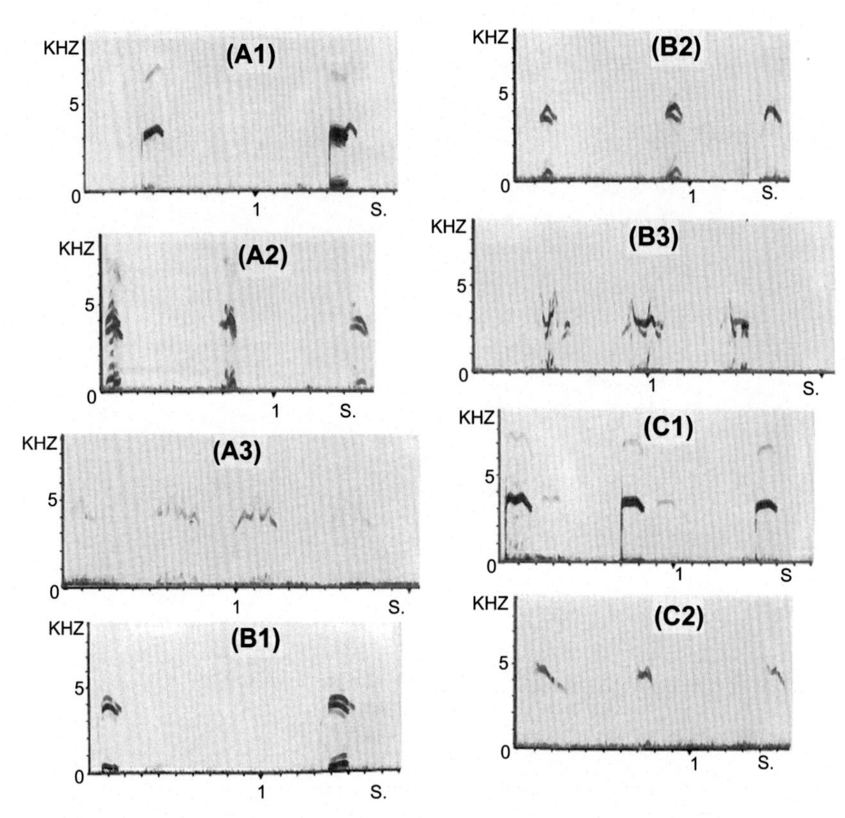


**FIGURE 4.6** Song of Adélie penguin chicks. (A–D) = four chicks. Stage (1) 4 days old. Stage (2) 10 days old (just before emancipation). Stage (3) 22 days old just before moulting. *From Jouventin, P., Roux, P., 1979. Le chant du manchot adelie (*Pygoscelis adeliae*). Role dans la reconnaissance individuelle et comparaison avec le manchot empereur non territorial. Oiseau et La Revue Francaise d'Ornithologie 49, 31–37.*

constant and may be distinguished from those of other chicks. Chick exchange experiments (i.e., cross-fostering) are very successful during the first week, but then become more and more difficult as soon as chicks start moving away from the nest, and ended in total failure once chicks were semiindependent. Song development was followed in the same chick from hatching to departure into the sea (Jouventin and Roux, 1979). Up to the time of leaving the nest (Fig. 4.7A–C), there was little change in the chick's songs. Afterward, however, chick songs become more and more differentiated (Fig. 4.7D–F) and developed into the adult song during feather molt (Fig. 4.7G–J). Syllables of the adult song originate from modulations of the chick's song (Fig. 4.7H).

The song of the Chinstrap penguin sounds chopped (Fig. 4.8A, A′). The frequency range of the song can reach a maximum of 4000 Hz, with frequencies usually around 2000 Hz. The song lasts 5–10 s (based on 10

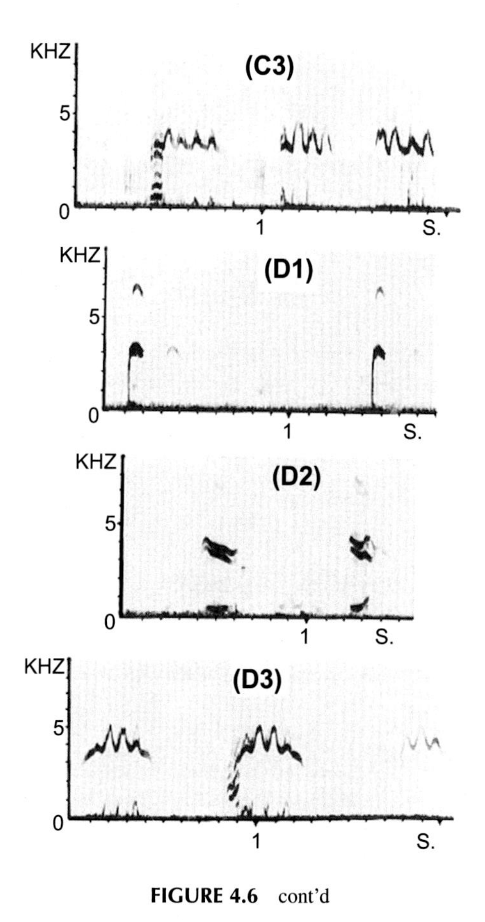


**FIGURE 4.6** cont'd

individual songs from South Georgia). Bowing, rare in Adélie penguins, is common in this species and associated with a song that resembles hissing (Fig. 4.8B). This song is often heard in agonistic situations, and it may have an appeasement function. It reaches 2500 Hz and lasts about 1.5 s. The mutual display song between partners or parents and chicks (Fig. 4.8C, C′, C″) is very different from that of the Adélie penguin and sounds less chopped than the Chinstrap penguin's ecstatic display song (20 songs of different individuals from South Georgia, provided by B. Despin). On recordings, variation among individuals is again greater than variation in repeated songs within individuals.

Bowing is the commonest gesture in the Gentoo penguin. It may be repeated up to 20 times in succession and is accompanied by a barely audible song (Fig. 4.9A), similar to the song given during bowing in Chinstrap penguins. As for all vocal signals of adult Gentoo penguins, this song sounds chopped. The contact song (Fig. 4.9E) is low-pitched. Agonistic songs are

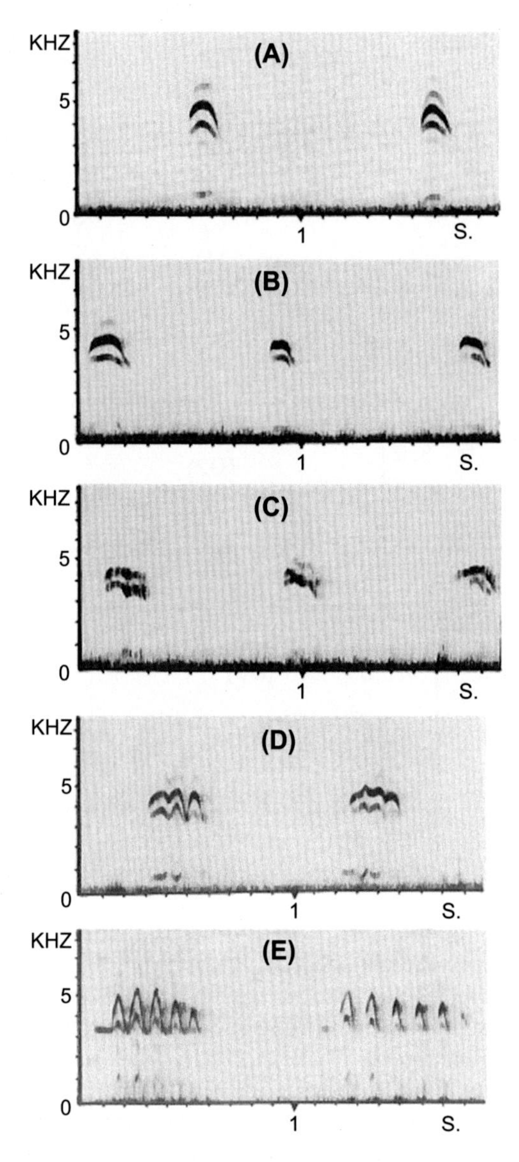


**FIGURE 4.7** Ontogenesis of the Adélie penguin chick song (from hatching to departure for sea). Song sonograms of the same chick at age: (A) 3 days, (B) 9 days, (C) 11 days, (D) 13 days (beginning of emancipation), (E) 15 days, (F) 19 days, (G) 29 days (beginning of molt), (H) 42 days, (I) 46 days, and (J) 49 days (end of molt). *From Jouventin, P., Roux, P., 1979. Le chant du manchot adelie (*Pygoscelis adeliae*). Role dans la reconnaissance individuelle et comparaison avec le manchot empereur non territorial. Oiseau et La Revue Francaise d'Ornithologie 49, 31–37.*

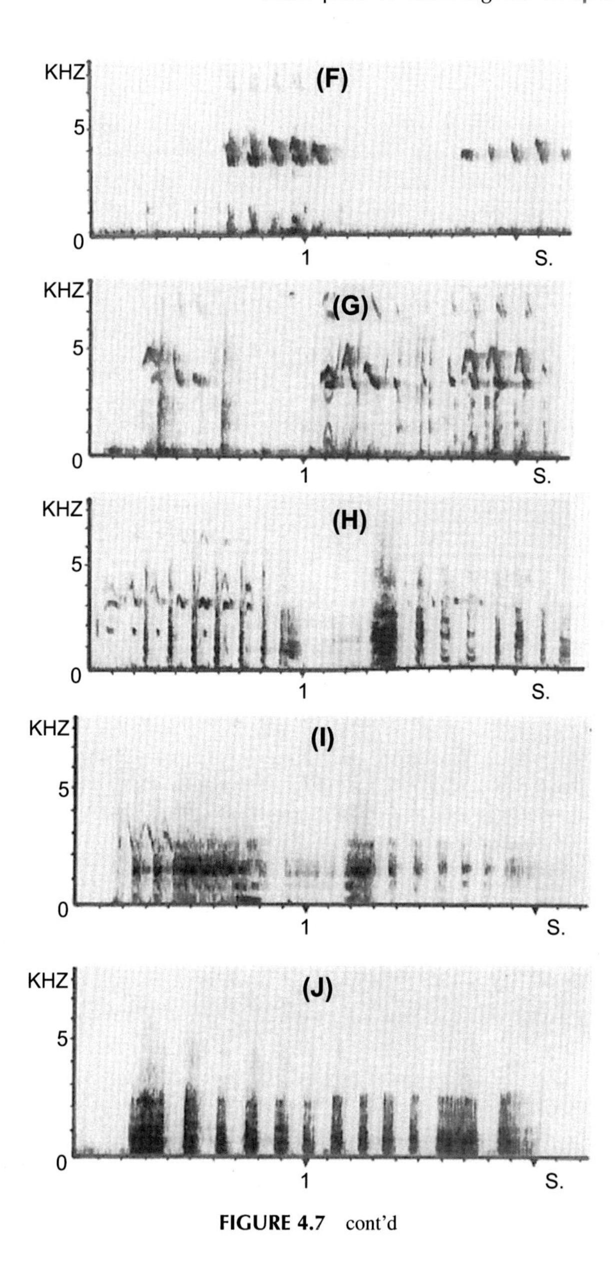


**FIGURE 4.7** cont'd

polymorphic grunts that suggest abbreviations of the display song and for this reason they are not shown. Ecstatic and mutual display songs of Gentoo penguins are closely related. Comparing 20 birds, mutual display songs are less irregular, with shorter phrases, than ecstatic display songs. They are highly

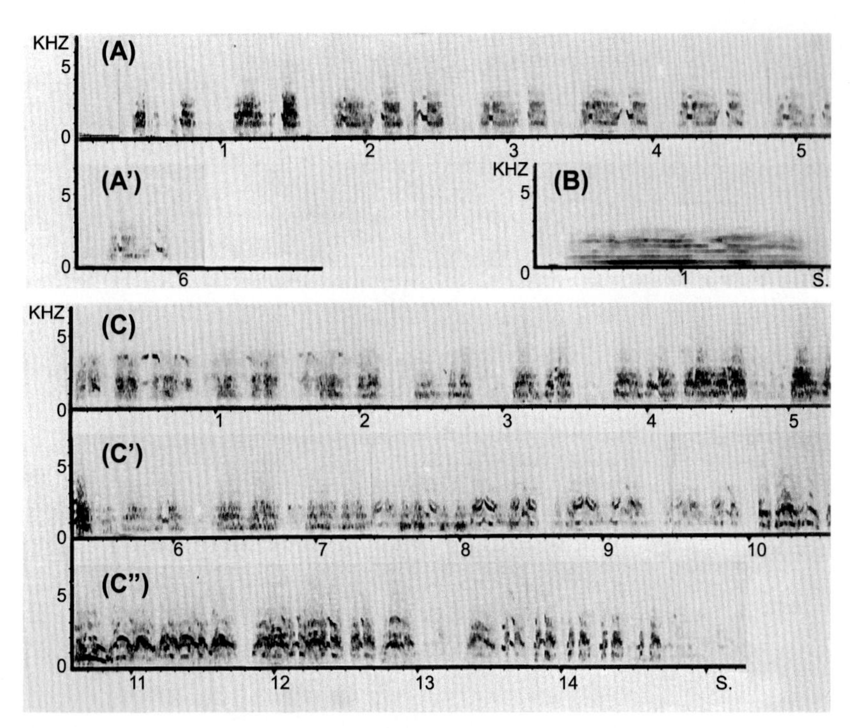


**FIGURE 4.8** Sonograms of the mating songs of the Chinstrap penguin. (A, A′) ectatic display song; (B) song uttered after bowing; and (C, C′, C″) mutual display duet. *From Jouventin, P., 1982. Visual and vocal signals in penguins, their evolution and adaptive characters. Advances in Ethology, **24**, 1–149.*

contagious; one song stimulates other birds to sing, which can spread throughout the whole colony. The structure of these songs (Fig. 4.9A–C) is very different from those in the other two *Pygoscelis* species. There are no syllables of varying length combined into similar phrases as in Adélie penguins or into slightly chopped songs as in Chinstrap penguins. Instead, the song is a highly regular series of almost identical syllables separated by inspirations (the unchopped pattern in sonograms). Song variation among individuals (Fig. 4.9A vs. B and C) was again greater than within individuals (Fig. 4.9B and C), thus giving a physical basis for individual recognition by the partner and chicks. Gentoo chick songs consist of a modulated whistle, a variation in time or frequency associated with amplitude modulations that is relatively stable in the same individual and differs between individuals. Fig. 4.9F shows two almost identical songs of one chick and (Fig. 4.9G) the different song pattern of another, again giving a physical basis for individual recognition, this time by parents. When molting their juvenile plumage (Fig. 4.9H), chicks develop the adult song.

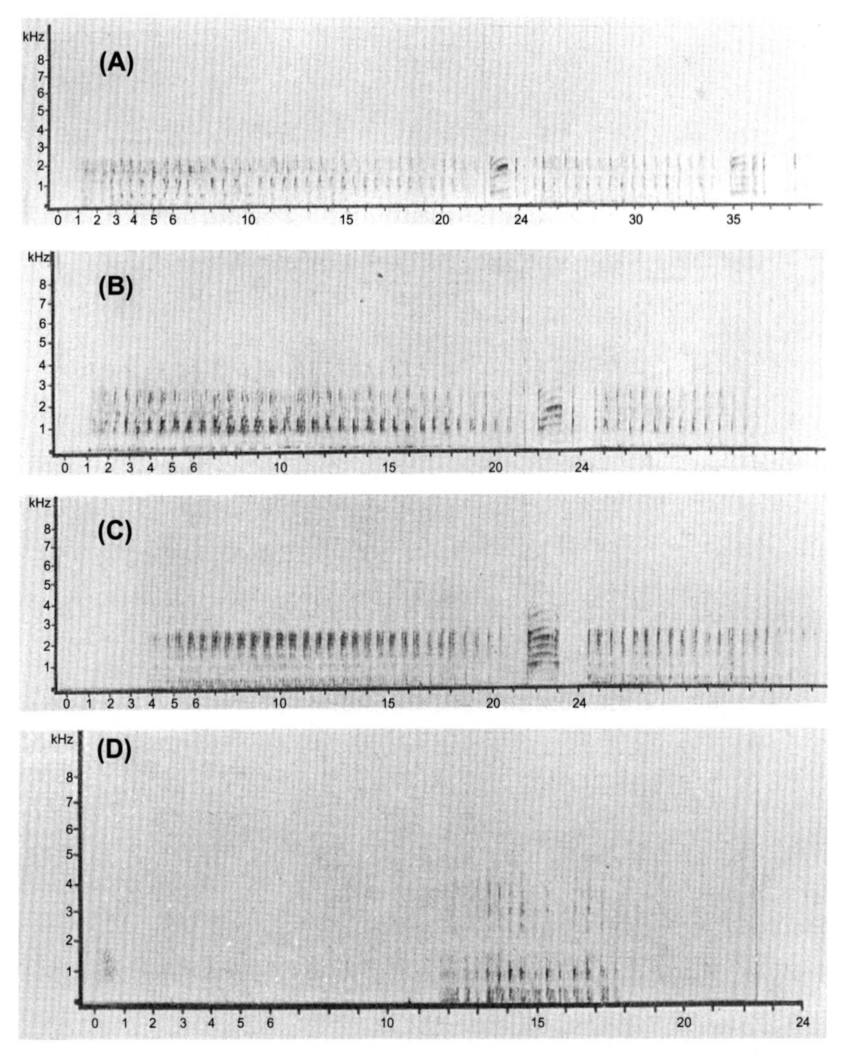


**FIGURE 4.9** Songs of the Gentoo penguin. (A) nuptial song; (B, C) two songs of the same individual; (D) song uttered during bowing; (E) contact song; (F) two songs of one molting chick; (G) song of another molting chick; and (H) chick song at the end of molting showing genesis of adult song. *From Jouventin, P., 1982. Visual and vocal signals in penguins, their evolution and adaptive characters. Advances in Ethology,* ***24****, 1–149.*

## Banded Penguins (Genus *Spheniscus*)

Paradoxically, although residing solely on "inhabited" coasts, and often kept in zoological parks, the Banded penguins were among the least studied for their vocal repertoire. But recently, Favaro et al. (2014) studied Jackass penguins in a zoological park. Our contact songs of Jackass penguins were recorded in the

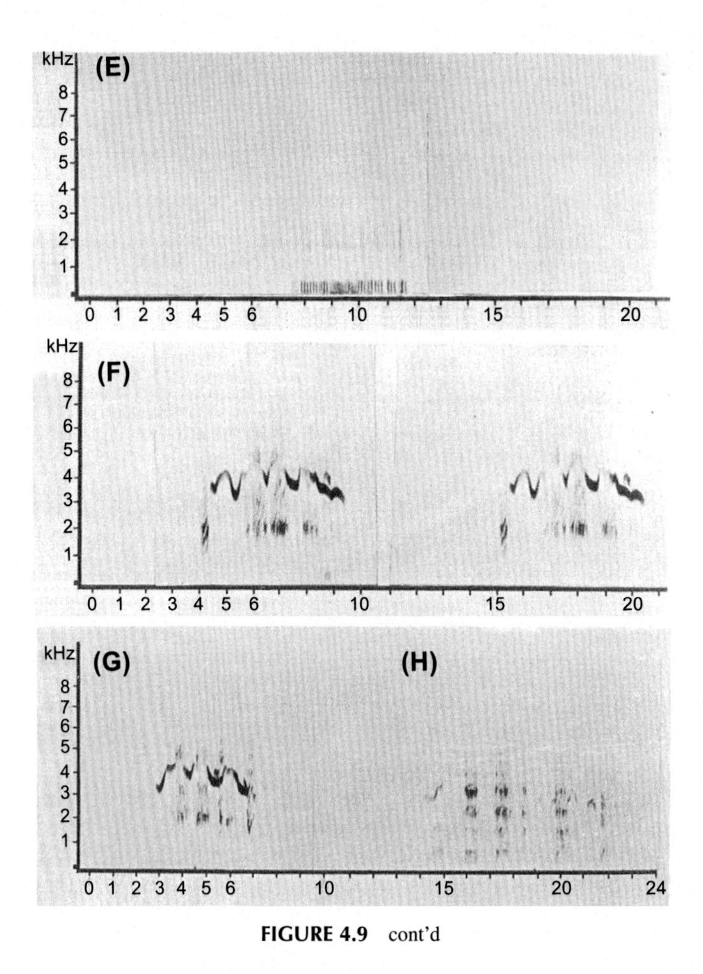


**FIGURE 4.9** cont'd

field, and Magellanic and Humboldt penguins were recorded in a zoological park (Jouventin, 1982). The ecstatic display song of the Jackass penguin that breeds in South Africa is very like the braying of a jackass, hence its name. This very peculiar song is a sequence of high-pitched modulated inspirations with longer and louder expirations (Fig. 4.10A and A′). It is highly contagious, as in the other species of Banded penguins, and up to several dozen individuals in a breeding colony can be heard to utter it in chorus. In the field, where singers are numerous and easy to compare, repeated songs of an individual looked similar, but differences among individuals were apparent.

The ecstatic display songs of the Magellan and Peruvian penguin appear more similar to those of other penguins; that is, they consist of a more uniform sequence of short phrases. The Magellanic penguin contact song (Fig. 4.10B) is monosyllabic. The intensity of this song increases then decreases rapidly, and the song is brief and shows a specific pattern. Although related in their general form, these contact songs show some species-specific characters. They are used to gather group members at sea. Mutual display songs are similar to

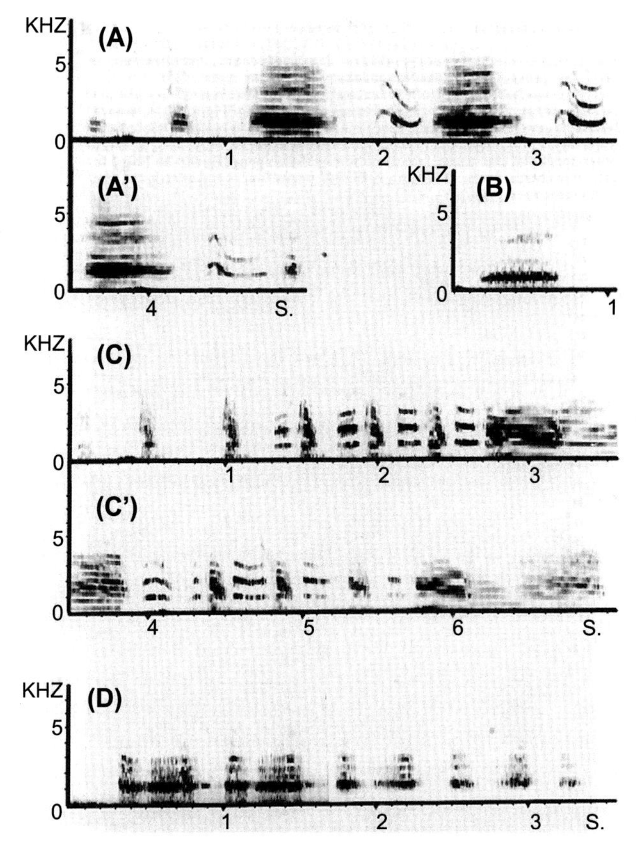


**FIGURE 4.10** Songs of Banded penguins. (A, A′) Ecstatic display song of the Jackass penguin. (B) Contact song of the Magellan penguin. (C, C′) Mutual display song of the Jackass penguin. (D) Mutual display song of the Magellan penguin. *From Jouventin, P., 1982. Visual and vocal signals in penguins, their evolution and adaptive characters. Advances in Ethology,* ***24****, 1–149.*

those of the ecstatic display, although their structure is more complex. Fig. 4.10C shows the sonogram of a mutual display song of the Jackass penguin. The song is composed of a series of six short expirations separated by four inspirations, followed by a very long expiration (which continues onto the next line), then a new series of short in- and expirations ending in a long expiration. Although the Magellan penguin mutual display song (Fig. 4.10D) resembles other nesting penguins in its temporal pattern, its expirations are similar to those of the Jackass penguin.

## Little Penguins (Genus *Eudyptula*)

The general form of the Little penguin's song recalls that of Gentoo penguins: phrases of uniform syllables with a short, modulated inspiration between them.

Just before sunrise, a continuous concert of Little penguin songs can be heard at Phillip Island, Australia, where Jouventin (1982) recorded them. Ecstatic (Fig. 4.11I) and mutual display (Fig. 4.11J) songs are similar, although the ecstatic song has a more homogeneous structure. Syllable form and intervals between syllables are more regular in Little penguins than in other penguin species. Various songs of one individual are almost indistinguishable (Fig. 4.11A–D), but between individuals the rhythm is completely different (Fig. 4.11E–H). In Little penguins, the agonistic song (Fig. 4.11K) is a grunt. Some highly differing agonistic songs are shown in Fig. 4.11L, to give an idea of the great variability, depending on the situation and motivational state of the animal. From recordings in the field (Jouventin, 1982), the contact song (Fig. 4.11M) is monosyllabic and differs between *Eudyptula minor minor* recorded in Dunedin, New Zealand, and *E. minor novaehollandiae* recorded at Phillip Island in Australia (Fig. 4.11N). This suggests that phylogenetic analysis, perhaps using DNA techniques, should indicate a split between the populations (i.e., a lack of gene flow between the populations over a considerable period of time).

As in Gentoo penguins, the chick songs of Little penguins show greater differences among than within individuals, both just before (Fig. 4.11O) and during chick emancipation (Fig. 4.11P). When molting (Fig. 4.11Q), the chicks develop the adult song as in other species. Statistical analysis of song characteristics in Little penguins confirms the sonogram study. Variation in length is similar between 15 songs of one individual and in songs of 30 different animals; thus, this song trait seems not to be used in individual recognition (Table 4.2). Main and maximum frequencies vary much more in a population sample of 30 individuals (69.6%) than in a single bird (16.9%), yielding a ratio of 5.99. As well, syllable length (parameter relative to sequence or syllable period) yields a similar value of 4.25 (23.4%). The syllable length varied only 0.0015 s in a sequence of 30 syllables of one song. Since measurements would have been within the range of individual variation, the period was calculated by dividing 1 s of the song by the number of its syllables. The difference between the extreme values of the period (Table 4.2) computed according to this method is 0.03236 s – 0.0299 s (for 15 songs of one individual). By this method, the variation in song period for one individual is only 0.00246 s at most.

Little penguins are completely nocturnal, living around their burrow and strongly maintaining this area as a territory. Their peculiar use of song for maintenance of the territory made them ideal subjects for recognition of individuals of the same species, rather than individual recognition experiments. As expected, they react strongly to conspecific songs played back near their burrows. In Fig. 4.12, a Little penguin is shown coming out of its hole near the loudspeaker. It attempted to out-sing the recording, even when the recorded song was its own. Table 4.3 summarizes the preliminary experiments

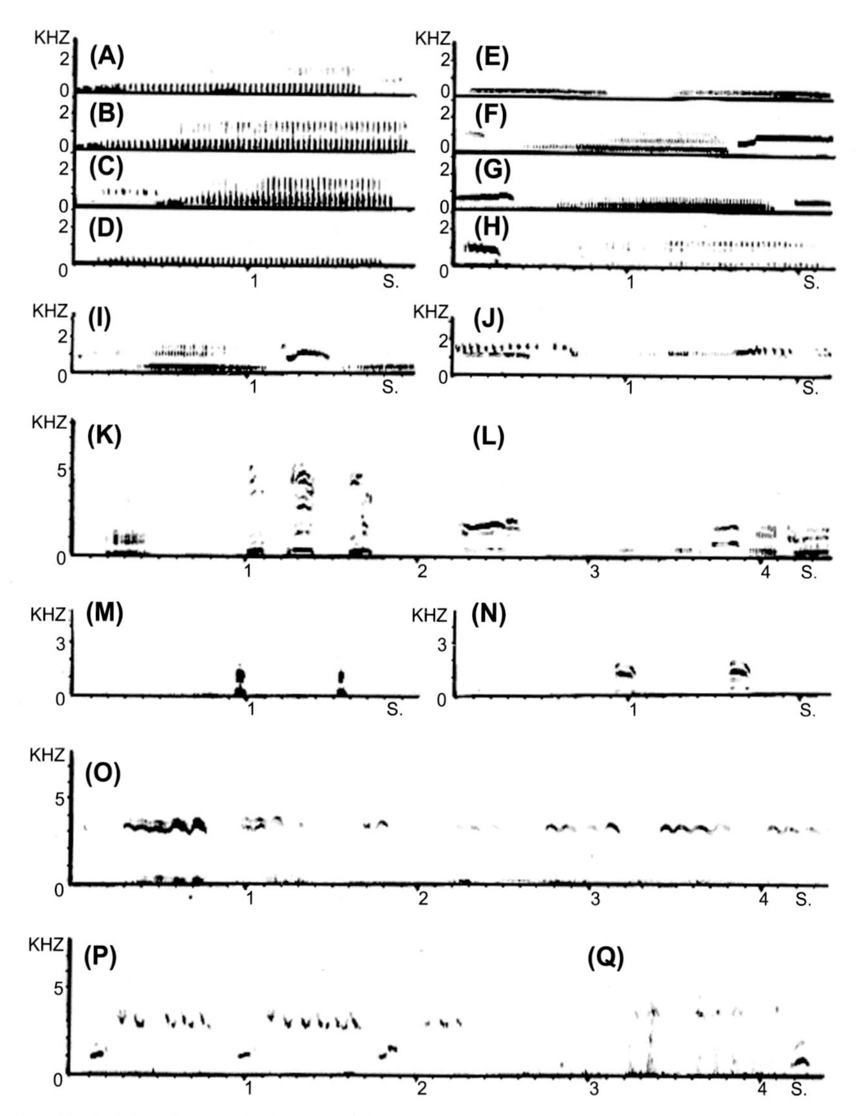


**FIGURE 4.11** Songs of Little penguins. (A–D) Four songs of the same Little penguin (*Eudyptula minor minor*). (E–H) Songs of four different individuals. (I) Whole ecstatic display song consisting of two short inhaled songs and two rhythmically fixed exhaled songs. (J) For a duo, the rhythm is much more variable. (K) Typical agonistic song. (L) Series of agonistic songs highly variable in form. (M) Two contact songs (*E. m. minor*). (N) Two contact songs (*E. m. novaehollandiae*). (O) Several songs of one emancipated chick. (P) Three songs of another emancipated chick. (Q) Song of a molting chick showing the future adult song sequence. *From Jouventin, P., 1982. Visual and vocal signals in penguins, their evolution and adaptive characters. Advances in Ethology, **24**, 1–149.*

**TABLE 4.2 For Little Penguins, Comparisons of the Principal Song Characteristics From Songs of 30 Different Individuals and 15 Songs From One Individual**

| | | Main Frequency in Hz | Maximum Frequency in Hz | Length of Song in 1/100 s | Period in 1/100 s |
|---|---|---|---|---|---|
| 30 songs of different individuals | Mean (n = 30) | 601 | 2402 | 151 | 2.515 |
| | Extreme values | 200–1950 | 700–6000 | 82–200 | 1.639–4.166 |
| | Coefficient of variation | 79.41% | 59.82% | 22.91% | 23.45% |
| 15 songs of same individual | Mean (n = 15) | 265 | 1578 | 143 | 3.198 |
| | Extreme values | 200–350 | 1000–1700 | 110–200 | 2.990–3.236 |
| | Coefficient of variation | 20.93% | 12.42% | 21.12% | 5.50% |

From Jouventin, P., 1982. Visual and vocal signals in penguins, their evolution and adaptive characters. Advances in Ethology, **24**, 1–149.

**FIGURE 4.12** Little penguin leaving the burrow and approaching loudspeaker and singing back in territorial defense against a song playback.

**TABLE 4.3 Preliminary Playback Experiments on Little Penguins**

| | | Response of Animal | | |
|---|---|---|---|---|
| Pitch of signal | High-pitched sound eliminated (+ than 1000 Hz) | + | + | + |
| | Low-pitched sound eliminated (– than 500 Hz) | + | + | + |
| Intensity of signal | Low (50 dB) | + | | |
| | High (70 dB) | + | + | + |
| Distance of recorder from burrow | 0–3 m | + | + | + |
| | 6–10 m | + | | |
| Signal divided into portions of | Several seconds | + | + | + |
| | 1 s | 0 | | |
| Rhythm of signal | ×2 | 0 | | |
| | /2 | 0 | | |

From Jouventin, P., 1982. Visual and vocal signals in penguins, their evolution and adaptive characters. Advances in Ethology, **24**, 1–149.

on Little penguins undertaken by Jouventin (1982) during November 1975 at Phillip Island (n = 8 in each test). Even after elimination of high and low frequencies of the song, the animal's response remained more or less the same, though influenced by intensity or distance of the sound source as by a more or less distant rival. Modification of the structure of the signal, by doubling or halving the speed of emission, rendered it unrecognizable to the bird. Nor did the bird respond to 1-s portions of a song, whereas 5-s portions of the same song were answered.

Sound frequencies are highly variable in the songs of Little penguins (mean of coefficients of variation for main and maximum frequencies of 30 individual songs was 69.6%). At the same time, song frequency (16.9%) and syllable length (23.4%) characterize the individual. As we will see in Chapter 5, and using a computer, we conducted more sophisticated experiments on Little penguins in southern Australia (Jouventin and Aubin, 2000). The coding process was very simple and highly redundant, based on repetitions of identical units of information at two levels and based on song syllables of the inhalation and exhalation of the breath. The information was enough to encode territorial information about the nest location as well as complementary individual information and was very similar to the song of a burrowing petrel. Petrels are faced with the same problem of localization and identification in the dark of night, but with topographical cues (Jouventin and Aubin, 2000).

## Crested Penguins (Genus *Eudyptes*)

The group of crested penguins has been subjected to intense speciation and has given rise to a series of populations and a variable number of species (see Chapter 6). As it is the only genus previously studied in its entirety (Warham, 1975; Jouventin, 1982), we limit our description to the two most problematic groups of species or subspecies.

Jouventin (1982) studied numerous individual Macaroni penguins that were temporarily tagged and found that parents feed their own chicks. Macaroni penguins lay two eggs, but the first egg, which is smaller, was generally abandoned at the onset of incubation. As in all penguins, partners recognized each other by song. If captured when returning to the nest, beak taped closed and released, an incoming bird is greeted by its nesting partner with a bow and a song and headshaking. But since the incoming bird cannot respond, its partner then pecks at it and chases it from the nest. Optical recognition thus appears again to be unimportant to individual recognition. Later when the chick is emancipated, both parents fish at sea and return at intervals to feed their chick. An incoming parent sings at the site of its previous nest, as in Adélie penguins, to attract its chick to the meeting site of the nest. If the chick has the beak taped shut, it still runs to the parent, but the chick is pecked at and excluded from getting a meal because it cannot identify itself by giving the chick song. As soon as the tape is moved to the top of the beak and the chick can sing, it is fed. This indicates that the parent was not disturbed by the tape, but by the chick's lack of the vocal signature.

We measured the proportion of birds singing when Macaroni penguin partners returned from the sea, using the same procedures that were used for Adélie penguins (Jouventin, 1982). The results depended on the specifics of the nesting situation (Fig. 4.13). First, when the incoming partner found a completely empty nest without chick or partner (top left), song frequency was low (11.5%, n = 96). However, song frequency increased considerably if the

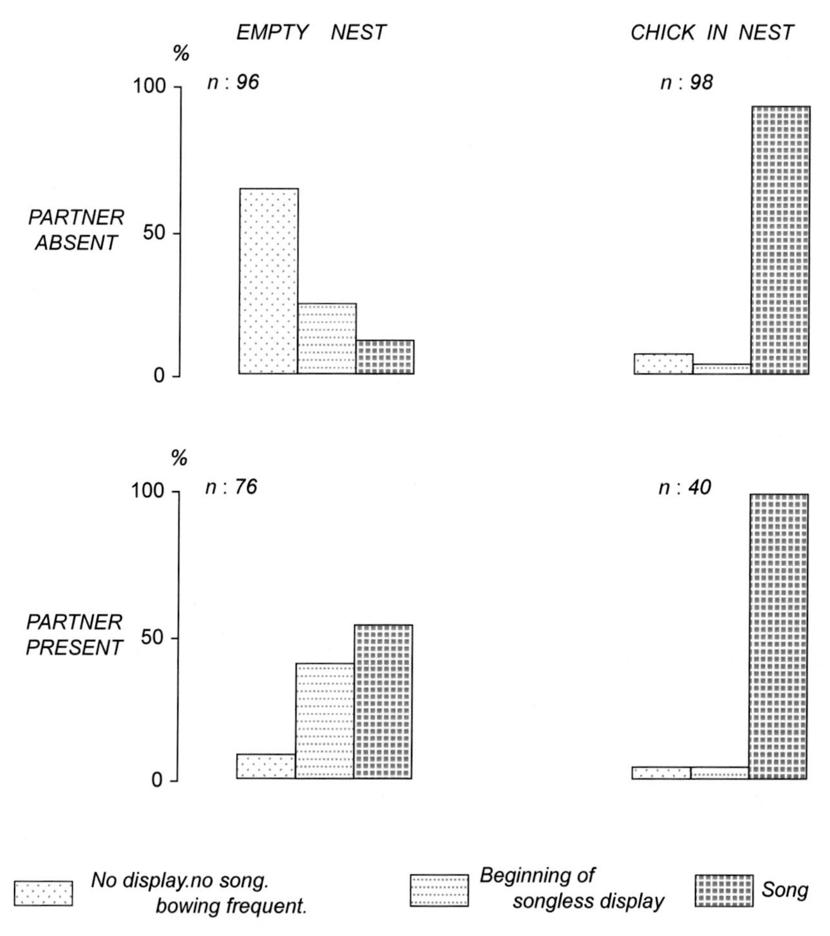


**FIGURE 4.13** Role played by singing in individual recognition. Top, a Macaroni penguin nesting alone is removed from its nest. Upon its return, songs are rare when the nest is empty (left) but greatly increase when a chick is present (right). Below left, songs are much more frequent if the mate is on the nest when the partner returns. Below right, the presence of the partner does not influence the number of songs if the chick is in the nest. *From Jouventin, P., 1982. Visual and vocal signals in penguins, their evolution and adaptive characters. Advances in Ethology,* ***24****, 1–149.*

partner was absent but the chick was in the nest (top right, 92%, n = 98); ($\chi_1^2 = 122.3$, n = 194, $P < .0001$), given that the chick was old enough to respond by singing to the incoming parent. Next, if the chick was removed from the nest but the partner was present (bottom left), song frequency by the incoming partner was much higher (52.5%, n = 76) than when the partner was absent, and mutual recognition of the partners occurred ($\chi_3^2 = 32.5$, n = 172, $P < .0001$). Finally, if the incoming partner found both mate and chick on the nest (bottom right), singing was normal (95%, n = 40). The same experiment is even more conclusive if the incoming partner's bill is taped shut (Fig. 4.14). In Fig. 4.14A, left, the three possible outcomes are depicted: the incoming bird is accepted by its partner on the nest (28% of the trials), it is chased with beak blows (22%) or it is first pecked and then tolerated (ambivalent attitudes, 50%). But if the nest also contains a chick (Fig. 4.14A, right), the incoming partner is less often accepted (percentage of expulsions 70%, n = 17). After removal of the adhesive tape, pairs soon engaged in mutual display and the nest owner stopped pecking. As shown in Fig. 4.14B, the sex of the nesting bird strongly influences the results; that is, males are much more aggressive toward an incoming partner that cannot sing than females are to a mute approaching male.

These results should be interpreted with caution because the condition of taping a beak shut is artificial and some gestures of the penguins are difficult to score. As well, the reaction depends on nuances of the incoming partner's behavior; if it insists, it was usually eventually tolerated near the nest. Such an "ambivalent reaction" of the partner sitting on the nest might be scored as either defense or acceptance. Then again, females were usually less aggressive and also more intimidated by the human researcher than males. If a female was left on the nest and did not hinder the male from settling close by, even though she might have been reluctant, the result was scored as an acceptance. Moreover, the mere fact of a potential partner reaching the territory often reduced the vigor of attacks by the sitting partner. As described, acceptance was not influenced by the adhesive tape as such but by producing an experimental effect: the absence of singing. This was confirmed in that some captured females returned to the nest too frightened to sing and were pecked at by their males. As for all penguins, however, the entire song is not necessary for recognition. Indeed, fortuitous observations showed that even a 1-s portion of the song was sufficient for acceptance of an incoming partner; so the song is highly redundant as time passes. Macaroni penguin chicks can be cross-fostered during the first week after hatching, but if exchanged at later stages the foreign chick is increasingly pecked. Recognition is thus a progressive process as in the equally strongly territorial Adélie penguins.

Throughout our observations of Macaroni penguins, the nest owner first began to sing whenever an adult passed by, whether the intruder was the partner or an alien bird. Pecking ensued only after initial singing by the sitting

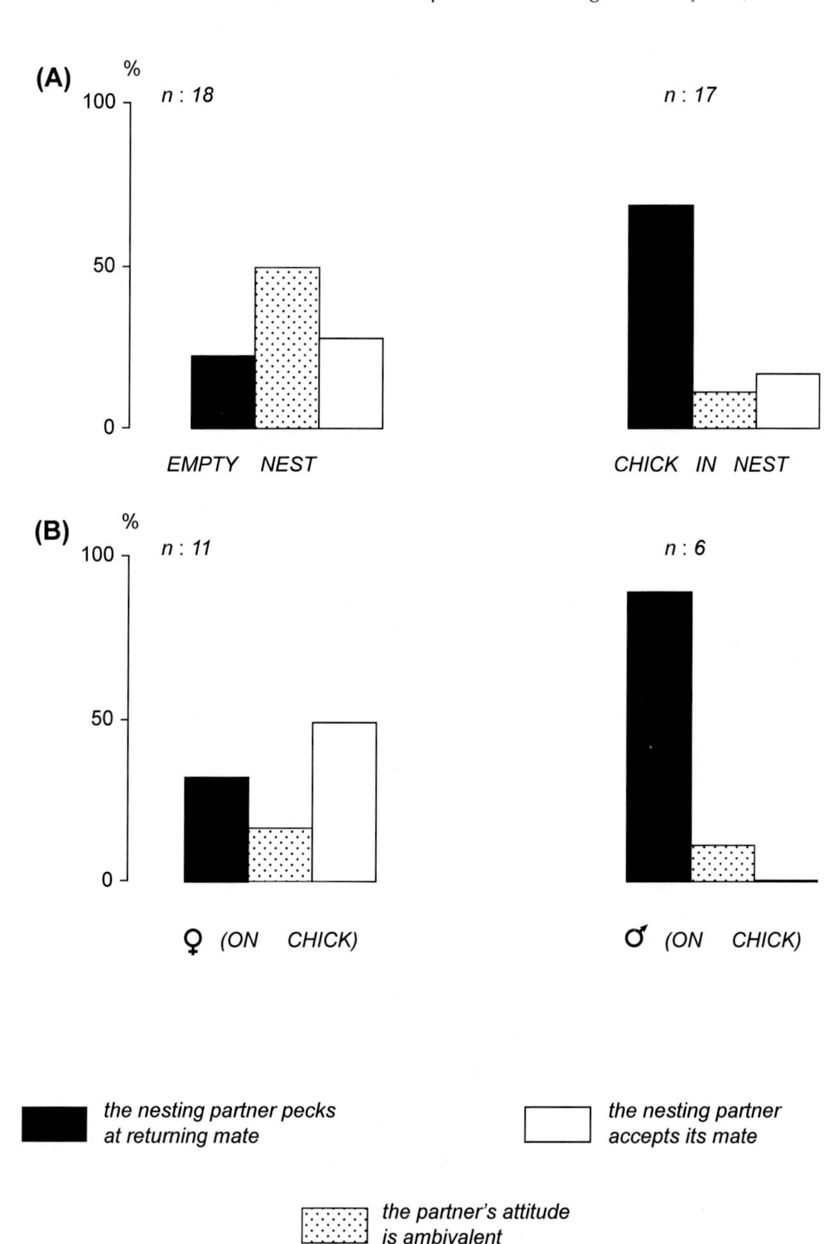


**FIGURE 4.14** Individual recognition and territorial aggressiveness. The beak of the nest-returning Macaroni is closed with adhesive tape to prevent singing. The nesting partner does not recognize its mate and attacks it more readily if a chick is present in the nest (A). Territorial aggressiveness is shown to be stronger in males (B). *From Jouventin, P., 1982. Visual and vocal signals in penguins, their evolution and adaptive characters. Advances in Ethology,* ***24****, 1–149.*

bird. Normally the incoming bird answers if it is the partner, inhibiting an attack of the nesting partner by song in reply, thus initiating a mutual display. As in Adélie penguins, the recognition system worked only after previous sifting of information. The incoming partner finds the nest occupied by a bird that indicated ownership by its song. In turn, the sitting partner met a bird whose song proclaimed it to be the returning partner. The risk of confusion is minimal. Parent–chick recognition in Macaroni penguins was also quite similar to that in Adélie penguins. The parent returned to its abandoned nest and sang; the chick came to the parent, singing. The parent pushed other food-seeking chicks aside and fed its own chick. In these cases, the territory no longer has a protective function but is solely a meeting place that simplified the problem of individual recognition.

Macaroni penguins reproduce on all of the islands of the sub-Antarctic zone. On Macquarie Island, however, we also find Schlegel penguins. It is currently uncertain whether these two forms should be categorized as separate species or as subspecies of a single species. In Fig. 4.15, agonistic songs of Macaroni penguins (Fig. 4.15A and B) are compared to songs of Schlegel

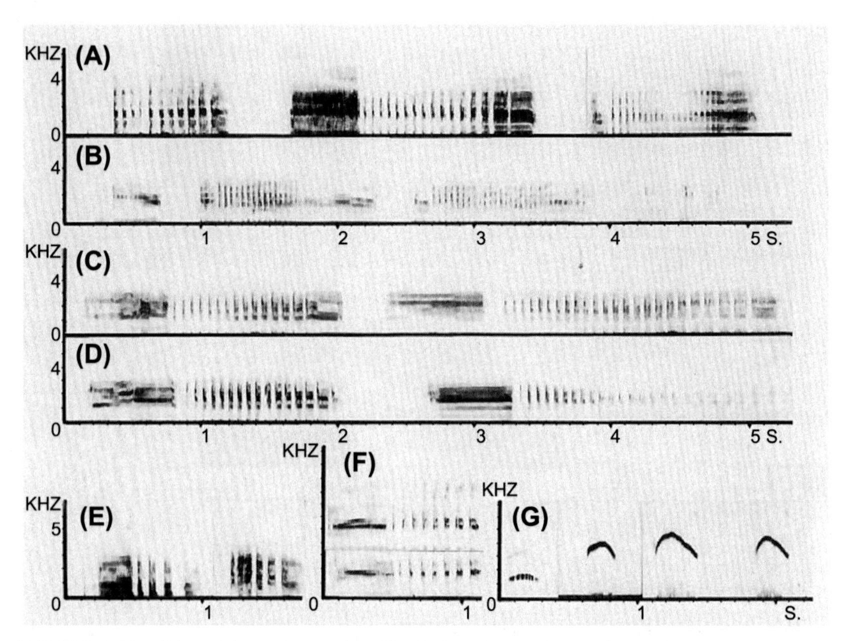


**FIGURE 4.15** Song of Crested penguins. Similarity of nuptial songs in the Macaroni (A and B) and Schlegel (C and D) penguins. (E) Two agonistic songs of the Schlegel penguin. (F) Superimposition of two songs of one individual Macaroni starting at the last syllable. (G) Songs of three Macaroni chicks before emancipation, the last two songs are uttered by one chick and already show the incipient adult song sequence. *From Jouventin, P., 1982. Visual and vocal signals in penguins, their evolution and adaptive characters. Advances in Ethology,* ***24****, 1–149.*

penguins (Fig. 4.15C and D). The difference is minimal to the ear as to the eye, and this agrees with the classification of these forms as subspecies or simply populations. Ecstatic and mutual display songs of Macaroni and Schlegel penguins also seem to be very similar. The general structure of a long song of Macaroni and Schlegel penguins consists of several phrases of varying number and length. At the beginning of the song, a phrase generally starts with a long syllable followed by a series of shorter syllables. In the middle of the song, short syllables are framed by two long ones. Finally, the phrase ends in a long syllable. Thus, it is impossible to deal with the problem of individual recognition without considering the position of a phrase in the song. Fig. 4.15F shows the superimposition of two very short songs, revealing sequence and frequency similarities.

The two agonistic songs of Schlegel penguins (Fig. 4.15E) are identical to those of Macaroni penguins (not pictured). They suggest an abbreviation of the display song and can be highly variable as in other penguin species. As contact songs are mostly uttered at sea, it is very difficult to record them. But they seem to be short, monosyllabic, and quite low-pitched with a chopped structure. The chick song is simple as long as the chick remains in the nest (Fig. 4.15G) and then differentiates after the chick leaves the nest. In crested penguins, the adult song apparently begins to develop from the chick song at a relatively early stage. Statistical analysis of Macaroni penguin songs confirmed these observations (Jouventin, 1982). A preliminary analysis revealed that phrase period lengths did not influence partner recognition in this species, intervals varied only slightly, and syllable length, although fundamental for studying individual recognition in other species, were too variable within the same song to be used for individual recognition.

Comparison of songs of both sexes of Macaroni penguins ($n = 8$) revealed no clear-cut vocal dimorphism (Jouventin, 1982). The only character showing slight differences was phrase length, with a mean of 0.0161 s (0.0120−0.220 s) for females and 0.009 s (0.0058−0.0145 s) for males. In the wild, many songs of males were clearly distinguishable as lower-pitched and containing longer phrases. The importance of phrase position within the song is revealed when the first and last phrases of an individual's songs were compared ($n = 5$). There was a stronger resemblance between the first phrases of two individuals than between the first and last phrases of each of the individuals. The last phrases of two individuals were also much more similar than the beginning and end of the same song. Analyses clearly showed that syllable length was similar in phrases occupying the same place in the song. This recognition system of the Macaroni penguin is thus more complex than that of previous nesting penguins, as we will see in the next chapter in a more sophisticated experimental study (Searby et al., 2004). It is worth remembering that the position of the long syllable acts as a landmark, inserting a "bookmark" in the phrase, even if only part of the song has been heard.

The taxonomical status of Rockhopper penguins was until recently uncertain because there are several populations that breed on several different islands, a common phenomenon in penguins. All the Rockhopper penguin populations were similar in their body measurements. Such measurements were an important trait for species definition before molecular data became the foundation of the systematics of birds. The Rockhopper penguin populations, however, were different in their nuptial displays, as described by Jouventin (1982). After comparing the vocal and visual displays of the different Rockhopper penguin populations, Jouventin concluded that there were at least two distinct species spread among the sub-Antarctic islands. This was later confirmed by DNA analyses (Jouventin et al., 2006; de Dinechin et al., 2009), as we will see in Chapter 6.

The historical associations of the Rockhopper penguin populations turned out to be complex, as revealed by comparison of phylogenetic and ethological studies of this once-supposed single species. In a comparison of vocal signals from several islands (Jouventin, 1982), Rockhopper penguin songs (Fig. 4.16)

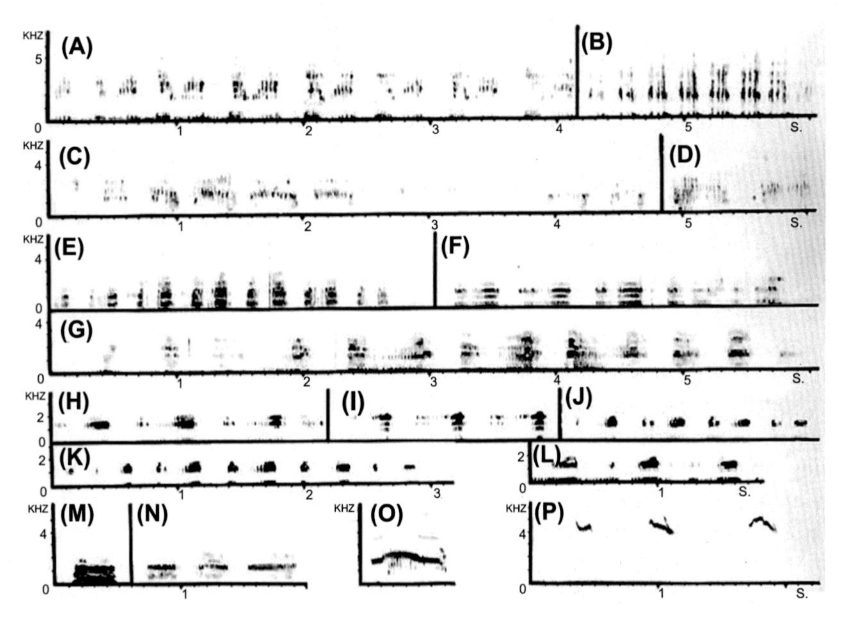


**FIGURE 4.16** Songs of Rockhopper penguins from different localities. (A) East Island, Crozet Archipelago. (B) Courbet Peninsula, Kerguelen Island. (C, D) Staten Island, Tierra del Fuego. (E–G) Gough Island, South Atlantic Ocean. (H–J) St. Paul Island, Indian Ocean. (K) Amsterdam Island, Indian Ocean. (L) Rockhopper penguin duo (St. Paul Island). (M) Agonistic song (St. Paul Island). (N) Agonistic songs (Gough Island). (O) Contact song (Staten Island). (P) Songs of three chicks before emancipation (Possession Island, Crozet Archipelago). *From Jouventin, P., 1982. Visual and vocal signals in penguins, their evolution and adaptive characters. Advances in Ethology,* ***24****, 1–149.*

change completely from the first to the last displayed sonogram. There is no resemblance between the most extreme songs (e.g., Fig. 4.16A, B, and H–K); that is, between songs of the Crozet or Kerguelen populations and those of St. Paul and Amsterdam Islands, all in the southern Indian Ocean. Furthermore, songs from Staten Island (Fig. 4.16C and D) and Argentina are related to the first (Fig. 4.16A and B), whereas those of Gough Island (Fig. 4.16E–G) are more similar to those of St. Paul and Amsterdam, the last displayed songs (Fig. 4.16H–K). Thus, characteristics of the songs divide the populations into northern and southern groups, rather than among more distant ocean basins. The ecstatic display song of the Northern rockhopper penguin was clearly similar to the mutual display song, as shown when comparing the duet (Fig. 4.16L) to the ecstatic display songs that preceded it (Fig. 4.16H–J). Agonistic songs resemble grunts derived from the display song (Fig. 4.16M and N). The contact song is again difficult to record, as it is rarely heard on land, but a single sonogram was recorded at Staten Island. Contact songs heard at the Crozet Archipelago were high-pitched and monosyllabic. Finally, chick song sonograms are quite amorphous but differentiate during emancipation (Fig. 4.16P) as in other species. Vocal dissimilarities between populations of Rockhopper penguins are apparent in the few sonograms described, and are clearly evident, although not based on the distances between islands. If we compare populations that are the most different in song characteristics, the songs appear less similar than those of some species belonging to different genera, for example Gentoo and Little penguins. Almost no superimposition of extreme values occurs for three parameters: phrase period and main and maximum frequencies. This is not due to individual variation, as the large sample size reveals obvious differences between populations ($n = 23$ for Crozet, $n = 28$ for St. Paul).

According to results of Jouventin (1982) and Searby and Jouventin (2005), the song structure of Rockhopper penguins is much simpler than in Macaroni penguins. Sequence (i.e., the position of the phrase in the song) plays almost no role. Intraindividual coefficients of variation are equivalent in different birds, but they differ greatly from coefficients of variation for a population. The song, then, consists of a succession of almost identical phrases. The variability of repeated songs within a single individual is much lower than that among individuals in the population (Jouventin, 1982). Interpartner or parent–chick identification can be revealed in the same way as for Macaroni penguins (Fig. 4.17). This sequence shows that vocal and not optical recognition is facilitated when parents meet their chick at the nesting territory.

## Yellow-Eyed Penguin (Genus *Megadyptes*)

The ecstatic display song of the Yellow-eyed penguin is similar to the mutual display, except that the ecstatic display is more regular. Its structure (Fig. 4.18, fourth and fifth lines) slightly recalls that of the Eastern rockhopper penguin of

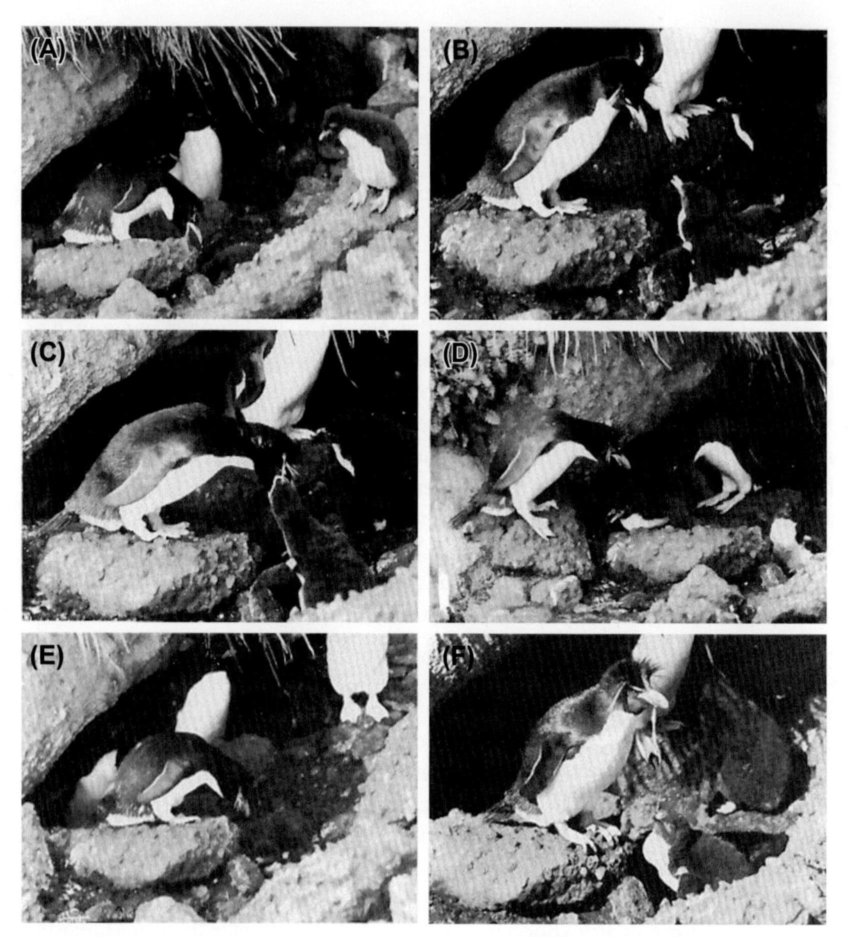


**FIGURE 4.17** Parent—chick individual recognition in Rockhopper penguins. (A) Returning from the sea and sings on the nest. (B) Singing chick comes running from the right. (C) Chick is immediately fed. (D) The chick is trapped and its beak closed with adhesive tape; it can be seen, right, in the photograph; a foreign chick begs in its place but is chased away once it utters its song. (E) The beak taped chick approaches to be fed but is also chased away as it cannot sing. (F) The tape is moved to the top of the beak; the chick can now sing, it is accepted and fed. *From Jouventin, P., 1982. Visual and vocal signals in penguins, their evolution and adaptive characters. Advances in Ethology,* ***24****, 1—149.*

the Crozet and Kerguelen Archipelagos. Song phrases became stable rather rapidly and consisted of short syllables with a large frequency band. The maximum frequency was around 6000 Hz, the main frequency about 1500 Hz, and the period was very short (around 0.030 s, phrase length about 0.016 s, each phrase comprises about five syllables).

The mutual display song was high-pitched, complex, and variable. Fig. 4.18 (first three lines) pictures a duet in which the female took part only rarely. Syllables became very short toward the middle of the song and phrases more closely resembled a variation in frequency with time, rather than a vertical sequence (8 songs of two couples). Yellow-eyed penguins are relatively quiet and breed in small colonies. Agonistic songs exhibited considerable variation in all aspects. The chick song corresponded exactly to

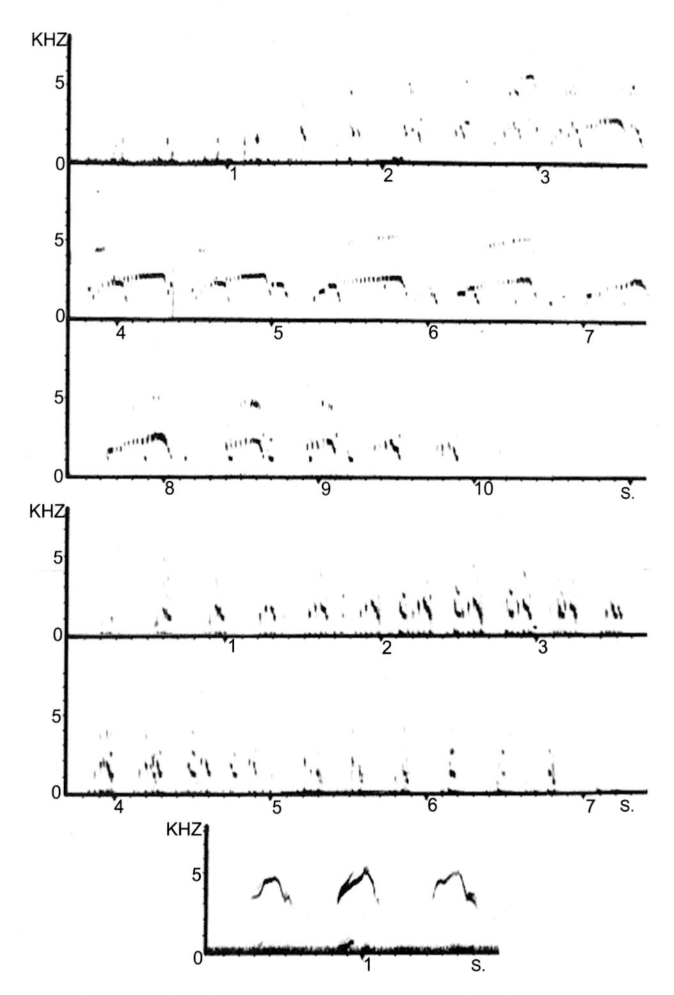


**FIGURE 4.18** The song of the Yellow-eyed penguin. The top three lines show the duet of a pair, the female's contribution mainly toward the end. Next two lines: song of a penguin alone on its nest. Below, three successive songs of one unemancipated chick. *From Jouventin, P., 1982. Visual and vocal signals in penguins, their evolution and adaptive characters. Advances in Ethology* ***24****, 1–149.*

previous descriptions of chick songs of penguins. The resemblance between three repeated songs of a chick just before emancipation was evident (Fig. 4.18, last line).

## LARGE NONNESTING PENGUINS

The Emperor penguin contact song was first described as "trumpeting" by Wilson (1907) and the sonogram of the song was later examined by Prévost (1961, Plate XIII). It is easy to record because, differently from nesting penguins, it is used not only at sea but on sea-ice when Emperor penguins are moving in group to the feeding zone. The frequency of this song ranges from about 500 to 6000 Hz, and the highest intensity is about 2000 Hz. It lasts about 1 s and is not interrupted by pauses. When the song is uttered, the beak of the bird is pointed upward, and the song can be heard over a kilometer away. As such, birds can be attracted by imitations from a human researcher. It is thus the general form of the song that stimulates a response and not the distribution of frequency bands, which are different in human imitations (Jouventin, 1971b). The frequency band of the agonistic song was more or less identical to that of the contact song, but it looked like a distorted abbreviation of the song (as in other species) but with hatched phrases (Jouventin, 1972a,b). Although it may take on diverse forms according to the context, from a complete display song down to an amorphous grunt, the most typical song was uttered with the "horizontal head circling motion" often shown during crucial periods in the reproductive cycle. The display song was first published as a sonogram by Prévost (1961) and later analyzed by Jouventin (1971b, 1972b) and Jouventin et al. (1979, 1980).

Emperor penguins are the only vertebrates that reproduce during the Antarctic winter. They withstand snow and the extreme cold by huddling together. Using individually tagged Emperor penguins, it was well established that pairs are faithful all through the reproductive cycle (Prévost, 1961). Thus, mates must be able to recognize one another within the enormous crowd of the penguin colony. There are no spatial landmarks to help them locate their partner or chick. Obstruction of olfactory ducts in penguins of four other species did not interfere with partner recognition and at far distances there was no visual identification. Nevertheless, experiments showed that wherever separated, pairs rejoined without singing because partners had not lost sight of each other ($n = 50$, Jouventin et al., 1979). When partners did lose sight of each other, however, they could not distinguish their partner from neighbors, except by song. These results were difficult to obtain because partners tended to keep each other in sight as long as possible and only sang to reunite as a last resort.

Emperor penguin display songs are particularly adapted for partner recognition and from inspection of a sonogram resemble a "bar code" from a commercial product. The sonogram forms a recognizable individual signature when heard by the partner (and even to the acute human ear). As shown in Fig. 4.19, there is considerable variation between the songs of different individuals, but repeated songs of a single individual are nearly identical, even from one year to the next (Jouventin et al., 1979). There is, moreover, a sexual dimorphism in song that is easily perceptible to human ears (Prévost, 1961). Female songs are characterized by many short syllables and have about double the number of syllables as male songs (Table 4.4 from Jouventin, 1982).

Conditioning experiments to determine the auditory capacity of Emperor penguins showed that their perception of sound pitch has a range similar to our own, from 30 to over 12,500 Hz (Jouventin et al., 1979). Temporal resolution occurs at less than 15 ms, roughly three times as acute as in human hearing (50 ms). Elimination of the frequency parameter by a level recorder leaves only amplitude modulations (Jouventin and Aubin, 2002). The song sequence remains the same but syllables now appear as peaks and silences as valleys, as will be further discussed in comparing the recognition system of Emperor penguins with those of other seabirds. Values measured from sonograms can be compared for different songs of the same individuals, songs of different individuals, and for both males and females (Jouventin et al., 1979). Syllable length, invariant in previous nesting penguins, was a determining parameter in the Emperor penguin. Signal duration and the number and length of different syllables were practically invariant in songs within an individual. The mean coefficient of variation for signal duration was only 3.4%. On the other hand, maximum and main frequencies are highly variable (mean of the two coefficients of variation was 22.6%). This rough analysis confirms the impression (from Jouventin, 1972) that the most stable individual character is the sequence, whereas frequencies may vary to a great extent. And the determining role of syllable sequence in vocal identification was later experimentally verified (Jouventin et al., 1979).

Comparison of coefficients of variation from different individuals (21 males and 34 females, n = 55) revealed a mean for sequence parameters 14.5 times greater (50.7%) than for repeated songs from the same individual, whereas the mean coefficient of variation for frequency parameters was only doubled (43.3%; from Jouventin, 1982). There were also sex differences in songs. Although males and females look almost alike, they could be distinguished, even by inexperienced human observers, by listening to the number of syllables in their songs (overlap between the sexes occurred only once in 59 cases). Further, the sexes were easily recognized by the general form of their songs. The distribution and length of syllables followed a type of syntax (Jouventin et al., 1979). Female songs consisted of two patterns, both short, containing several short syllables framed by two longer ones. Rarely, there was only one pattern (15% of 60 female songs). The male song was more complex. Syllable length slowly decreased, then abruptly increased followed by a rapid

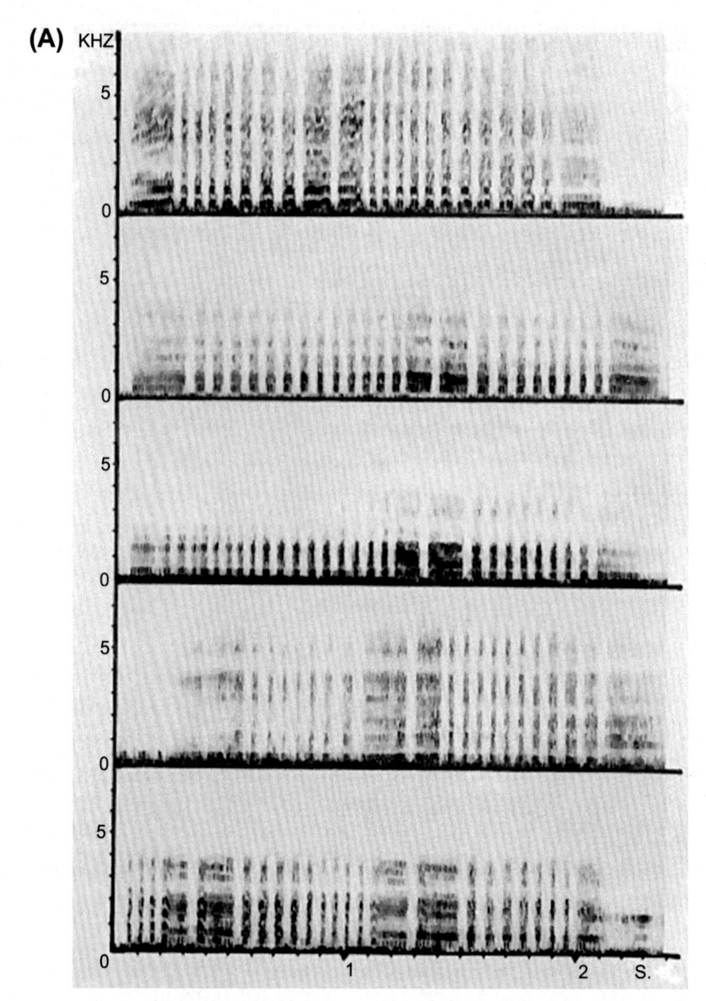


**FIGURE 4.19** The song of the Emperor penguin. (A) Variation in songs of five females; (B) top, different songs of three males; below, five similar songs of one female. Vocal dimorphism of the sexes is based on sequence of phrase (notice twice the number of syllables for females in the same time period). *From Jouventin, P., 1982. Visual and vocal signals in penguins, their evolution and adaptive characters. Advances in Ethology,* ***24****, 1–149.*

decrease and finally a slow increase (Fig. 4.19). Within such a structure, the number of possible combinations is considerable, although less than if the elements and intervals were highly variable. As mentioned, syllable sequence is 14.5 times less variable within than between individuals.

Yet the fine structure of the sequence would not be perceivable if two penguins attempted to sing side by side. In this case, the territory of the nest is enough in seabirds to discriminate the partner from neighbors. But in large

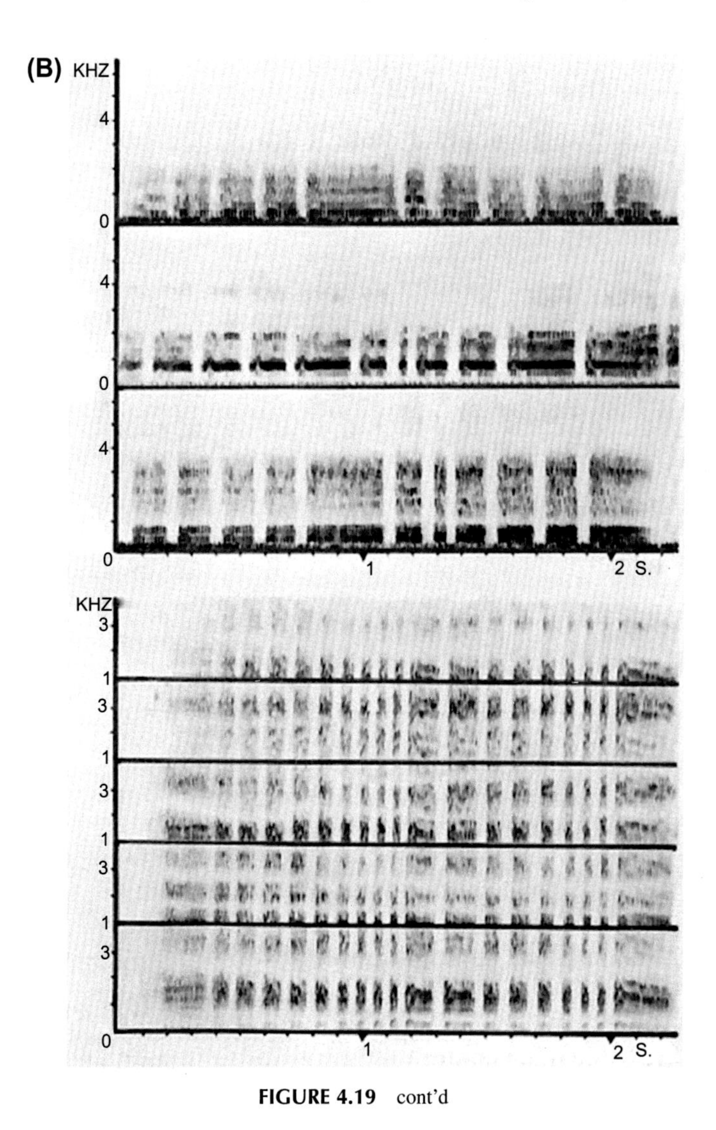


FIGURE 4.19 cont'd

nonnesting penguins, we observed a prelaying silence during pair formation, in part in the King and strictly in the Emperor penguin. That is, during the mating phase of the breeding cycle, Emperor penguins (but not King penguins) could not sing when they heard another singer nearby. One bird always stopped upon hearing its neighbor. This innate behavior functions as a "courtesy rule," preventing jamming and poor communication when there is no nest to isolate singers. Jouventin (1971b, p. 54) often managed to inhibit a stimulated bird from singing by imitating the song himself. When a song recording was played back to an Emperor penguin that was dropping its head to sing, he or she

**TABLE 4.4** For Emperor Penguins, Comparison of General Song Characteristics of Males and Females, Showing Vocal Dimorphism and Interindividual Variation for Each Sex

| | | Main Frequency in Hz | Maximum Frequency in Hz | Length of Song in 1/100 s | Number of Syllables | Length of Syllables (1/100 s) | | | |
|---|---|---|---|---|---|---|---|---|---|
| | | | | | | 1 | 2 | 3 | 4 |
| 7 songs of same male | Mean | 1185.71 | 4306 | 200.14 | 11 | 7.5 | 10.57 | 12.5 | 12.21 |
| | Extreme values | 500–2000 | 4000–5000 | 200–204 | 11–11 | 7.5 –7.5 | 10–11 | 12.5 –12.5 | 12 –12.5 |
| | Coefficient of variation | 54.96% | 10.40% | 0.18% | 0% | 0% | 5.05% | 0% | 2.18% |
| 8 songs of another male | Mean | 9625 | 3877.14 | 200.37 | 11 | 11.87 | 12.12 | 13.5 | 12.81 |
| | Extreme values | 600–1300 | 3300–4600 | 199–202 | 11–11 | 11–12 | 12–13 | 13–14 | 12–13 |
| | Coefficient of variation | 20.73% | 11.42% | 0.58% | 0% | 2.97% | 2.91% | 3.9% | 2.9% |
| 7 songs of same female | Mean | 1114.28 | 4385.71 | 145.42 | 17 | 20 | 4 | 4 | 4 |
| | Extreme values | 800–1500 | 3500–6000 | 144–150 | 17–17 | 20–20 | 4–4 | 4–4 | 4–4 |
| | Coefficient of variation | 24.53% | 18.02% | 1.47% | 0% | 0% | 0% | 0% | 0% |

| | | Main Frequency in Hz | Maximum Frequency in Hz | Length of Song in 1/100 s | Number of Syllables | Length of Syllables (1/100 s) | | | |
|---|---|---|---|---|---|---|---|---|---|
| | | | | | | 1 | 2 | 3 | 4 |
| 9 songs of another female | Mean | 1012.5 | 4562.5 | 147.88 | 17 | 22.27 | 5.16 | 5.55 | 5 |
| | Extreme values | 900–1300 | 2900–5300 | 145–152 | 17–17 | 22–23 | 5–6 | 5–6 | 5–5 |
| | Coefficient of variation | 22.62% | 18.29% | 1.71% | 0% | 1.97% | 6.85% | 9.49% | 0% |
| Songs of 21 different males | Mean | 962.96 | 5583.9 | 209.28 | 10.66 | 11.16 | 14.33 | 14.97 | 17.38 |
| | Extreme values | 500–2500 | 3800–8000 | 135–265 | 7–12 | 2–23 | 8–28 | 10–43 | 8–47 |
| | Coefficient of variation | 64.81% | 32.14% | 11.54% | 10.40% | 52.86% | 29.17% | 45.62% | 60.52% |
| Songs of 34 different females | Mean | 1077 | 5524.32 | 182.09 | 20.5 | 16.16 | 6.08 | 5.36 | 5.05 |
| | Extreme values | 450–3000 | 4000–8000 | 126–232 | 11–25 | 2.5–40 | 2.5–19 | 2–8 | 2–7.5 |
| | Coefficient of variation | 50.15% | 26.27% | 14.08% | 17.41% | 57.30% | 47.03% | 30.78% | 29.32% |

From Jouventin, P., 1982. Visual and vocal signals in penguins, their evolution and adaptive characters. Advances in Ethology, **24**, 1–149.

stopped if the speaker was less than 7 m away. Our experiments (Fig. 4.20; from Jouventin et al., 1979) thus disclosed a proximity threshold within which these birds avoided song overlap.

Perhaps even more remarkable is the long-lasting silence that Emperor penguins exhibit after pair formation, and that continues through the long incubation period, when males stand alone with the egg. When there is no nest to isolate incubators, breeding King and especially Emperor penguins stay mute. They stop singing to prevent the attraction of additional potential partners. Of course, once paired, Emperor penguins have little incentive to cause disturbances by singing. These silences are difficult to investigate, as one needs known couples that can be followed for several weeks. Yet indirect support was gathered by daily counts of songs during the period of copulation (Fig. 2.15), which were greatly reduced from the previous pairing period and the subsequent laying period (Guillotin and Jouventin, 1979). Emperor penguins give only one series of songs during pair formation, as well as when relieving their partner. This song gives both sex and individual identity. Briefly, the prelaying silence is an adaptation to the absence of the nesting territory (Isenmann and Jouventin, 1970). In contrast to nesting penguins, whose chicks give a nondifferentiated song during their first days or weeks of life, the chick songs of large penguins are already complex from hatching. This is particularly so for Emperor penguins. They were slightly modified during the first weeks of development and soon remained unaltered as long as

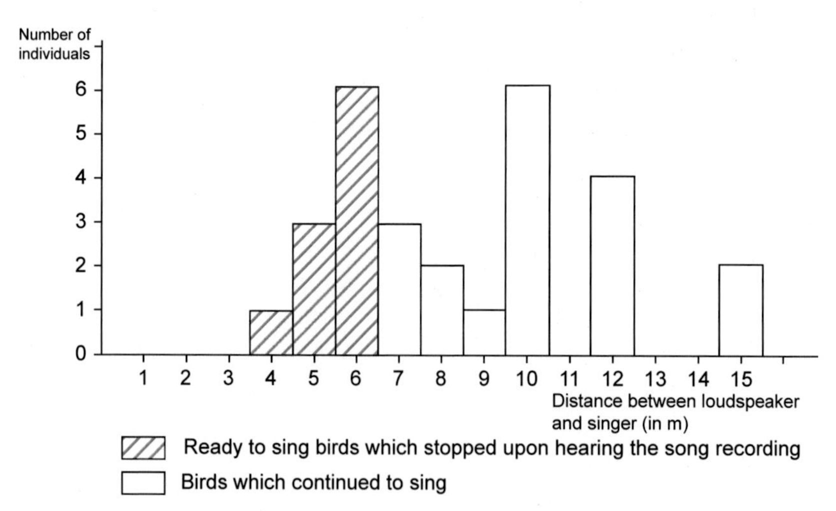


**FIGURE 4.20** Innate courtesy rule: play-backs by a loudspeaker among bachelors singing confirm observations that Emperor penguins less than 7 m apart abstain from singing together when a neighbor is present, to eliminate signal jamming. *From Jouventin, P., Guillotin, M., Cornet, A., 1979. Calls of the Emperor penguin and their adaptive significance. Behaviour 70, 230–250.*

chicks were dependent on their parents (Jouventin et al., 1979). This early differentiation is probably another adaptation to lack of territoriality in King and moreover in Emperor penguins, at a time when there is considerable risk of offspring loss.

The King penguin's contact song is very similar to that of Emperor penguins but slightly shorter and lower-pitched (Fig. 4.21A). It has a frequency range from 250 to 5000 Hz, maximal intensity around 1000 Hz, and a duration of 0.4–0.8 s. This song is mostly uttered by single birds, but beak pointing upward, and response songs can also be induced by vocal imitations by researchers. The agonistic song resembles a shortened display song, again as in Emperor penguins but with its own characteristic pattern. It is generally uttered by a bird before it pecks at another but it is not associated with a ritualized posture. Unlike the Emperor, King penguins have two versions of the same display songs: a short one ("short call") and a long one ("long call" according to Stonehouse, 1960). To quantify this distinction, Jouventin (1982) timed the songs of 50 unpaired birds that were displaying on the beach. The mean duration of songs was $2.62 \pm 0.80$ (range = 1.5–5) s, and the mean for the same number of paired birds in their territories was $3.47 \pm 0.84$ (range = 2–5.4) s. This difference is highly significant ($t = 5.25$, $P < .0001$). Short and long songs of the same individual, however, are not clearly distinguishable as Stonehouse (1960) believed. Some of songs are identical in length (Fig. 4.21B and C), but all length gradations are possible and variations are fairly continuous. In Fig. 4.21D, four different songs of one bird have been superimposed, showing that their length varies greatly depending on circumstances, although the structure remains unaltered. As shown in Fig. 4.21E and F, short songs of the same individual may also vary considerably, as the main part of the song is common to both songs (after the dashed line in Fig. 4.21E and before the dashed line in Fig. 4.21F are the same).

King penguins have no nest, but stand on and defend a small patch of ground. They only move after fighting with congeners, flooding due to storms, from approaching avian predators, or beaching elephant seals. Mates are faithful all through the reproductive cycle, and partners know each other by their long songs. Solitary birds on the beach searching for a mate indicate their sex by short songs ("short call" of Stonehouse 1960). We were able to distinguish the sexes by their songs in King penguins (Table 4.5), and it is even easier in Emperor penguins (Fig. 4.22; Prévost, 1961; Jouventin et al., 1979). Identification of sex requires differentiating between only two possibilities, while individual recognition requires discrimination among thousands of possibilities. Thus, sexual discrimination requires much less information. It is thus understandable that more complete information given by the long song ("long call" of Stonehouse 1960) is provided by partners on their rearing sites in the King penguin. Only partial information, given by the "short call," need be conveyed by mate-searching birds on the beach. Songs have to be as short

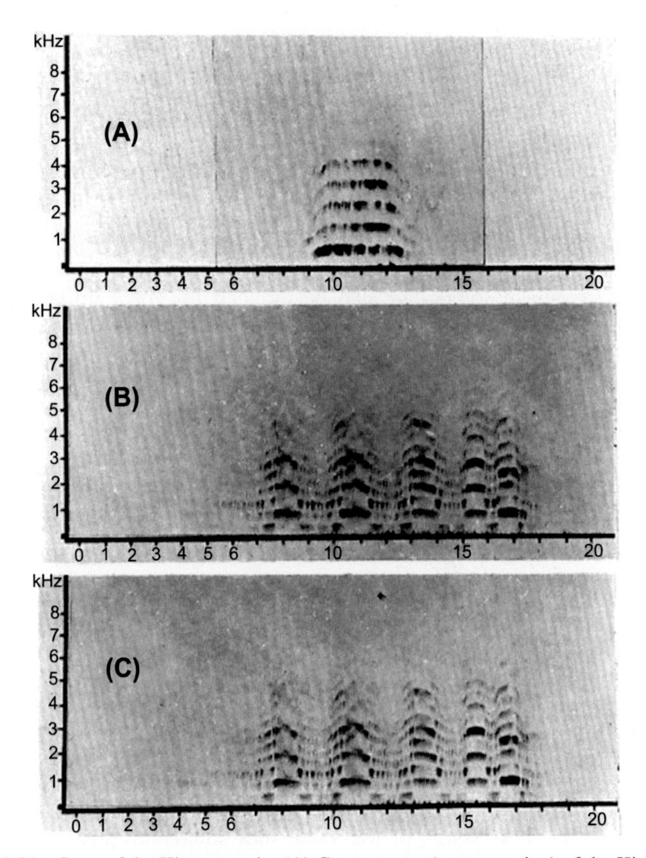


**FIGURE 4.21** Song of the King penguin. (A) Contact song (or trumpeting) of the King penguin. (B and C) Short identical songs of one bird. (D) Four superimposed lengthening sonograms of the same individual. (E and F) Two long songs of one bird, with identical middle parts. *From Jouventin, P., 1982. Visual and vocal signals in penguins, their evolution and adaptive characters. Advances in Ethology,* ***24****, 1–149.*

as possible, so that singers can find a mate; but finding two mates as in trios can lead to breeding delays or failure.

King penguin songs (Fig. 4.22) compared to those of Emperor penguins (Fig. 4.19) have a smaller frequency range from 0 to 5 kHz but more variation in sound intensity. However, the general form of the signal is similar in the two species. A very fine-structure analysis of Emperor penguin sonograms showed that during intervals there was never absolute silence, but a very low sound distinct from background noise. The general structure of the King penguin song was even more pronounced than in the Emperor, it was a series of patterns repeated up to five times. Syllable length, measured between amplitude maxima, was nearly constant within individuals (Fig. 4.22), and interindividual variation in syllable sequence was greater

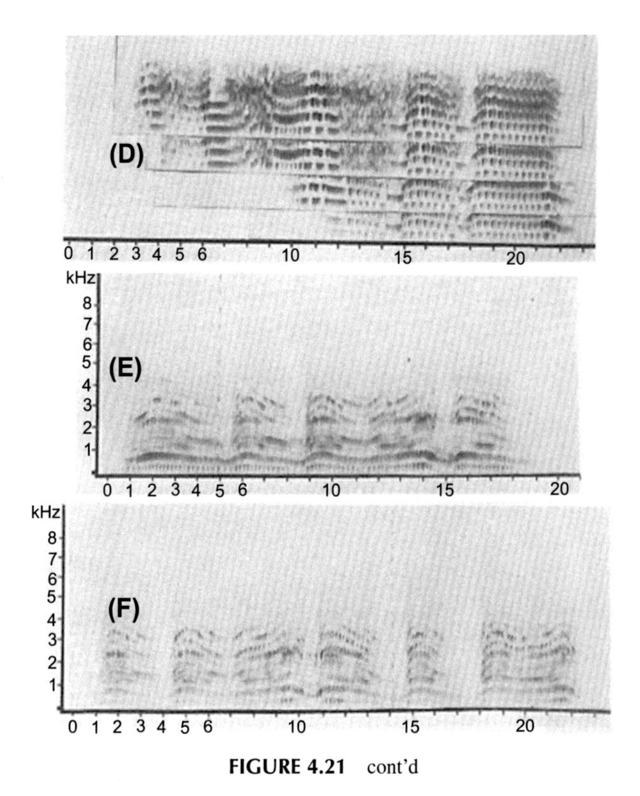


**FIGURE 4.21** cont'd

than intraindividual variation. The mean coefficients of variation for the parameters related to temporization (syllable length) was 4.6% for several songs of one individual (24 songs of two females and one male), whereas songs of different individuals yielded a coefficient of variation of 37.4% (n = 28, 16 males and 12 females). Thus, the ratio (8.1) between sequence variability of this population sample and that of individuals was considerable but lower than in Emperor penguins (14.5). On the other hand, the coefficients of variation computed for parameters related to frequency (main and maximum frequency bands) yielded a value of 1.9% for individuals and 19.8% for the population, a ratio of 10.5. In King penguins, frequency parameters were as constant as syllable length, in contrast to Emperor penguins (where the ratio of coefficients of variation was 1.9).

Vocal sexual dimorphism was evident in sonograms (Fig. 4.22), as well as in measurements of sonogram characteristics (Table 4.5). While female songs comprised a mean of 12 syllables, male songs averaged only 9.3. However, because of strong overlap, this criterion alone was insufficient to identify the sex of individual birds. Female songs were also higher-pitched (mean main frequency: 2937 Hz for females and 2400 Hz for males; mean maximum

**TABLE 4.5** For King Penguins, Comparison of Main Call Characteristics, Showing Vocal Dimorphism and Interindividual Variation for Each Sex

| | | Mean Frequency in Hz | Maximum Frequency in Hz | Length of Song in 1/100 s | Number of Syllables |
|---|---|---|---|---|---|
| 9 songs of same female | Mean (n = 9) | 2916 | 6000 | 276 | 10 |
| | Extreme values | 2750–3250 | 6000–6000 | 261–279 | 10–10 |
| | Coefficient of variation | 7.42% | 0% | 1.66% | 0% |
| 9 songs of another female | Mean (n = 9) | 3000 | 6666 | 331 | 9.83 |
| | Extreme values | 3000–3000 | 6500–7000 | 270–380 | 10–13 |
| | Coefficient of variation | 0% | 3.87% | 10.60% | 49.94% |
| Songs of 12 different females | Mean (n = 12) | 2937 | 6781 | 333 | 12 |
| | Extreme values | 2750–3500 | 6000–7000 | 21.2–445 | 8–18 |
| | Coefficient of variation | 17.47% | 9.12% | 21.05% | 21.75% |
| Songs of 16 different males | Mean (n = 16) | 21.00 | 6200 | 369 | 9.3 |
| | Extreme values | 1000–3800 | 4000–8000 | 220–520 | 6–13 |
| | Coefficient of variation | 33.08% | 19.82% | 27.38% | 27.31% |
| 6 songs of same male | Mean (n = 6) | 1500 | 7000 | 480 | 11 |
| | Extreme values | 1500–1500 | 7000–7000 | 480–480 | 11–11 |
| | Coefficient of variation | 0% | 0% | 0% | 0% |

From Derenne, P., Jouventin, P., Mougin, J.-L., 1979. The King penguin call *Aptenodytes patagonica* and its evolutionary significance. Gerfaut 69, 211–224.

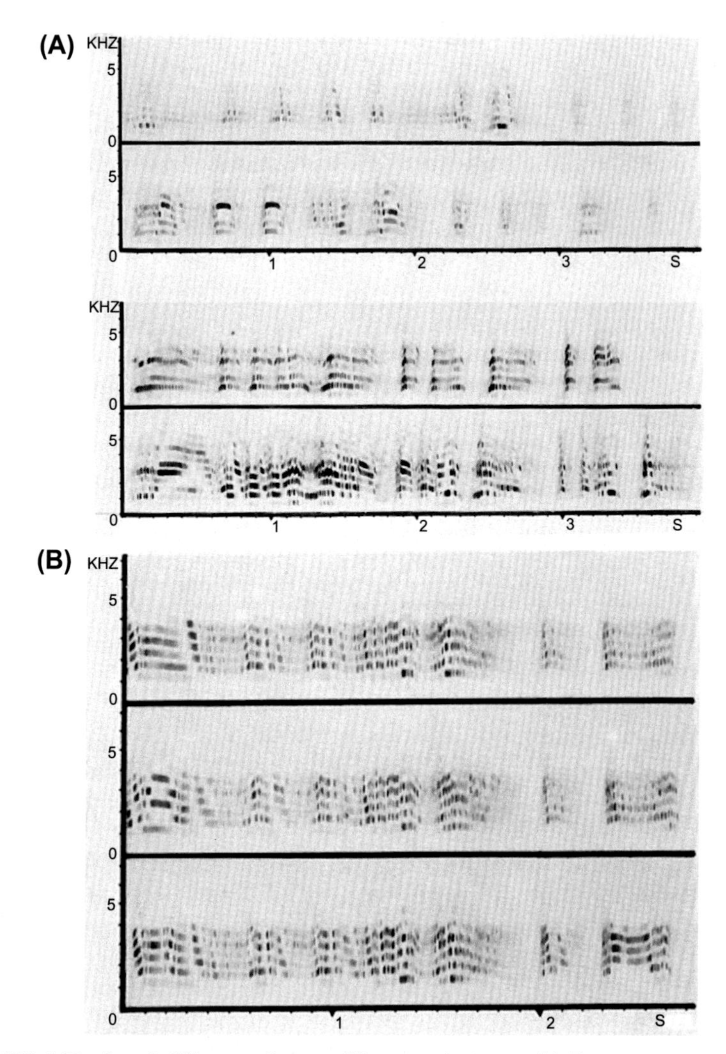


**FIGURE 4.22** Sexual differences between King penguin songs. (A) Top, sonograms of two male songs; and below, of two females with always more syllables. (B) Three songs of one bird; although very similar, a slight variation (particularly in the frequency bands) is shown in the beginning and end of each song. *From Derenne, P., Jouventin, P., Mougin, J.-L., 1979. The King penguin call Aptenodytes-Patagonica and its evolutionary significance. Gerfaut 69, 211–224.*

frequency: 6781 Hz for females and 6200 Hz for males) although there was some overlap for frequency. Vocal sexual dimorphism shared the same basis in Emperor and King penguins, but in the latter species frequencies varied no more than did syllable sequence. Consequently, even though singing King penguins could mostly be sexed by the human ear, this was not as reliable as in

Emperor penguins. Experiments on individual recognition were difficult to undertake in Emperor penguins owing both to the cold and lack of a nest for individual pairs. It was easier to assess individual recognition with King penguins because although they had no nest, their movements during incubation and brooding are limited to a small area or zone and much more restricted than in Emperor penguins. So, did King penguins recognize each other by voice only? Fifteen nonbrooding pairs that firmly settled in a breeding spot (territory) in the colony were separated but left within sight of one another. These pairs immediately rejoined afterward without singing, as in Emperor penguins. Fifteen pairs were individually paint-marked, and then their partners were placed out of sight. These pairs crossed one another, passing many times without taking notice of each other. Two pairs did not rejoin. In five cases, one partner sang outside their mobile territory and was then joined by its mate. In four cases, both partners returned to their territory; one of them sang, which allowed the other to find it. In only one case, partners met on their territory within the breeding colony without singing. The last three pairs disappeared into the crowded colony.

For the chicks of *Aptenodytes*, the vocal repertoire is limited to one song, becoming higher-pitched and more variable when the chick is frightened. Chick songs of Emperor and King penguins were very similar in amplitude and were frequency-modulated whistles lasting half a second at the most, with dominant frequencies at about 2000 Hz (Fig. 4.23; from Jouventin, 1982). The frequency band in Emperor penguins often extends over 3 kHz (1–4 kHz, for example), and in King penguins only over 1.5 kHz. The song of Emperor penguin chicks is generally longer, with more amplitude and frequency variations (Fig. 4.24). In both species, chick songs are emitted in series of threes. Harmonics occur above the most marked fundamental. Normally, songs of individual chicks are almost identical and very different from others. In both species (Stonehouse, 1960; Prévost, 1961), parents feed and brood their own chick, contrary to the communal brooding that was suggested by early observers (Wilson, 1907). In other penguins, individual recognition is facilitated by the nest, but in King penguins returning from the sea to approach their territory zone, and in Emperor penguins, parents return to the area where they last fed their chicks. The adult utters its song, which attracts the chick. Frequent interparent or parent–chick duets form the basis of individual recognition.

## Recording and Playback Procedures

Calls were recorded with a Sony TCD Pro II DAT (frequency response flat within the range 20–20,000 Hz) and an omnidirectional Sennheiser MKH 815T microphone (frequency response 100–20,000 Hz $\pm$ 1 dB) mounted on a 4 m pole, so that birds could be approached without disturbance. The distance between the beak of the recorded bird and the microphone was $\sim$1m.

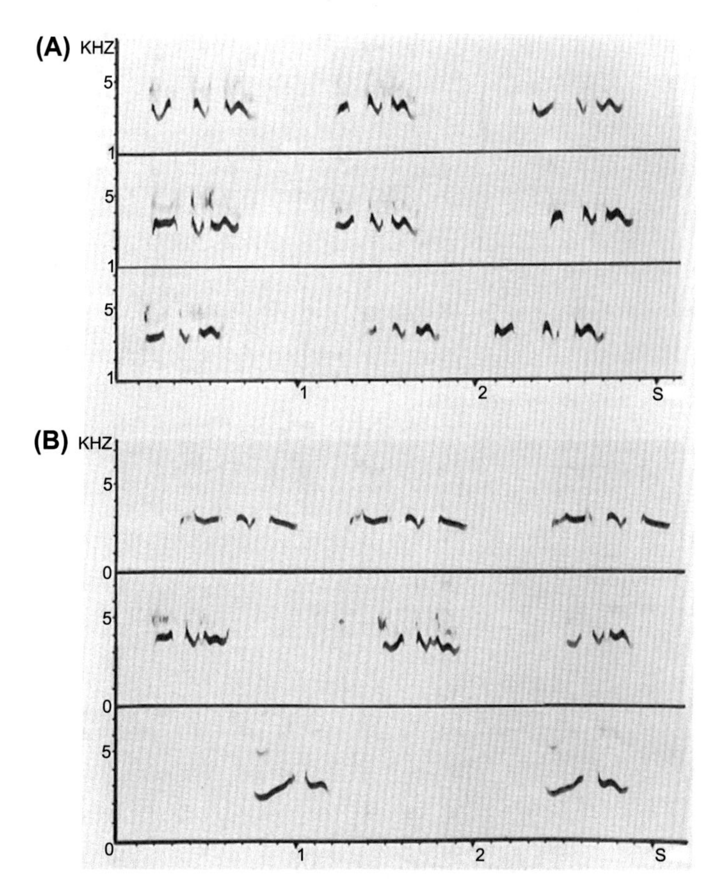


**FIGURE 4.23** (A) Nine very similar songs of the same King penguin chick. (B) Three different penguin chicks. *From Jouventin, P., 1982. Visual and vocal signals in penguins, their evolution and adaptive characters. Advances in Ethology,* ***24****, 1–149.*

Experimental signals were broadcast with the previous tape recorder connected to a to a PSP-2 E.A.A. preamplifier and a 20 W self-powered amplifier built in the laboratory, equipped with an Audax loudspeaker (frequency response 100–5600 Hz ± 2 dB) (Fig. 4.25). For propagation tests, signals were rerecorded by means of an omnidirectional Sennheiser MKH 815T microphone connected to another Sony TCD10 Pro II DAT. For sound pressure level measurements (SPL in dB), we used a Bruël & Kjaer Sound Level Meter type 2235 (linear scale, slow setting) equipped with a 1 in. condenser microphone type 4176 (frequency response 2.6–18,500 Hz ± 2 dB). Calls were analyzed and signals synthesized using mainly the Syntana software built in the laboratory (Aubin, 1994).

Signals were digitized through 12- or 16-bit acquisition cards equipped with an antialiasing filter (low-pass filter, i120 dB/octave) at a sampling rate of

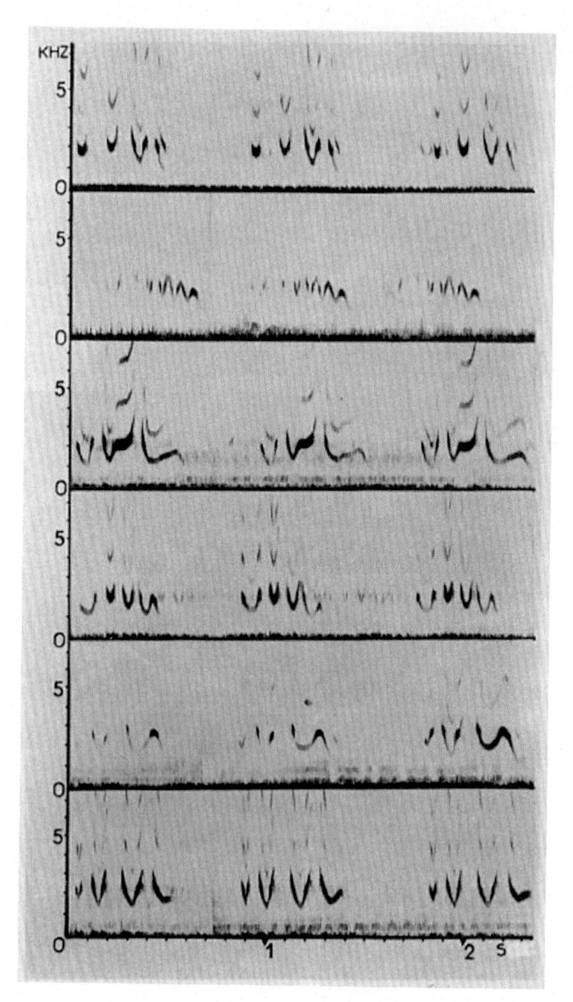


**FIGURE 4.24** Songs of six Emperor penguin chicks: the three songs of each chick are the same, but differ among the six chicks. *From Jouventin, P., 1982. Visual and vocal signals in penguins, their evolution and adaptive characters. Advances in Ethology* ***24****, 1–149.*

12–48 kHz (depending on the call studied) and stored on the hard disk of a computer. Calls were then analyzed and signals synthesized mainly using Syntana software. Signals were examined in the amplitude-versus-frequency domain by spectrum analysis (fast Fourier transform (FFT) calculation) and in the amplitude-versus-time domain by envelope analysis (analytic signal calculation). To follow the time evolution of the frequency, we used the Hilbert (Papoulis, 1977) or zero-crossing calculations, which provide a representation of the instantaneous frequency. Fundamental frequencies were detected and measured using the Cepstrum calculation defined as the power spectrum of the

**FIGURE 4.25** Marked King penguin chick (with temporary numbered plastic tape in foreground) singing back (with head up and beak open) to a playback of its parent's song that is being broadcast through the loudspeaker. Other chicks were not reactive to the song broadcast.

logarithm of the power spectrum (Noll, 1967). Experimental signals were built either by constructive synthesis (i.e., by computer synthesis starting from scratch) or by destructive synthesis (i.e., by modifying natural calls). For the constructive synthesis, natural calls were modified in the temporal, frequency, and amplitude domains. Amplitude and frequency modulations (respectively, AM and FM) were modified or removed using the Hilbert transform calculation (Brémond and Aubin, 1992; Mbu-Nyamsi et al., 1994). For modifications of the harmonic structure, natural calls were filtered by low-, high-, or band-pass digital filters, by applying optimal filtering with FFT (Press et al., 1988; Mbu-Nyamsi et al., 1994). The window size of the FFT was 4096 points. Natural calls were also shifted up or down in frequency by picking a data record (from a natural call) through a square window, applying short-term overlapping (50%) followed by a linear shift up or down of each spectrum, and finally by a short-term overlapping inverse FFT (Randall and Tech, 1987). As before, the window size was 4096 points. To modify call or syllable durations, we truncated the sounds. To prevent spectral artifacts arising from gaps in amplitude, an envelope was applied (by multiplication) to the data set in the time domain to smooth all the edges.

Chapter 5

# Experiments on Vocal Signals

## EARLY EXPERIMENTS ON VOCAL SIGNATURES

In the earliest studies of individual recognition via vocal signatures, parent–chick separation experiments were undertaken in Emperor penguins (Jouventin et al., 1979) and King penguins (Derenne et al., 1979). A chick singing in concealment generally evokes a strong reaction from its parents. A chick colored red or blue (n = 5, in each species) was accepted, although the parent appeared to perceive the alteration. The parent stoped and sang a duet several times with the chick, probably to verify the chick's identity. King penguin chicks colored with dye and enclosed by a fence with three other chicks were readily accepted by their parents (n = 5). Tests and field observations seemed to indicate a few mechanisms of parent–chick recognition. Whenever a chick is stressed, particularly when cold or hungry, it sings. A lost chick will answers any adult, but immediately responds to its parents, which recognize it and peck at nonoffspring chicks. Parents make the final choice of acceptance or aggression. Visually, the chick's general body form is important, but actual appearance is secondary. Song and an associated submissive posture by the chick are decisive for recognition.

Playbacks of chick songs in King penguins (Table 5.1) always attracted the chick's parents (n = 12). Even half of the song suffices (n = 8). Predictably, longer fragments resulted in a stronger reaction. If the speed of emission of the chick song was halved (n = 8) or doubled (n = 8) no reaction is visible, indicating that, as in adults, modifications of song structure impeded recognition. In choice experiments with a singing chick that was visible to a parent, and the recording of the chick's song played from a loudspeaker, the parent recovered its chick and ignored the loudspeaker (n = 5). The sight of the chick aided in its recovery by the parent, especially if the recording was less clear than the natural song. Most likely both voice and posture were important to acceptance of the chick. Some experiments were inconclusive, the tested birds evidently distinguished between the real chick song and the recording, which elicited only head movements toward the loudspeaker (Fig. 4.25). Moreover, the choice between a chick with its beak taped shut and a recording of its song (n = 7) was an insoluble problem; the parents went from one to the other. However, several chicks, after being successively

**Why Penguins Communicate. http://dx.doi.org/10.1016/B978-0-12-811178-9.00005-6**

**TABLE 5.1 The First Playback Experiments on Parent–Chick Recognition in King Penguins**

| | Parent's Responses | Conclusions |
|---|---|---|
| Emission of chick song | In each case the parent rushes forward toward the loudspeaker (n = 12) | Evident vocal individual recognition |
| Emission of chick song fragments | The parent does not react in two cases, slightly reacts in two cases and reacts strongly in 4 cases (n = 8) | Part of the chick song is sufficient for recognition, the more evident, the longer the fragment |
| The parent must choose either the singing chick or it's song recording | The parent recovers its chick without taking notice of the loudspeaker (n = 5) | **1.** The sight of the chick aids in its recovery<br>or<br>**2.** The recording is less effective than the natural song |
| The parent must choose either the beak-tapped chick or the recording | In two cases, the parent recovers its chick while executing intention movements toward the loudspeaker. In three cases, it recovers its chick then abandons it for the loudspeaker. In two cases, it rushes from the chick to the loudspeaker and cannot choose between the two (n = 7) | The further the chick is separated from its song, the more the parent hesitates |

From Derenne, P., Jouventin, P., Mougin, J.-L., 1979. The King penguin call Aptenodytes patagonicus and its evolutionary significance. Le Gerfaut 69, 211–224.

adopted and abandoned by a parent, decided to beg to the loudspeaker for adoption, before going back to their parents.

The necessity of good partner recognition can be discerned from when vocal communication is lacking. In the Emperor penguin, after singing and pairing, partners stay together but stay mute, even if moving. This is because if one partner sings, other prospectors come searching for a mate. Because Emperor penguins lack a nesting territory where they can exclude intruders, trios form that can contribute to a breeding failure, as we often saw in the field. In the same species, neighbors do not sing when another bird sings nearby. As noted previously, we observed that Emperor penguins stopped singing completely when another bird began to sing (Jouventin, 1971a), and then we did tests to confirm this experimentally (Jouventin et al., 1979). Observing from a distance, we recorded a song that we played back, just before an Emperor penguin close to the loudspeaker was ready to sing. The bird lowered its head just before singing, then at the start of the recording it stopped instantaneously (i.e., the penguin did not sing). This response occurred for any bird ready to sing, until the playback was about 7 m away, but not farther (Fig. 4.20). The biological significance of this innate rule of courtesy, unique to the penguin family and in the most mobile species during breeding, obviously prevents the jamming of songs between neighbors. Thus, ineffective communication is highly possible during mating in Emperor penguins, due to the lack of topographical cues, and can even lead to disruption of reproduction.

The first experiment conducted by Derenne et al. (1979) showed that King penguin pairs avoided singing unless strictly necessary, corresponding to the prelaying silence in Emperor penguins. Newly formed King penguin pairs rarely sing. Moreover, King penguin mates sing simultaneously within a short distance, which never occurs in Emperor penguins. The second experiment showed that even well-acquainted King penguin partners do not recognize each other visually, and singing serves as a confirmation of identity. As in Emperor penguins, only one King penguin partner needs to sing to allow pair reunion, if the meeting is not disturbed. At a later reproductive stage, brooders returning from the sea were captured and their beaks closed with adhesive tape (Derenne et al., 1979). Six out of ten birds fled toward the sea and had to be caught, and the tape removed. Four others reached their territories and remained several meters away from their partners, delaying an approach until three out of the four partners standing with the egg began to sing. One of returning birds was pecked by the incubating partner and departed; the others, although unable to answer because of the tape, were finally tolerated by the brooding partner. In conclusion, singing is not indispensable for entering the territory, but it seems necessary for normal reliefs, since no "egg exchange" occurred for the partners that could not sing.

The results of our preliminary playback experiments with the partner's song in King penguins were that incubating birds became agitated, looked around, and/or sang in response (Derenne et al., 1979). Only a few responses of incubating partners were feeble or negative ($n = 60$). To their own songs, incubating partners ($n = 10$) reacted much more ambiguously. For controls, the songs of three birds of the brooder's sex and of birds of opposite sex ($n = 20$) evoked no response, suggesting that only the partner's song counted. The first half of the partner's song ($n = 20$) elicited fewer responses than the complete song, the first three quarters of the song ($n = 10$) was slightly more stimulating, and only the second half of the song ($n = 20$) induced relatively few positive responses. The complete song of the partner was more stimulating than any partial songs. After elimination of frequencies higher than 4 kHz (n = 10) or lower than 1 kHz (n = 10), there were still many positive responses. High and low frequencies played an important role in song recognition, but they were not indispensable. When the speed of the song was doubled, or halved, responses were generally negative (n = 20), probably because of changes in temporal structure. These preliminary playback tests were important because the speed modifications changed several physical parameters. These tests, however, were not as accurate as later experiments, in which we modified the recorded song with computer software. Nonetheless, the early experiments initiated our series of studies on vocal signatures, especially in the King penguin, and opened the way to more sophisticated experiments.

## HOW TO FIND A PARTNER IN A NOISY CROWD

We know that oceanic birds have to feed at sea and reproduce on land, where terrestrial sites free of predators are limited. Consequently, almost all seabirds are colonial, particularly penguins. Due to this sharing of feeding and breeding by parents, a female cannot reproduce alone, as occurs in some avian and most mammalian species that exhibit parental care. Rather, it is necessary for both mates to take turns at incubation, and then later at brooding. Consequently, all seabirds are monogamous. Thus, penguin pairs have to be reunited several times during the breeding cycle, as part of the process of relieving their partner on the nest. The problem for a foraging partner that is returning to a coast or an island is as follows: first, to find the right location of the nesting colony, then to find the nest locality, and finally to find its partner on the nest (Dobson and Jouventin, 2003a). In these tasks, visual topological cues are important, but for the final step, vocal cues are essential for identification of the land-based partner. This principle even holds true when colonies are flat, as often occurs in Adélie penguins (Fig. 5.1). To reduce confusion about identity among birds, they use auditory perception and identify their partner by their song; and in the same way, their partner identifies them.

Finding the partner or chick is particularly difficult in a dense seabird colony and even harder if partners and chicks move about the colony, as occurs in the two largest species of penguins (Fig. 5.2). Penguin partners and chicks identify each other by their voices, rather than visually. This simplifies the experimental problem of understanding how the vocal signal works, because they use only one mode of communication for individual recognition (Jouventin et al., 1979; Jouventin and Roux, 1979; Derenne et al., 1979).

Penguins are also particularly interesting to study because of their wide variety of breeding habitats. They use the largest diversity of types of breeding sites known for colonial-breeding species. Some species nest in a burrow (dispersed in Little penguins, genus *Eudyptula*, or sometimes nesting in huge colonies like some species of banded penguins, genus *Spheniscus*), others

**FIGURE 5.1** A typical colony of Adélie penguins, with each pair breeding on a nest of pebbles. In small colonies like this one, the partner or chick is easy to find at the "rendezvous" nest site by visual inspection, when returning from the sea; but parents and chick nonetheless have to identify each other from their songs.

**FIGURE 5.2** A colony of King penguins, with a singer advertising for a mate by extending its head skyward (ecstatic posture) and singing (short song). Pairs mate without a nest, and then must find a place to lay and stand with their egg in the breeding colony.

among rocks (crested penguins, genus *Eudyptes*), on flat areas among thousands other pairs (Adélie and Chinstrap penguins, genus *Pygoscelis*), on grass plateaux, with the colony shifting to a new spot each year (sub-Antarctic Gentoo penguins, also *Pygoscelis*). Finally, species without a nest, incubate and brood on their feet while moving about beaches or sea ice (the large King and Emperor penguins, respectively, genus *Aptenodytes*). Thus, penguins are the only bird family in which we can compare about 20 species that have a variety of ecological conditions for nesting and an absolute requirement that parents cooperatively raise one or two chicks. Consequently, penguin species engage in a series of increasing difficult tasks while finding their partner and identifying each other during reliefs from incubation shifts and later while feeding the young chick(s). In two words, penguins provide a unique "natural experiment" for the study of individual acoustic recognition.

The lack of a nest in the two large penguins explains why new pairs are almost mute, to prevent attracting new interfering adults that are searching for a partner. But when they have to shift from one location to another in or about the colony, they need to keep in contact with their partner without singing. This is accomplished by keeping their partner in view, and this connection is easier if they have a special posture that is easy to see. Thus, in each species of large penguin, we observed that new pairs have a special posture, the waddling gait or attraction walk, to distinguish the partner from birds that might be milling about (Fig. 5.3).

Observations demonstrated that all penguins were able to recognize their partner or chick solely by the voice (Derenne et al., 1979; Jouventin and Roux, 1979; Jouventin et al., 1979; Jouventin, 1982; Waas, 1988; Robisson et al., 1989; Speirs and Davis, 1991; Lengagne et al., 1997; Jouventin and Aubin, 2000). We developed experimental methods to test penguin species for their ability to use a "vocal signature" to correctly identify their partner and chick. We discovered that each species had an accurate system of recognition that was adapted to each nesting problem. Initially, we found that it was possible to capture the attention of a bird on the nest by broadcasting its partner's display song with a loudspeaker, while the partner was away at sea (Fig. 4.12). But with time, adults learned to distinguish this artificial song from the natural one. So we needed to continuously change to new birds that had not been tested before. It was often more effective to broadcast parental songs to naïve chicks, because hungry chicks were easier to attract and more motivated to reply.

According to Darwin's (1859) theory of natural selection, parents should rear their own offspring. To do so, they have to identify their chick(s) and, when the chick is too young or in the egg stage, identify their breeding partner. Only nuptial songs were given during mating, incubation, brooding, and feeding during the crèche stage. Vocal signatures uttered during mating contain more information than other signals, both in temporal and frequency parameters of the sound in all penguins, as shown by comparisons of

**FIGURE 5.3** New pairs show a waddling gait ("attraction walk") to maintain contact and to identify their partner without singing; singing is avoided because nonnesting birds, as here in King penguins, stay mute after mating to prevent attracting unmated birds and thus trio formation.

among-individual variation versus within-individual variation (Jouventin, 1982). These nuptial songs gave several sorts of information: (1) the species of the singer; (2) the sex of the singer, particularly in the two large *Aptenodytes* species; and (3) individual recognition, the main focus of our experiments and the most difficult type of information to demonstrate.

We were consequently fascinated by the opportunity to conduct experiments that could reveal the evolution of differences in communication abilities among species, thus producing a natural comparative experiment in acoustic recognition. We conducted sound trials in the field from 1969 to 2004, using several kinds of tests and a variety of different techniques. Six different species of penguins were experimentally studied, including both nesting and nonnesting penguins, and hundreds of birds of each species. Our goal was to identify the significant physical parameters of the vocal system of individual recognition for each species. A seabird colony, and particularly a penguin colony, is an extreme environment from an acoustic point of view. This is only partly due to the loud background noise of waves and nearly constant winds, as well as the songs of sometimes thousands of penguins. These problems are augmented by the physical properties of sound propagation in the biotic community of the penguin colony. The penguin bodies themselves act like a screen that blocks sound. Depending on the density of the penguin colony, this factor can be relatively more or less important to how far a song will carry to a partner or chick.

Faced with the problem of finding a particular individual in a penguin colony, among several hundreds or thousands of conspecifics, the display call alone seems at first glance inadequate to secure individualization. Nevertheless, penguins succeeded, performing acoustic identification of the partner or chick, and usually within a few minutes. But how well can each species perform, for example, when pair mates are perhaps 20 or 30 m apart, and apparently unable to hear the song of their partner in the noisy crowd of their colony? How much of the song do they need to hear for recognition, and can they recognize a partner when they hear only a small part of the song in this loud environment? Answering these questions might indicate how efficient penguins of different species are at identifying their partner or chick, given their particular ecological circumstances. To address these questions, we used computer software that manipulated previously recorded songs. The experimental songs contained modifications of specific song parameters. Then the modified songs of a partner or parent were played to "targeted" penguin individuals from a loudspeaker. The data produced were the reactions of the targeted penguins to the modified songs. If they replied, the physical parameter modified in the song had no influence on individual recognition. But if they did not reply, then the modified parameter was an important component of recognition of the vocal signature. In this way, we could "ask the penguin" if it was able to identify their partner or parent. This inquiry was obviously much more efficient than simply examining and comparing the characteristics of songs on sonograms among the different species, as we did before.

The goal of the experimental study was to examine several species that reflect an increasing degree of difficulty in recognizing their partner or parent. The primary division was an ecological one between penguins that have topographical cues (i.e., species with a burrow or open nest) and those that do not (the two large nonnesting species). As we demonstrated, vocal signals are essential to successful breeding in penguins. Songs initially attract unmated individuals during the courtship process, so that pairing can begin, and the vocal signature is essential for recognition of the breeding partner. In a sense, the initial songs sung by an available unpaired bird is the invitation to learn the song of a potential partner. In all of the penguin species, successful reproduction is impossible without close cooperation between partners during the breeding season. Additionally, partners only cooccur during brief periods over a season of successful reproduction. While pairs stay together from mate choice (pair formation) to egg laying, one partner then leaves to forage at sea. Early in the breeding season, they take turns in egg care and in feeding and protecting young chicks. During this period, they are together only long enough to ensure individual recognition by duets and conduct the transfer of responsibilities, and then the previously sitting partner leaves for the sea. While initially the nest site provides information about where the partner and offspring can be found, accurate and definitive information is only provided by the voice. Later, as both parents forage and bring back food to the developing young, chicks often congregate in a small or large crèche. Whether on their own or in a crèche, these chicks must be recognized by their parents if they are to receive essential sustenance. This recognition is accomplished from identification of a vocal signal from which parents and offspring can "sign off" on their relationship (Fig. 5.4).

The importance of the penguin voice was easily observed in experiments on visual signals, and these same experiments showed how visual and vocal signals were integrated into a multimodal system of information about both the quality and identify of mates (Partan and Marler, 1999; Higham and Hebets, 2013). Multimodal systems of information transfer are those in which more than one sensory modality is used in communication (Fig. 5.5). In penguins, auditory and visual signals are dominant, but tactile and olfactory senses can be used for gaining information about the environment. Perception of conspecific individuals occurs primarily, and probably solely, by the visual

**FIGURE 5.4** Parents returning from the sea to feed their chick must find the right one by vocal identification, as some chicks try to obtain extra feedings. This King penguin parent has three chicks too many!

**FIGURE 5.5** Male Adélie penguin on his nest giving the "ecstatic display," both singing and flapping his flippers. Visual and vocal signals are often linked and reinforce each other, a bachelor male on its nest singing and flapping is easy to locate by bachelorette females among less energetic and sometimes sleeping penguins on nests.

sense, to roughly perceive where the breeding place is in the colony, and then more precisely and with certainty by auditory sense to confirm the identity of the supposed partner and chick on the nest. These two senses convey alternative forms of information, and so we can study them separately. The vocal individual recognition system of penguins contrasts with that used by many mammalian species that use olfaction. Olfaction is used by many bird species that also use both visual and vocal cues. We examined olfactory communication in several species of seabirds (Jouventin, 1977; Jouventin and Robin, 1984; Lequette et al., 1989; Verheyden and Jouventin, 1994; Bonadonna et al., 2003a,b). For penguins, however, it appeared that individual identification was only vocal.

In large penguins that are able to move during breeding, specifically the Emperor and King penguins, we found the most complex and complete vocal signals. These species have no nest and care for the egg and young chick on their feet. Individuals are thus somewhat mobile during care of the developing offspring, though less so for the King than the Emperor penguin (making individuals harder to locate in the breeding colony for the latter species). King penguins are easy to manipulate during their courtships and pairing on the beach, without disturbance within the adjacent or nearby breeding colony. We identified the pairing status of duos of birds that were standing together; that is, we asked whether duos that had at least initially become partners were in fact behaviorally paired together. We determined this by walking between the two birds, so that visual contact between them was broken (Derenne et al., 1979; Jouventin, 1982; Keddar et al., 2015c). Such pairs often subsequently walked by one another without recognition, indicating that vision was not sufficient for mate identification. But if they were behaviorally paired, one partner would eventually sing. If the other "partner" had learned the song of the vocal bird, it would immediately make an abrupt turn toward the singer and waddle to it in a beeline, frequently singing in return and then both birds walked toward one another. This reunion showed that color ornaments or other outward appearances, as well as odors, were not sufficient for partners to identify each other by alternative sensory modalities; identification was accomplished by a vocal signal. When standing in close contact, vision may provide a quick assurance of the partner's presence, but only the voice ensures individual recognition. Earlier experiments in which the auditory canal or beak were closed also made it impossible for partners to find each other (Jouventin, 1982). They needed the voice. Thus, penguin songs have a very specific and essential purpose. They not only say "here I am" but also "I am your partner because my name is Mr. or Mrs…"

Traits that are directly linked to breeding, especially if they are essential to successful reproduction, are especially favored by natural selection (Endler, 1986). Additionally, we can think of traits that aid survival, like a quicker escape from a predator or a better digestive system, as aiding reproduction. In all of the penguin species, voice and the recognition of the partner are essential for successful reproduction. No seabird can produce surviving offspring without cooperation between parents. And this cooperation occurs in a crowded environment, since terrestrial habitat is limited in the southern hemisphere where penguins live. All penguin species are accordingly colonial breeders. Thus, if individual recognition of the partner was not extremely accurate, many penguins might make mistakes, and aid a bird that was not their partner or feed a chick that was not their offspring. In effect, they would commit their energy and resources to misdirected parental care, in which they help to raise offspring that are not their own (Hoogland and Sherman, 1976; Searby and Jouventin, 2005). This would result in favoring another bird's genetic inheritance rather than their own. Natural selection works strongly against such mistakes by favoring recognition of partners, as partners aid one another in passing on their genetic inheritance. Natural selection also strongly favors recognition of offspring, so that misdirected care seldom occurs.

The directing thesis of the experimental research program was that the location of the nest is a great aid to reproduction in seabirds other than the two largest species of penguins. To make sure that they are raising their own egg and chick, penguins need information that will guide them to the right spot where the mate is sitting or standing with the egg. The location of the nest, even if just a pile of pebbles, gives this information to penguins, as in many other species of birds (Jouventin, 1982; Lengagne et al., 1997). And this cue is more informative when a breeding colony is less tightly packed, presenting fewer choices to make in finding the right partner at a given location. We can observe complete identification of the breeding partner in all of the penguin species, but it is especially easy in the nesting penguins. A partner returns from foraging at sea to relieve the sitting partner during incubation or brooding. The approach is first made to the nest location, and then a series of songs are given. The partners recognize each other from their songs and trade places and duties, and the hungry sitting partner can then depart to the sea for much-needed foraging.

Unlike humans and other social mammals, penguins do not have to recognize a lot of individuals by voice, but only the partner, and later on their chick(s). The problem is discriminating the song of the partner and chicks from all the other songs nearby. Thus, confusion may sometimes occur when songs are too similar in a dense breeding colony. Nonetheless, the combination of location and song usually gives a sure identification in most penguin species. For the two nonnesting penguin species (Emperor and King Penguins), however, locations can change and confusion increases (e.g., Stonehouse, 1960; Isenmann and Jouventin, 1970; Jouventin, 1982). Among all birds, the situation of these species deserves special attention because they are so unusual in not having a nest, and the permanent breeding location that comes with the nest. Since most birds have a nest, it is difficult to evaluate how important it is to avian life. But in species without a nest, we can see just how important a nest is from the benefits these species lose, and the adjustments that they must make.

In Emperor penguins, the breeding colony is on sea ice in winter, and the colony may move about quite a bit as blizzards and temperature conditions make some microhabitat locations more favorable than others (Stonehouse, 1960; Prévost, 1961). In addition, individual parents, standing with their eggs or young chicks, move about the colony as they jockey for positions of shelter in the lee of the other birds, and in this way huddles are pushed along by blizzards. Formerly, Prévost (1961) reported that the Emperor penguin's colony moves toward more stable substratum of ice, closer to the coasts of islands. He supposed that movements of huddles were due to movements of the sea ice, which can be broken up by a storm. Jouventin (1971a) followed huddles during blizzards, with special attention to the direction and magnitude of the winds (Fig. 1.30). He observed that huddles were moved about only during a blizzard and were always pushed in the direction away from the wind. The movements of the penguins were due to the high freezing power of the blizzard, since breeders were unable to stay a long time along the leading edge of the huddle, where the force of the wind was greatest. Individual penguins, with their egg or young chick on their feet, moved from the windward to the lee side of the huddle, where they were more protected from the blizzard. Thus, the huddle moved continuously until the blizzard was over. Then the breeders would shift back to their traditional position on the sea ice (Fig. 1.30). Consequently, the Emperor penguin cannot use topographical cues to find its partner or chick.

Although the male Emperor penguin takes the first, long-standing shift with the egg, assuming a full incubation shift during more than 2 months, his female partner must relieve him, so that he can return to the sea to forage. In King penguins, again it is the male that takes the first long-standing shift with the egg, but this lasts for only about 2–3 weeks, and then his female partner relieves him (Weimerskirch et al., 1992) (see the breeding schedules of Emperor and King penguins in Fig. 1.17). If there is no disturbance, the male King penguin waits for his female partner within a zone in the colony, not moving more than a few meters from where the female left him, until her return (Jouventin, 1982; Lengagne et al., 1999a). So the female can return to the spot where she left her partner, and he likely has not moved as far as in the Emperor penguin. In all cases of relief, the song is used for mutual identification of partners before the egg is transferred, and the lack of landmarks, given in other penguins by the location of the nest, increases the importance of the vocal signature for these two species of large penguins.

Relieving the partner is essential for successful breeding in penguins and is an important element of the high degree of cooperation that occurs between mates during incubation. Equally important is the next stage of reproduction, after the egg is hatched. The young are initially naked and helpless and must be brooded by one parent or the other. Later, the young stand in the colony, often in a group called a crèche, but are only semi-independent. They retain chick plumage that is not suitable for swimming, or even entering the sea. So the still-developing chicks have to be fed by both parents and the chick(s) wait in the breeding colony for the parents to return to the nest location (or, if possible, the previous feeding location for the chick, for the two largest species). Meanwhile, both parents forage at sea and return to stuff the chicks full of regurgitated fish and invertebrate foods such as squid. During this time, the importance of partner recognition wanes, and a new identification, that of parents and chicks becomes paramount. This stage involves learning of the chick's song by both parents, but also of both parental songs by the chick(s).

During the stage of chick semiindependence, nesting penguins have the location of the nest to provide a rendezvous location. For the nonnesting King penguin, chicks form crèches to be warm and protected. The crèche is usually within the breeding colony, running through like a thick thread. Parents can find their chick in a "zone" in the colony where several hundreds of chicks stand together, and the parent sings to attract the chick to come and feed (Jouventin, 1982; Dobson and Jouventin, 2003a). Once chicks are in a crèche, they are semiindependent because they wait for their parents to bring food. But they are also subject to greater movements than when they were on the feet of their parents. Chicks move both to and about the crèche and may be forced to move to avoid the often-hostile adults near them, as well as away from potential predators. The recognition system of the voice is needed and must develop between offspring and the parents, or parents cannot find their chicks and the chicks will starve. The importance of songs as vocal signatures for mutual individual recognition between parents and chicks can be shown by the observed fact that multiple chicks frequently beg for food from adults when they return from the sea to feed their chick (Dobson and Jouventin, 2003a). These parents identify the chicks only acoustically and not by sight. Most often, parents chase off all but their own chick before giving food (Fig. 5.4). To conclude, to find and feed its own chick, among several scores or hundreds of chicks trying to lure the parent to obtain extra food, is a challenge. Without a nest location to serve as a rendezvous, this task is much more difficult, and it can only be accomplished using the right voice in a noisy crowd.

## TESTING INDIVIDUALS

Bird calls and songs have long been a favorite listening hobby for ornithologists, both amateur and professional. The usual way of studying bird songs is to record them using a tape recorder, so that the structure of the sound can be represented in a sonogram (a graph of sound frequency, shown over time so that temporal changes are revealed). The songs are then modified electronically and played back to the birds, and subsequent bird behaviors are recorded. These playbacks are easy to conduct because we modify only one physical parameter that might elicit species recognition. The manipulation is then tested for several birds. To study individual recognition, a new song is recorded and modified for each partner or chick.

Individual recognition by the voice has rarely been studied in detail (Beecher et al., 1994). In the late 1970s, we began a long-term study on vocal signatures across the entire family of penguins. Studies of three species pointed to similarities and differences in partner recognition via individual songs, comparing ecological conditions of reunion for species with and without nests: the nesting Adélie penguin (Jouventin and Roux, 1979), the nonnesting King penguins (Derenne et al., 1979), and the nonnesting Emperor penguins (Jouventin et al., 1979). Through observations, followed by experiments in the field, each of the species exhibited individual recognition between partners in breeding pairs and between chicks and parents.

Songs were played back via a loudspeaker to show recognition by an individual's partner and chick. Further, in each species, we found that individuals were very consistent in their songs (see Chapter 4), but different individuals had different songs. So the raw material for a vocal signature was present, and often only part of the song, a second or two, was sufficient to produce recognition. We also found that redundancy was necessary to communicate efficiently in such a loud environment as a colony of penguins, thus explaining why adults and chicks repeat their songs several times when they sing. At the same time, examination of early sonograms showed plainly that each species had a unique aural pattern of the song. But which of the specific elements of the song carried the "signature" that would tell a bird which of scores or hundreds of other penguins was their partner or parent? To answer this question, it was necessary to go further with the use of computer programs that could modify elements of songs for carefully designed experiments in which modified songs were "played back" to partners and chicks. The natural song was recorded and modified in different ways: modifications of amplitude and frequency modulations, modifications of frequency and temporal parameters, modification of "the two-voice system" that we will describe, and other changes (Aubin and Jouventin, 2002a). Table 5.2 summarizes the main results to our numerous playback experiments on five species, three nesting and two nonnesting penguins.

Playing back a generic, species-typical song often elicited a territorial response. This experiment was easy because a single unfamiliar song of the same species was tested on numerous birds in the field. Understanding the systems by which individual recognition takes place requires testing individual birds with the songs of partners and unknown birds, as well as using modified songs (Falls, 1982; Dhondt and Lambrechts, 1992). Tests of individuals were performed on marked birds with a personalized modification of the song. However, calls and songs in nature were given in an aural environment, and this had to be taken into account. For penguins, this means a windy and noisy breeding colony (Lengagne et al., 1999a). Recorded songs were played back so that the response of listeners could be evaluated in the natural environment. The sounds of the song were changed, with increased adjustments until the songs no longer elicited a behavioral response from the tested individual. Playbacks can also reveal the contexts and functions of songs (e.g., Searby et al., 2004). This was done carefully for four species of penguins which use a nest and two that have no nest, and for species that breed in high- and low-density colonies (Jouventin et al., 1999a; Aubin et al., 2000; Lengagne et al., 1999b, 2000; Aubin and Jouventin, 2002b). Of course, the various species of penguins also breed in different environmental conditions, from sparse colonies on relatively flat ground, to noisy crowded colonies on cliff faces or sandy beaches, to winter sheets of sea ice with frequent blizzards. Playbacks in these conditions can show how difficult it is to communicate individually under different environmental conditions and with noisy background sounds.

As described in the recording and playback procedures in Chapter 4, penguin songs were recorded in the field using omnidirectional microphone mounted on a 4-m pole (Fig. 5.6). The pole allowed recording with the human observer at a distance from the target individuals, to avoid upsetting the penguins. The microphone was held 1 m from the penguin, and songs were captured on NAGRA IV band tape-recorders then on a Sony TCD Pro

**TABLE 5.2 Responses of Penguins to Experimental Signals Corresponding to Modified Natural Songs in View to Compare Vocal Signatures in five Species**

| | Species | | | | |
|---|---|---|---|---|---|
| **Signals** | **EP** | **KP** | **MP** | **AP** | **GP** |
| **Modulation** | | | | | |
| Without AM | – | ++ | – | – | – |
| Without FM | + | – | + | + | + |
| Reversed | + | – | ++ | ++ | ++ |
| **Frequency Domain** | | | | | |
| Fo | nt | + | nt | – | – |
| Low-pass | ++ | ++ | ++ | ++ | – |
| High-pass | + | – | ++ | – | – |
| Shift ±25 | nt | nt | ++ | + | ++ |
| Shift ±50 | ++ | ++ | ++ | – | – |
| Shift ±75 | ++ | + | ++ | – | – |
| Shift ±100 | – | – | + | nt | nt |
| **Temporal Domain** | | | | | |
| HalfCall | + | ++ | ++ | ++ | nt |
| OneSyl | – | ++ | + | ++ | ++ |
| HalfSyl | nt | ++ | – | – | – |
| QuarterSyl | nt | – | nt | nt | nt |
| **Two-Voices Phenomenon** | | | | | |
| One voice suppressed | – | – | nt | nt | nt |

Only the adult King penguin (and chicks of all penguin species) uses the frequency modulation (FM) for individual recognition. In the frequency domain, Adélie and Gentoo penguins code the identity mainly on timbre analysis (spectral profile and precise frequency values of the harmonics). In the temporal domain, the redundancy is high in all penguins, thus improving the probability of receiving a message in a noisy environment. The two-voice system occurs only in nonnesting penguins.
++, strong response; +, moderate response; –, no response; nt, not tested. *EP, KP, MP, AP, GP,* respectively, Emperor, King, Macaroni, Adelie, and Gentoo penguins. (Results from: Jouventin, 1982, Aubin et al. 2000, Hildebrand et al., unpublished data, for the Emperor penguin; Jouventin et al., 1999a,b, Lengagne et al. 2000, for the King penguin; Jouventin and Aubin, 2001, for the Adelie and Gentoo penguins; Searby et al. 2004, for the Macaroni penguin)
Modified from Aubin, T., Jouventin, P., 2002a. How to vocally identify kin in a crowd: the penguin model. Adv. Study Behav. 31 (31), 243–277 and Aubin, T., Jouventin, P., 2002b. Localisation of an acoustic signal in a noisy environment: the display call of the King penguin Aptenodytes patagonicus. J. Exp. Biol. 205, 3793–3798.

**FIGURE 5.6** (A) Penguins are tame, but their colonies are loud. Ornithologists usually use a parabolic reflector to record songs, but this does not work well in a noisy background and modifies sounds. (A) Here P. Jouventin uses a pole, to obtain a much better recording. (B) Recording the song of a King penguin chick.

II DAT digital audiotape recorder. Songs were then transferred to a computer hard drive for both analysis and manipulation using Syntana computer software (Aubin, 1994).

The idea behind manipulated songs was basically to remove or change the song characteristics, so that they could be played back to the birds using the tape recorder attached to an Audax loudspeaker. During playbacks, target birds were examined for behavioral responses, such as singing in return, standing up-alert, or moving towards the loudspeaker. Altered songs were played twice with a short delay between the pair of songs, as occurs naturally under field conditions. By their actions, the penguins could then indicate what they could recognize by responding or not to the playbacks. Behavioral responses were then ranked from no reaction to a very strong reaction, on a scale and via behaviors that were appropriate to the circumstances of each species tested. Since some birds may have been more motivated to respond than others, several birds were tested under similar circumstances for each experiment (i.e., for each identical computer manipulation of the song), and the results analyzed statistically. This basic procedure is common to many studies of individual recognition in birds (e.g., Jouventin et al., 1979; Beletsky, 1983; Sieber, 1985; Jouventin and Aubin, 2002; Lovell and Lein, 2005; Clark et al., 2006; Buhrman-Deever et al., 2008; Jacot et al., 2010).

## BURROWING PENGUINS

The Little penguin is a burrowing species and thus has a specific location of the nest that is easy to differentiate from others. The nesting territories of Little penguins are widely dispersed enough that it is perhaps the least "colonial" of the penguin species. Little penguins exhibited strong territorial defense and used highly vocal "territorial songs" to defend their burrow and nest against intruders. Thus, these songs may serve the function of advertising behavioral dominance, as well as individual quality as a mating partner for the opposite sex, and finally for partner and chick recognition. Having a rendezvous site that is so easy to find, the individual recognition of the partner and with chicks by voice is not as accurate from the vocal signals alone as in other penguin species. Indeed, the song of Little penguins is both more variable and less sophisticated than for more colonial penguins (Fig. 5.7). Little penguins suffer less confusion, as reflected by the vocal signal, because the the presence of the burrow lessens confusion and aids individual recognition (Jouventin and Aubin, 2000; Nakagawa et al., 2001; Miyazaki and Waas, 2005).

In this territorial species, as for most birds, the voice is first used to maintain the territory around the burrow (territorial defense) and only secondarily for the final identification of a partner or chick (i.e., vocal recognition). So the sound pressure (loudness) of a normal song that is used to repel rivals is loud (about 90 dB from 1 m away), especially for such a small bird. Experimental changes to the songs show that as in other penguin species,

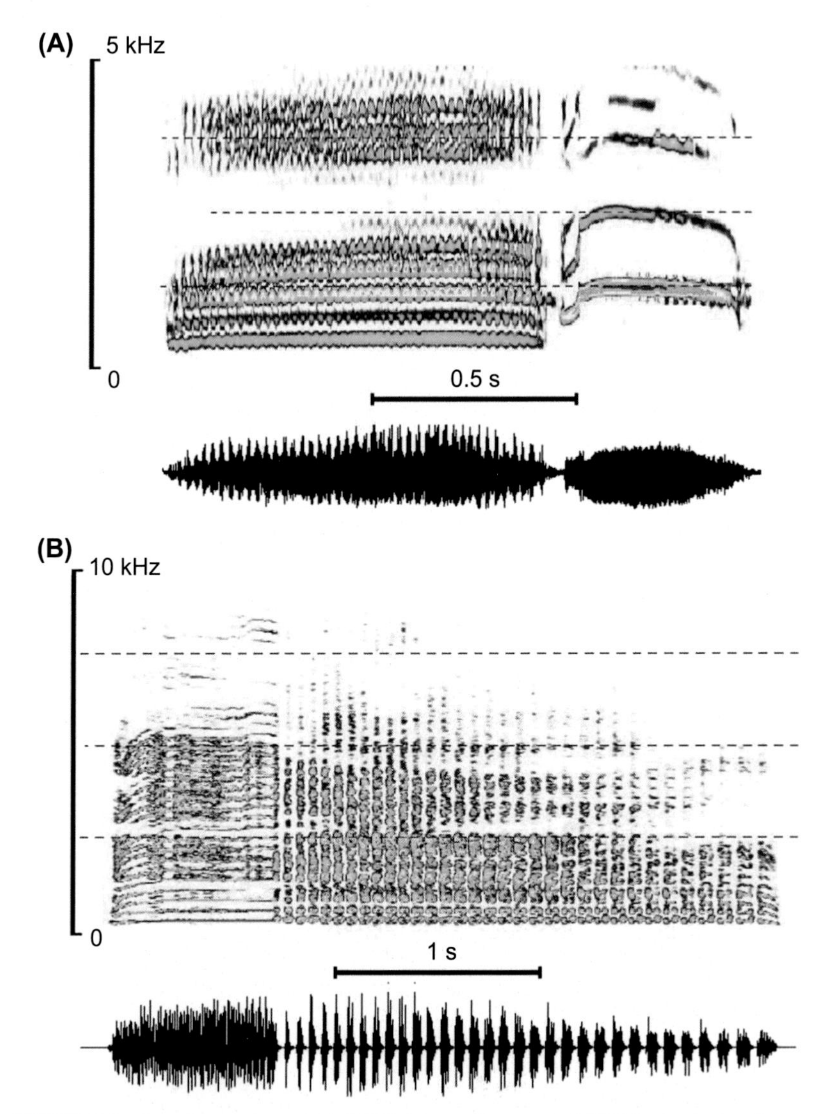


**FIGURE 5.7** Sound spectrograms of (A) Little penguin song and (B) Macaroni penguin song. Both comprise hatched phrases that are easy to locate by comparing sounds coming to each ear. Oscillograms, indicating the intensity of the sound in dB, are given under each sonogram. *From Aubin, T., Jouventin, P., 2002a. How to vocally identify kin in a crowd: the penguin model. Adv. Study Behav. 31 (31), 243–277.*

changes in the amplitude of Little penguin songs are relatively unimportant to song recognition (Fig. 4.12). Rather, it is the frequency of the sound that provides the information in the song. In addition, each syllable of the song has an exhale and inhale sub-syllable, and both are needed for the song to be recognized as that of a conspecific and responded to with a replying territorial song (Jouventin and Aubin, 2000).

The four banded penguins (*Spheniscus* species) are also burrowers, and song may facilitate both location of the territory and mate-chick recognition. While this seems likely, specific experiments on vocal signatures have not been conducted on species of banded penguins.

## NESTING PENGUINS

Gentoo penguins have relatively low-density breeding colonies because, different from other penguins on sub-Antarctic islands, they breed during winter when food is scarce (Bost and Jouventin, 1990, 1991). But the density and environmental circumstances of these colonies differ among populations. On sub-Antarctic islands, Gentoo penguins nest on gently sloping or flat expanses of meadowland above the shoreline and back from the beach, often between tufts of grass (Fig. 5.8). Individual recognition has been tested using modified songs in playback experiments (Jouventin and Aubin, 2002). Gentoo penguins use amplitude modulation (AM) to recognize songs. This means that as time passes, the strength of the song changes, with stronger sound emitted in "syllables" and gaps of relative silence between them. To study individual recognition, we first tested adults, but they unfortunately stopped responding to played back songs after a few experiments. These penguins quickly learned the difference between a song given by the electronic speaker and a real song.

**FIGURE 5.8** Gentoo penguin parent returning to the nest to find its two chicks, among many that respond to the adult's song. The chicks sing, hoping to be fed by mistake. The vocal signature of adult's chicks permits identification by the parent, so that its own chicks can be fed.

Thus, we shifted to testing the responses of chicks to playbacks of the parents' songs, and the chicks were easier to lure with broadcasted songs (Fig. 5.9; examples of modified parental songs from Adélie penguins, Jouventin and Aubin, 2002). The chicks were almost constantly hungry, so they responded much more readily to what they perceived as a nearby food-laden parent.

For the Gentoo penguin, sound pressure (loudness) of the parental song was about 83 dB (decibels), likely well above the background noise of the relatively sparse colony. The duration of the song phrase was anywhere from 0.7 to 1.9 s (for 30 birds), and the length of time of a syllable of the song gave information about the identity of the singer. Interestingly, reversing the song (playing it backwards) provided the same information, and at least one syllable (one maximum amplitude syllable of the song) was necessary for recognition. As well, the timbre of the song (the fundamental frequency and harmonics of the song) also gave the main information that was used in individual recognition. If the song was shifted up by 25 Hz (a relatively small shift in a musical note), the chick could still recognize it, but recognition was lost when we implemented greater shifts in frequency. Amplitude modulation (AM) was important for individual recognition of the song of the Gentoo penguin. Parental songs were not recognized if the AM was artificially removed and played back to the chick. On the Antarctic peninsula, Gentoo penguins breed in somewhat denser colonies than on sub-Antarctic islands, and their nuptial song is different (Jouventin, 1982). This suggested that genetic differentiation of these populations has occurred, and DNA analyses confirmed that these forms may be different species, as we will see later (Chapter 6; de Dinechin et al., 2012).

Adélie penguins live in more crowded breeding colonies than Gentoo penguins, often in rocky slopes along the shore of the Antarctic continent where the ocean is rich. Adélie penguin chicks were also motivated by hunger to respond to their parents, and thus we tested chicks for their responses to parental songs (Jouventin and Roux, 1979; Jouventin and Aubin, 2002). Adélie penguins are related to Gentoo penguins, but the former species is more adapted to high latitudes where positive temperatures are limited to 4 months of the year. So Adélie penguins have to breed quickly. These two nesting species have songs of similar structure, with song syllables that are repeated, with AM (Fig. 5.10). The breeding colonies of Adélie penguins are usually larger and denser than those of Gentoo penguins, and thus are noisier. The background noise level of an Adélie colony was about 57 dB (decibels), and the Adélie song was given at about 86 dB of sound pressure. So the problem of misidentification was increased over that of Gentoo penguins.

The duration of the Adélie song varied from 0.6 to 2.3 s, and at least one syllable was needed for individual chicks to recognize their parents. Repetition of the same syllable was consequently a guaranty against the confusion of the background noise. So we would expect greater discrimination possibilities in the Adélie songs, and they appear to have greater information content. Also, playing back the isolated fundamental frequency of the song was not sufficient

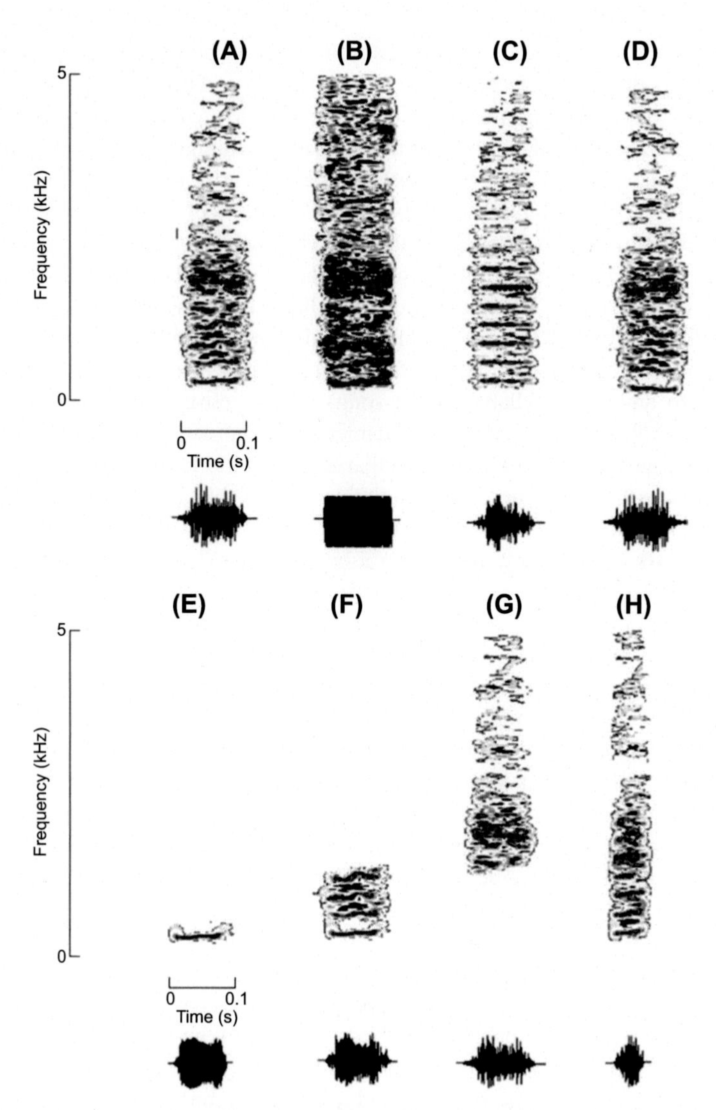


**FIGURE 5.9** Sound spectrograms and oscillograms of different modified parental songs played back to chicks: (A) natural (control), (B) without amplitude modulation, (C) without frequency modulation, (D) reversed, (E) fundamental frequency, (F) low-pass, (G) high-pass, (H) first half syllable signals. Only one species (Adélie penguin) and one syllable of the song are represented here, but we did systematic experiments in the field on six species of penguins. *From Jouventin, P., Aubin, T., 2002. Acoustic systems are adapted to breeding ecologies: individual recognition in nesting penguins. Anim. Behav. 64, 747–757.*

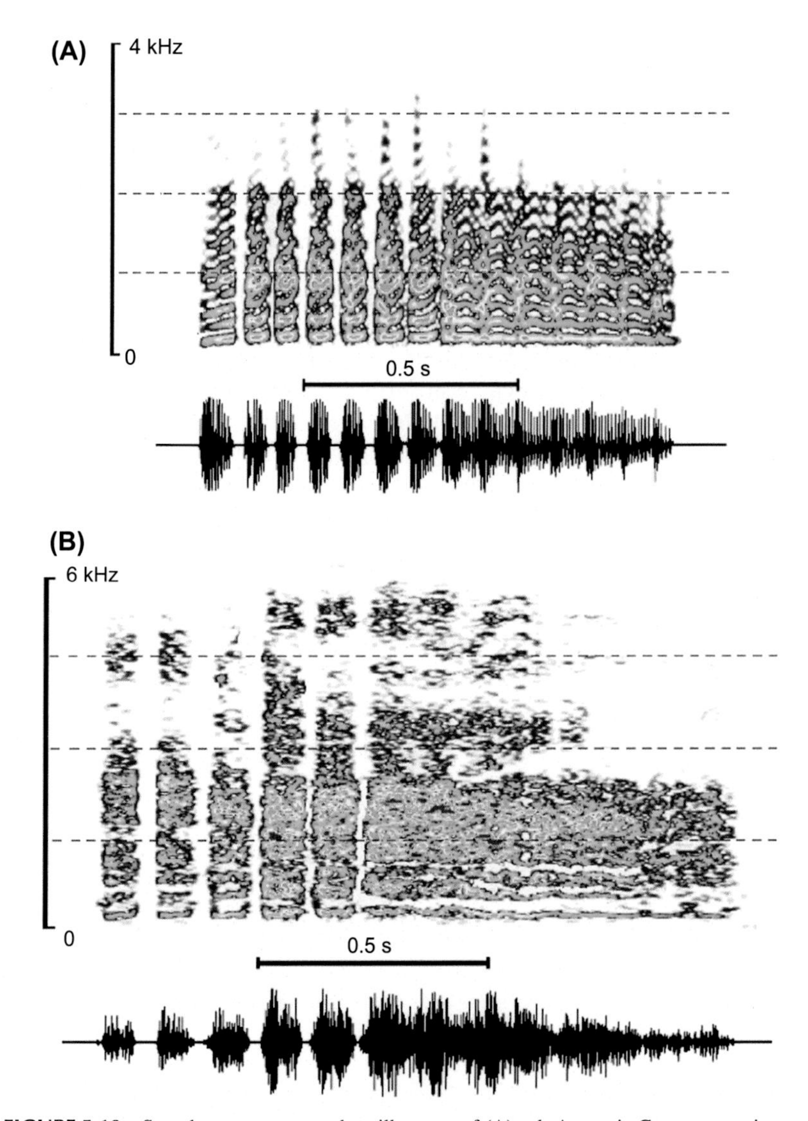


**FIGURE 5.10** Sound spectrograms and oscillograms of (A) sub-Antarctic Gentoo penguin song and (B) Adélie penguin song. The two mutual songs are hatched, which assists location of the singer on its nest in the colony, but the second one is louder, because the density and consequently the confusion is higher. *From Jouventin, P., Aubin, T., 2002. Acoustic systems are adapted to breeding ecologies: individual recognition in nesting penguins. Anim. Behav. 64, 747–757.*

for individual recognition, though upper (higher Hz) harmonics could be cut from the song, and individual recognition was still possible. If the song was shifted by 25 Hz, the chick could still recognize it, but again at greater shifts recognition was lost. Adélie chicks could hear adults when they were 7 m apart, if the song was given at the same loudness as the background noise of the colony (Fig. 5.11). This ability to pick out a known voice at a relatively low volume compared to the background noise is known as the "cocktail party effect," for comparison with similar effects in human party conversations, the only other species known to exhibit the phenomenon (Aubin and Jouventin, 1998).

Macaroni penguins are crested penguins that live in dense colonies on rocky shorelines of sub-Antarctic islands, often nesting on cliffs above the sea. To study individual recognition, we tested chicks to see if they could identify their parents. Macaroni penguins have a song with a long first syllable and then repeated short syllables. The two halves of the song are roughly equal in length. Oddly, only about 40% of playbacks of parental songs stimulated chick responses, while real songs virtually always stimulated a response in the chicks. Probably the artificial sound was not as accurate for these clever birds as the natural one. It was clear that the first half of the song did not carry information for individual recognition, whereas the second half of the song was recognized. This big difference in vocal identification between the two parts of the song was tested by substituting an alien song element for either the first or second part of the song and leaving the other half of the song from the

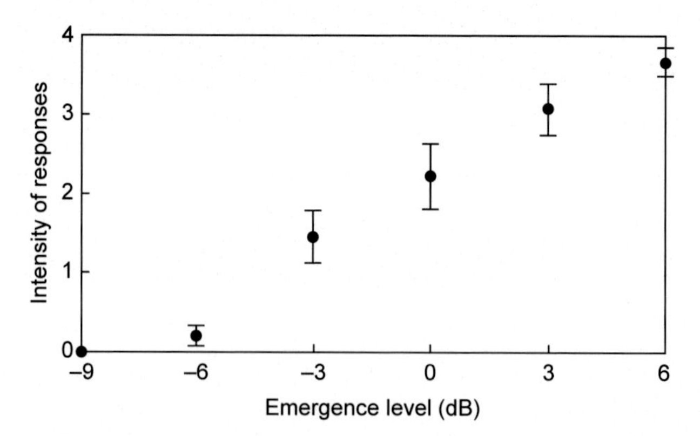


**FIGURE 5.11** Test of detection of the Adélie penguin song by chicks, with respect to a jamming situation. The response of chicks given the loudness of the background noise measures the ability of penguins to identify a song among others. The parental song is experimentally masked in the time and frequency domain by five extraneous songs of other adults to yield different emergence levels. The intensity of chick responses to playbacks is given according to six emergence levels of the parental song detected from the background noise. *From Jouventin, P., Aubin, T., 2002. Acoustic systems are adapted to breeding ecologies: individual recognition in nesting penguins. Anim. Behav. 64, 747–757.*

parent for playback to the chick (Fig. 5.12; Searby and Jouventin, 2005). Macaroni penguin chicks responded to the reversed second part of the song, and appeared to ignore the first part (coming afterward in the reverse playback) for individual recognition. Also, as in the previous two species, playing the song backwards still stimulated individual recognition by the chicks.

Macaroni penguins used AM to recognize the song (in the second syllable) and needed a complete short syllable for individual recognition. Cutting the song in half in the frequency range (much like listening to the bass and alto of a voice separately) still produced individual recognition of the parental song from just the lower and higher ranges. Thus, they could recognize the timbre of the song with only partial harmonic information. Finally, shifting the song by more than 25 Hz precluded individual recognition. Due to the focus on the second syllable of the song and better ability to pick out partial harmonics, macaroni penguins seemed more discriminating than Gentoo or Adélie penguins. The second part of the nuptial song was composed of the same short syllable, repeated up to 12 times to ensure individual recognition. Is it possible that the first part of the song has no biological function, and the second part is required for individual recognition? We hypothesize that the first part of the song is mainly used to attract the attention of potential mates or chicks before

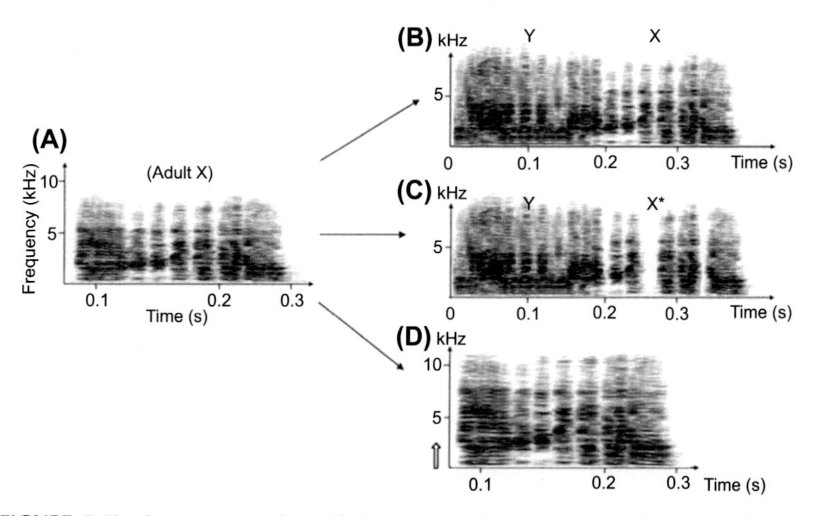


**FIGURE 5.12** Spectrograms of modified songs used in playback experiments to Macaroni penguin chicks to separate the use of the first and second part of the song. Each gray scale represents an amplitude class of 2.5 dB. (A) Natural signal (control), (B) composite signal: part one from adult Y, part two from adult X, (C) composite signal with modified rhythmic pattern in part two; X*, modified song of adult X, and (D) a frequency shifted signal (+50 Hz). *From Searby, A., Jouventin, P., 2005. The double vocal signature of crested penguins: is the identity coding system of Rockhopper penguins* Eudyptes chrysocome *due to phylogeny or ecology? J. Avian Biol. 36, 449–460.*

the identification part. Macaroni penguin colonies are as crowded as the previous *Pygoscelis* species that nest on flat areas where a singer is easy to see in an open environment. But a Macaroni penguin singer is often impossible to locate by observation (i.e., vision) alone, due the rocky landscape of a rock-slide or cliff face. This increased difficulty compared to other penguins perhaps explains why the vocal signature of the Macaroni is more complex than that of the species of *Pygoscelis*. The Macaroni penguin's song uses not only tempo but also AMs to be identified.

Rockhopper penguins are also crested penguins that live along rocky shorelines, but often among tussocks of grass where they nest in a burrow or between boulders. When they cooccur with Macaroni penguins, the latter species nests on higher and often steeper ground, above the Rockhopper penguins. Hungry Rockhopper penguin chicks were tested for individual recognition of their parents' songs (Searby and Jouventin, 2005). Playbacks of parental songs stimulated chick responses only about half the time, while real parental songs always stimulated chick responses, as in other species of penguins. Nonetheless, chicks responded to playbacks of parental songs significantly more often than songs of unfamiliar adults, so they appeared to clearly recognize the songs of parents. The information content of the song of the two crested penguin species was roughly equal (Fig. 5.13).

Rockhopper penguin songs also have a long first syllable, followed by repeated short syllables. By substituting an alien first syllable or second part of the song, the necessity of these different parts of the song was tested. As long as the first syllable or second part of the song was from the parental song,

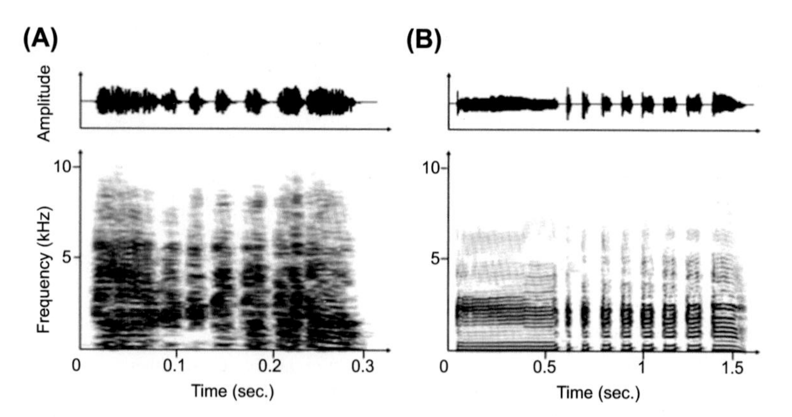


**FIGURE 5.13** Oscillograms and spectrograms of songs (vocal signatures) of adult Rockhopper (A) and Macaroni (B) penguins that were modified experimentally to test individual recognition. *From Searby, A., Jouventin, P., 2005. The double vocal signature of crested penguins: is the identity coding system of Rockhopper penguins* Eudyptes chrysocome *due to phylogeny or ecology? J. Avian Biol. 36, 449–460.*

chicks reacted with recognition. Thus, Rockhopper penguin chicks could use more parts of the song for recognition than could Macaroni penguins (that used only the second part of the song for recognition, see Fig. 5.12). Chicks were sensitive to the AM of the song and used the timbre of the voice to recognize their parent. They also could recognize harmonics to a higher degree. Cutting the song in half still produced individual recognition of the parental song separately in lower and higher frequency ranges. But Rockhopper penguin chicks could not recognize songs that were shifted in frequency by more than 25 Hz, as in the other nesting penguin species. For all the nesting species studied, parents need to hit the right "note" for their chicks to recognize them.

## CRESTED PENGUINS ARE SOPHISTICATED

Vocal signatures of Rockhopper penguins were analytically compared to Macaroni penguins (Searby and Jouventin, 2005). A methodology derived from the theory of information was used to determine which parameters of the song were likely to encode individual identity. Playbacks of modified songs in the field complemented the analyses, and parent–chick reunions were also compared between the two species. Summarizing the main conclusions, the results revealed a similar double signature system within the *Eudyptes* genus, which integrates information simultaneously from the temporal and spectral domains. This double encoding is made through the tempo given by the successive syllables of the song and its harmonic content. While it confirms the hypothesis that signatures are simpler in nest-building species, this result reveals differences in the efficacies of signatures within these species. This suggests that other parameters such as the mean distance at which recognition of parent and chick occurs should be considered to account for the encoding differences in the vocal signatures and the resulting efficacies of the different songs.

The mean distance between parent and chick at the parent's first song, and mean recognition delays were measured for the two species (Fig. 5.14). Parent–offspring recognition happened at a significantly shorter distance in Rockhopper penguins (mean Rockhopper distance = 1.7 m, Macaroni = 3.0 m; Mann–Whitney U-test: $U = 961$, $P < .05$, $N_1 = 30$, $N_2 = 50$), though the mean recognition delay was not significantly different between the two species (mean Rockhopper penguin delay = 42 s, Macaroni penguin delay = 48 s; Mann–Whitney U-test: $U = 882$, $P > .18$, $N_1 = 30$, $N_2 = 50$). This suggests that the mean search area covered by a returning parent singing for its chick might be smaller in Rockhopper penguins. As the densities were similar in the colonies, the number of adults that a chick compares to identify its parent should therefore be fewer in Rockhopper penguins. This was verified by comparing colonies of the two species, and as expected, adult Macaroni penguins that chicks

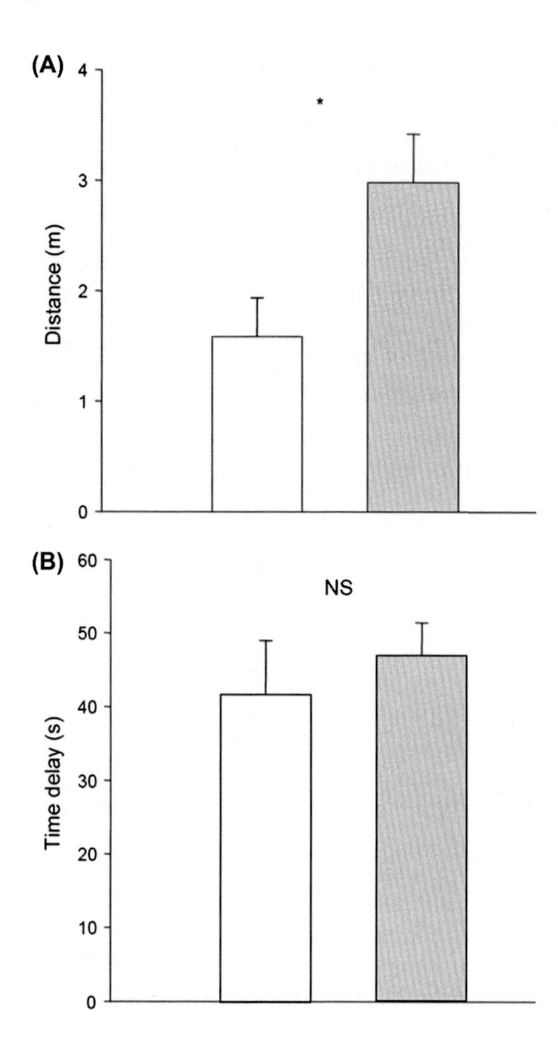


**FIGURE 5.14** Comparison of parent–chick reunions in Rockhopper penguins (*white bars*) and Macaroni penguins (*light gray bars*). (A) Mean recognition distance (in meters), the distance separating parent and chick when the parent began to sing. (B) Mean time delay (in seconds), the time elapsed between the first song of the adult and the beginning of feeding. *From Searby, A., Jouventin, P., 2005. The double vocal signature of crested penguins: is the identity coding system of Rockhopper penguins* Eudyptes chrysocome *due to phylogeny or ecology? J. Avian Biol. 36, 449–460.*

encountered were much more numerous. So the vocal signature of Rockhopper penguins appears to be somewhat less reliable that of Macaroni penguins. This is consistent with the reduced probability of confusion in recognition in Rockhopper penguins, suggesting stronger selection for efficacious recognition via the vocal signature in Macaroni penguins. Thus, even within a single genus, the complexity of the system of individual recognition appears to be adaptively linked to the nesting ecology of the penguins, that is, the difficulty of parent and chick meeting in the breeding colony.

The first studies of vocal signatures of penguins compared the complex vocal signatures found in nonnest-building penguins (Robisson, 1992; Jouventin et al., 1999a; Lengagne et al, 2000, 2001) with nest-building species from the genus *Pygoscelis* that had a simple signature system, based on the harmonic content of the song (Jouventin and Aubin, 2002). However, results from the genus *Eudyptes* also revealed significant variation within nest-building species (Fig. 5.15; Searby et al., 2004). The Macaroni penguin, belonging to the genus *Eudyptes*, uses an intermediate coding system that is more complex than the simple signature of *Pygoscelis* penguins. Our analysis of the Rockhopper penguin's vocal signature showed that this peculiar system is not restricted to the Macaroni penguin (Searby and Jouventin, 2005). Despite small variations in signatures, both *Eudyptes* species encode information in a similar redundant way, which is not found in nest-building penguins from the genus *Pygoscelis*. Rockhopper and Macaroni penguins' vocal signatures integrate information in

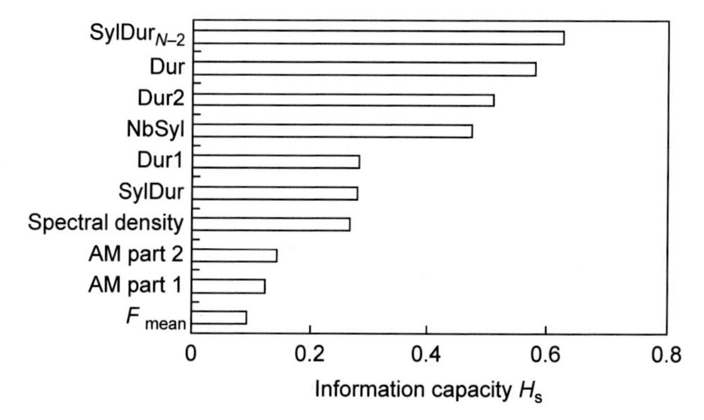


**FIGURE 5.15** Parameters of the song of the Macaroni penguin that facilitate discrimination between individuals, sorted by their information capacity value ($H_s$). SylDur, the mean duration of syllables in the second part of the song; SylDur$_{N-2}$, the mean duration of syllables not including the last two syllables; Dur, duration of song in (s); Dur1 and Dur2, the duration of first and second parts of the song; NbSyl, the total number of syllables; Spectral density, the average power spectrum of song; AM part 1 and AM part 2, the amplitude modulations in, respectively, the first and second parts of the song; $F_{mean}$, the mean value of the fundamental frequency. *From Searby, A., Jouventin, P., Aubin, T., 2004. Acoustic recognition in Macaroni penguins: an original signature system. Anim. Behav. 67, 615–625.*

both the time and frequency domains, and in this they are similar, though not as sophisticated as the large *Aptenodytes* species.

The time domain of the vocal signature refers to the timing of song syllables. For example, we can ask how far along we are in the song before recognition is achieved. Thus, if recognition requires half of a syllable, a shorter period would be required, other things being equal, than if a whole syllable were given. The same is true for comparing a single syllable to the full length of a single song. The time domain shows how the song changes over time, so it can be studied by cutting the song or even syllables into segments and looking for responses to playbacks by partners or chicks. The frequency domain refers to the range of frequencies of the song. The "spectrum" is the range of sound frequencies that a song entails, usually described as harmonics. Naturally, a range of frequencies are expressed by any sound, like a penguin song, at any given time. The range of frequencies that occur in a song can be thought of as the frequency domain, and this range is larger in some species than others. Frequencies can be studied by cutting the frequency range in half and expressing the highest or lowest, by shifting the frequency domain up or down, and by filtering all but the fundamental frequency (viz., the lowest frequency band) from the song. Once these manipulations are performed on a computer, the manipulated song can then be played back and responses of partners or chicks recorded.

Information on identity is usually repeated in both time and frequency domains, as each domain alone holds enough information to allow correct individual identification. But this is not the case for the large *Aptenodytes* penguins, for whom temporal and frequency parameters are strictly complementary, as both are necessary for recognition. In addition, the encoding of information in the time domain is achieved differently between *Aptenodytes* species and Macaroni penguins. In the first case, information is coded in the way amplitude or frequency varies over time, resulting in a sophisticated two-dimensional code. In the latter case, the temporal coding consists of a tempo given by the regular spacing of similar syllables. This system is thus simpler, as it is characterized by a single variable. In this way, the Rockhopper and Macaroni penguins' system is more similar to the simpler system based on one-dimensional variables found in the *Pygoscelis* genus. However, the ability to gather information simultaneously from the time and frequency domains is specific to the genus *Eudyptes*, resulting in a more complex system.

Unsurprisingly, the common factor in the signatures of nest-breeding species is the use of the harmonic content of the song to convey individual information. It is likely that the individuality of this pattern of frequencies simply reflects the specific sound-producing morphology of each animal, such as its tracheal resonance (Suthers, 1994). This might be an analogue of the formants found in mammal vocalizations, where some species are also known to communicate information such as individual differences (Fitch, 1997, 2000;

Owren et al., 1997; Reby et al., 1998, 2001; Phillips and Stirling, 2000; Charrier et al., 2003; McComb et al., 2003; Riede and Zuberbuhler, 2003; Searby and Jouventin, 2003). All *Eudyptes* signatures studied share a common rhythmic pattern due to abrupt and regular AMs (Warham, 1975; Jouventin, 1982). This similarity, along with our results, supports the hypothesis that there is a similar signature system within the whole genus *Eudyptes*. This signature system appears to be intermediate between the sophisticated system of nonnesting penguins of the genus *Aptenodytes* (Fig. 5.16) and the simpler system of nest-building penguins of the genus *Pygoscelis* (Fig. 5.10 from Aubin and Jouventin, 2002a).

## NONNESTING PENGUINS

Our main hypothesis is that the mutual song of the nesting penguins does not need to be highly complex because the nest location provides a stable rendezvous point that helps partners and chicks find each other. When a confident and aggressive penguin is at the nest, it is likely an owner. The adult or chick only has to verify the identity of a bird that is already close by and singing. But this is not the situation for the two largest penguins of the genus *Aptenodytes*. These species have no nest, but incubate and brood their young chick on their feet, moving about the breeding colony. Emperor penguin individuals make extensive movements throughout the breeding colony, and the colony itself may move about on the sea ice, due to weather conditions during the Antarctic winter (Jouventin, 1971b). The sub-Antarctic King penguin, on the other hand, breeds mainly in the spring and summer on sandy or pebble beaches, and individuals usually move only a few meters with the egg or young chick on their feet (Lengagne et al., 1999a). How do they find their partner or chick without the topographical cues given by the nest? To answer this question, experimental study of the vocal signatures of these species is also necessary, and in the field where the ecological context is influential and most evident (Fig. 5.17).

The greatest number of experiments on penguin songs have been conducted on King penguins. They breed on sandy beaches of sub-Antarctic islands, along the shore where they have relatively rapid access from the sea to their partner and chick. King penguin colonies are densely packed, and studied colonies vary from several thousand to hundreds of thousands of pairs. In such a crowded environment, there are extreme challenges for sound propagation from wind, from flipper flapping, from the noisy environment of thousands of singing penguins (Fig. 5.18; Aubin and Jouventin, 2002b), and from the bodies of breeding penguins that can mask the song between individuals at a distance (Fig. 5.19; Aubin and Jouventin, 1998). King penguins move during incubation and during brooding of the young and helpless chick,

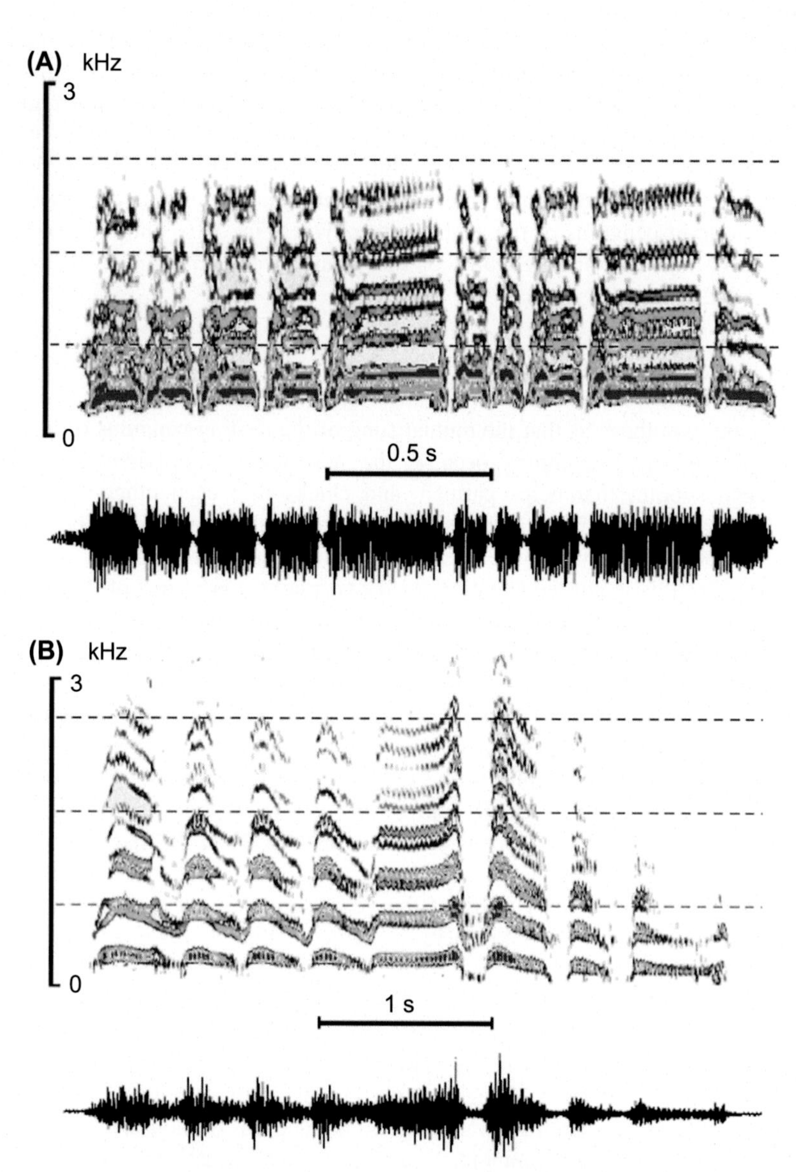


**FIGURE 5.16** Sound spectrograms and oscillograms of the more informative and sophisticated vocal signatures in penguins of: (A) Emperor and (B) King penguins. *From Aubin, T., Jouventin, P., 2002a. How to vocally identify kin in a crowd: the penguin model. Adv. Study Behav. 31 (31), 243–277.*

**FIGURE 5.17** On the left, P. Jouventin is playing back a King penguin song through the bodies of the penguins. On the right, two colleagues record the returning song to measure how the song passes through the intervening penguin bodies.

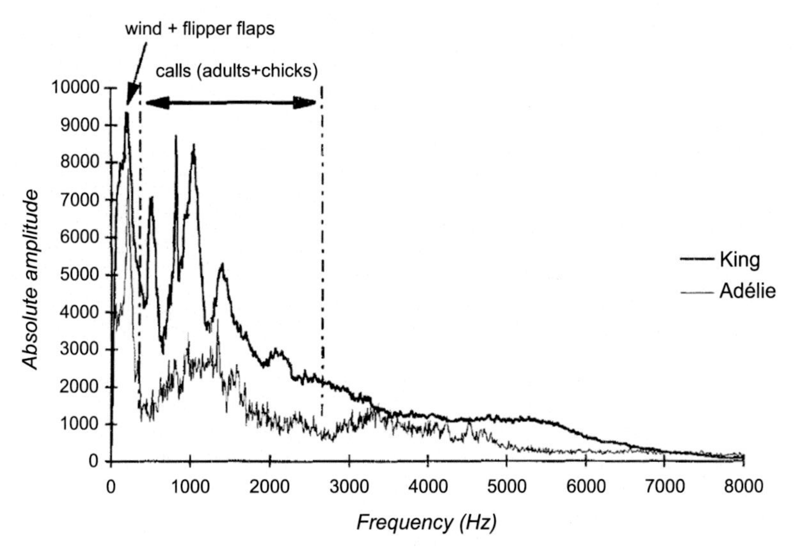


**FIGURE 5.18** Spectra of 4 min recording of the background noise in King and Adélie penguin colonies. Adult and chick songs represent 60% and 55% of the energy in the spectra of King and Adelie penguin colonies, respectively, showing the difficulty of vocal partner identification, particularly when penguins have few meeting points (King penguins with no nest). *From Aubin, T., Jouventin, P., 2002a. How to vocally identify kin in a crowd: the penguin model. Adv. Study Behav. 31 (31), 243–277.*

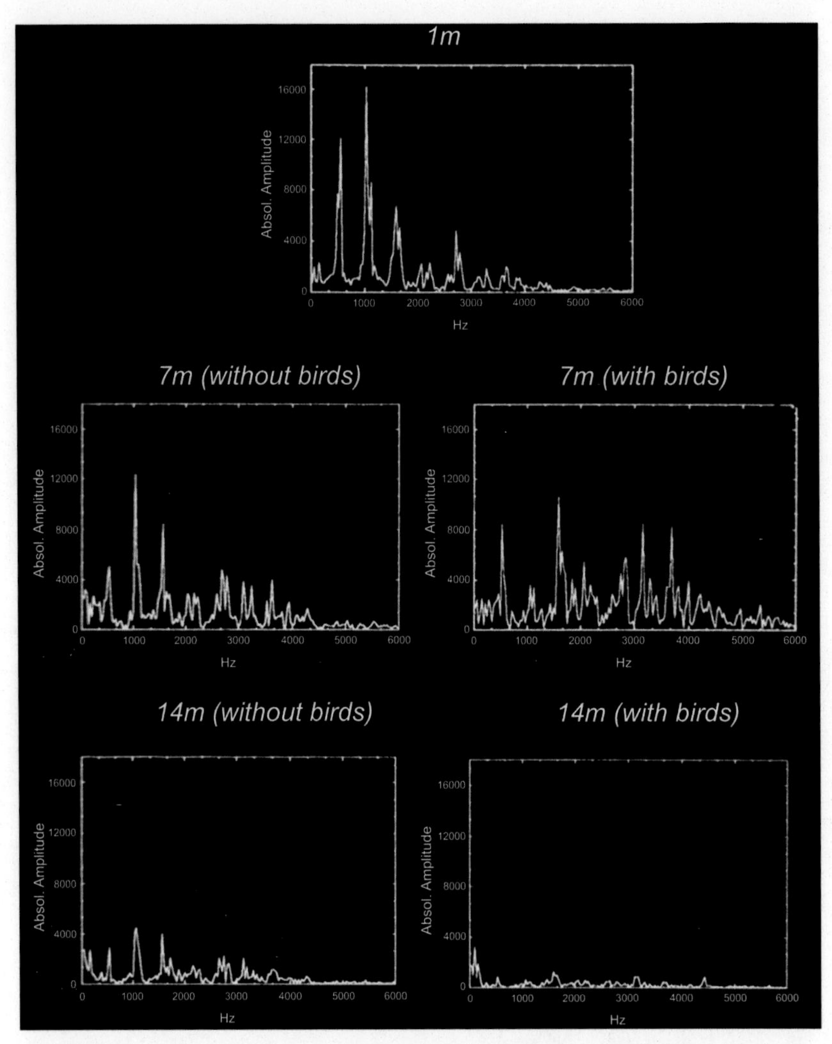


**FIGURE 5.19** Results of the Fig. 5.17 experiment according to sound pressure of the King penguin's song at different distances (1, 7, and 14 m) under different conditions (with or without intervening penguin bodies blocking the sound). After only 14 m little sound of the song remains, particularly when penguins mask the propagation. *From Aubin, T., Jouventin, P., 1998. Cocktail-party effect in King penguin colonies. Proc. R. Soc. B Biol. Sci. 265, 1665–1673.*

even without extreme weather events, such as the storms that are common on sub-Antarctic islands. Breeding penguin pairs moved an average of about 4–5 m during incubation and brooding due to disturbances, and some pairs moved up to 9 m or more during this period (Lengagne et al., 1999a). Late in brooding, most of the parents (about 54%–64%) changed location between one partner's turn at chick care and relief by the other partner. Subsequently, both parents forage at sea and return to the colony to feed their chick, while the chick waits in a crèche on a flat beach without landmarks. So, again, how can a breeding adult king penguin find its partner and chick in such a noisy and very crowded colony?

It might seem less difficult in a King penguin colony to find a partner because they move much less than the Emperor penguin. But Emperor penguins finish breeding in 1 year by using the winter for incubation and brooding. King penguins usually take about 14–16 months to finish a breeding cycle, and cannot successfully begin a new breeding cycle until the following year. Thus, even successful breeding King penguins have a 2-year breeding cycle, though many do not do even this well due to variation associated with individual body condition (Jouventin and Mauget, 1996; Fig. 1.27). Since there is no nest to aid individual recognition, King penguins must rely almost solely on the voice of their partner and chick. But the remarkable voice of the King penguin is aided by position in the colony and posture.

First, returning adult penguins approach a "zone" where they left their partner when they previously departed to feed at sea. During incubation, the partner usually moved only a meter or two (average = 1.3 ± 0.2 SE meters) while standing, and during brooding of the young chick the movement was about 3 m (average = 3.1 ± 0.2 SE meters; Lengagne et al., 1999a). Playbacks were used to estimate the distance at which a returning partner could be heard by the standing breeder, and recognition occurred within about 9 m (average = 8.8 ± 0.4 SE meters) and had a maximum distance of 12 m (Fig. 5.20; Aubin and Jouventin, 2002b). Thus, returning to the position that a partner left before foraging at sea should connect it with its mate most of the time (87% of tested birds heard and responded to the partner's song at 8 m distant). The approaching partner began to sing before reaching the partner, however, when about 8 m away (average = 8.3 ± 0.2 SE meters, maximum = 20 m). The first song given by the arriving partner stimulated the standing partner about 70% of the time, and about 548 standing birds were within range of the initial song. Later in the breeding cycle during the crèche stage, when both partners are feeding the independent chick, it is more difficult for a returning parent to find its chick. Only about half of parents find their chick immediately; and about a quarter of parents have to search for chicks broadly, taking nearly 3 times as long and giving more than twice the number of songs (Dobson and Jouventin, 2003a). Even so, mistakes in feeding were sometimes made, with 5%–10% of feedings given to chicks that were not genetic offspring of the returning adult.

Other features influenced these individual recognitions and reunions with a partner. A King penguin colony is so tightly packed that the bodies of other penguins limit the ability of sounds to carry between partners. The posture of a singing adult penguin, returning from the sea, is with head extended on high and thrown back, about 1 m off the ground. The initially incubating partner

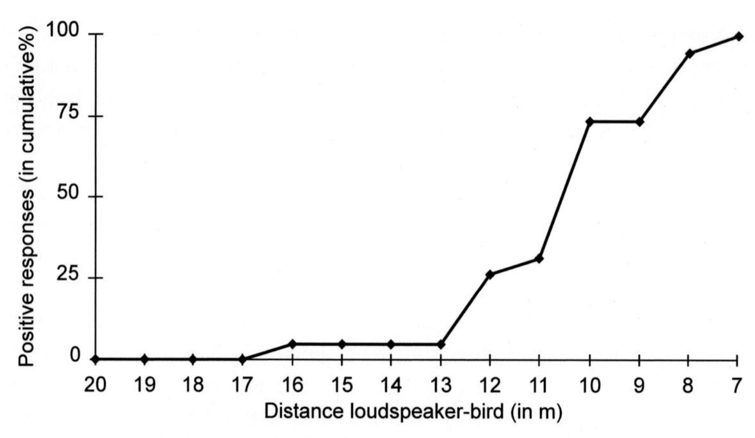


**FIGURE 5.20** For King penguins, maximal distance discrimination of the parental display song by the chick. During experiments (12 chicks tested), the chick-to-loudspeaker distance was progressively reduced from 20 m by approaching the loudspeaker in steps of 1 m. Parental song recognition disappears after only 11–13 m. *From Aubin, T., Jouventin, P., 2002b. Localisation of an acoustic signal in a noisy environment: the display call of the King penguin* Aptenodytes patagonicus. *J. Exp. Biol. 205, 3793–3798.*

sings in return from a lower position but still upright and about half a meter (an average of 45 cm) from the ground. Playbacks made from 10, 45, and 90 cm from the ground revealed that degradation of songs is much less at the two higher levels (Lengagne et al., 1999c). Thus, the height of the song, that is the position of the head, has a significant impact on whether it is heard by the partner. It is easy to understand why King penguins always sing with the head up. This body position not only aids acoustic transmission but also increases the visibility of the singer.

A further problem for partner and chick recognition is the high wind conditions of the breeding colonies. Sub-Antarctic King penguins breed on beaches of islands that occur between 45° and 55° south latitudes named the "roaring forties" and the "furious fifties." Here there is little land to impede the quasi-permanent winds as they whip around the globe, and thus wind is another impediment to hearing the sound of a partner or chick. The noise of singing birds in the colony increases when the winds raise, making discerning the signals in songs even more difficult. To compensate, partners sing even louder songs, but this contributes to the greater background noise of the colony. This was shown by calculation of the emergent entropy of the song, which decreases as the winds raise (Fig. 5.21; Lengagne et al., 1999b). The breeding colony, however, is not a "wall of sound," but an irregularly throbbing noise that includes low frequency sounds such as flipper flaps and wind. The noise of the breeding colony can be measured in units of loudness, or decibels (dB). Within a 4-min period, there are over 30 s of relative "silence" in which the sound level is at least 30 dB below its normal level of

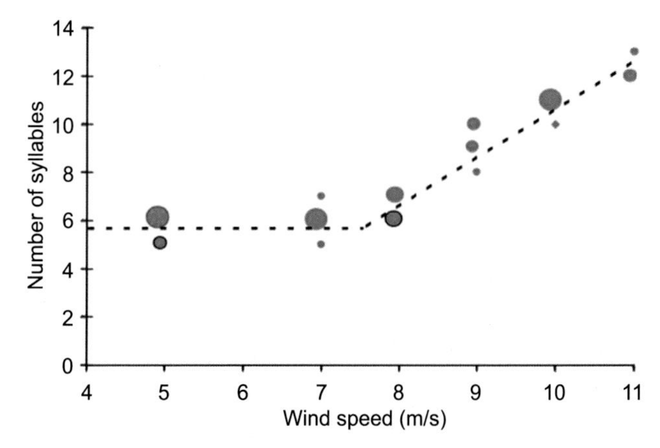


**FIGURE 5.21** For King penguins, enhancement of the number of syllables per song as wind speed increases. In windy conditions, birds maintain the efficiency of their communication by increasing the redundancy of the signal and singing longer. The circle size is proportional to the sample size (n = 3–6), and the *dashed line* represents least-squares linear models for wind speeds 0–7.5 and above 7.5 m/sec. *After Lengagne, T., Aubin, T., Lauga, J., and Jouventin, P., 1999b. How do King penguins (*Aptenodytes patagonicus*) apply the mathematical theory of information to communicate in windy conditions? Proc. R. Soc. B Biol. Sci. 266, 1623–1628.*

about 75 dB (Aubin and Jouventin, 1998). Thus, redundant syllables and repeated songs, as we saw above, are particularly effective at getting the signal of the song "through" to a partner. The birds increase both the number of songs and the number of individual syllables in their songs with increasing wind. Playback experiments showed that chicks recognize their parents in the same way that parents recognize each other (Jouventin et al., 1999a). Here too, increasing the number of songs and syllables (i.e., increased redundancy) makes it more likely that that the song will be active during moments of relative lull in the noise of the colony, again increasing the chance that the song will be heard.

The songs of King penguins have some special features that increase the likelihood of correct individual recognition. The sheer number of potential individuals to screen out songs makes finding the right partner seem an impossible task. For this, the King penguin recognizes a dominant facet of the song, frequency modulation. Unlike other penguin species that have been tested, AM can be removed from King penguin songs and played back to the partner with no loss of recognition. But if modulation of the frequency of the song is altered, recognition is stopped. Inspection of the song shows that each syllable has a rising and then more gradually falling frequency of sound. The rising frequency of sound in the song is different for different individuals, and this is used in individual recognition. In fact, the inflection of the first half of the syllable is all that is required, according to playback experiments, for the song to be recognized. The absolute frequency of the song is not so important, only its changes in frequency. The low frequencies in the song were also sufficient to produce recognition, and shifts in frequency of the song of up to 75 Hz did not alter the ability of partners to recognize each other.

King penguins increased individual recognition of the song with the use of a new characteristic, the difference between their "two voices." In humans, the origin of the voice is single, but it is double in birds. Birdsong is produced not by vocal cords, but in the syrinx, at the base of the bird's trachea. The syrinx has tympanic membranes that produce sound, and some occur at the opening of the two forking bronchia that descend into the lungs. From these paired structures, some birds, including Emperor and King penguins, produce lateralized songs or songs with "two voices" (Fig. 5.22; Aubin et al., 2000). This feature is well known to anatomists and ornithologists, but a biological function for the two voices was unknown until our discovery of its use in individual recognition in large penguins (Aubin et al., 2000). The double song is produced as a two-voiced signal, but each voice has a particular frequency. The difference in frequency between the two voices of the song varies from individual to individual.

Our playback experiments showed that both species of the nonnesting penguins use the two-voice system in individual recognition, but no similar use was found in nesting penguins. Computer software was used to remove the difference in frequency of the two voices, and when these modified songs were played back, recognition failed. The frequency level of the suppressed "voice," higher or lower, had no influence on the lack of recognition. The interaction of the two voices creates a "beat" in the song that produces information about identity. The beat is consistent within individuals but differs among them. The two-voice system plus the use of frequency modulation greatly increases the number of individuals that can be discriminated. In the King penguin, the two voices were fairly close together in frequency, differing on average by only about 25 Hz (Robisson, 1992). This second system of individual recognition is not so efficient by itself, when compared to frequency modulation, but it considerably increases the efficiency of the first sophisticated system, preventing confusion among thousands of shifting individuals.

When chicks are old enough, both parents forage at sea and bring fish and invertebrate foods back to the feed chicks. During this period, the digestive fluids of the parents change so that more of the food remains undigested for the chick (Gauthier-Clerc et al., 2000). Chicks congregate in crèches within the breeding colony. When a parent returns, the chick must recognize its parent so that it can approach and beg for food, or the transfer of resources will not be successful. Playbacks of parental songs to chicks revealed that even though AM and the fundamental frequency of parental songs were not used in individual recognition, they still helped the chicks localize the songs. That is, they helped the chick to discern the direction in which the parent stood (Aubin and Jouventin, 2002b).

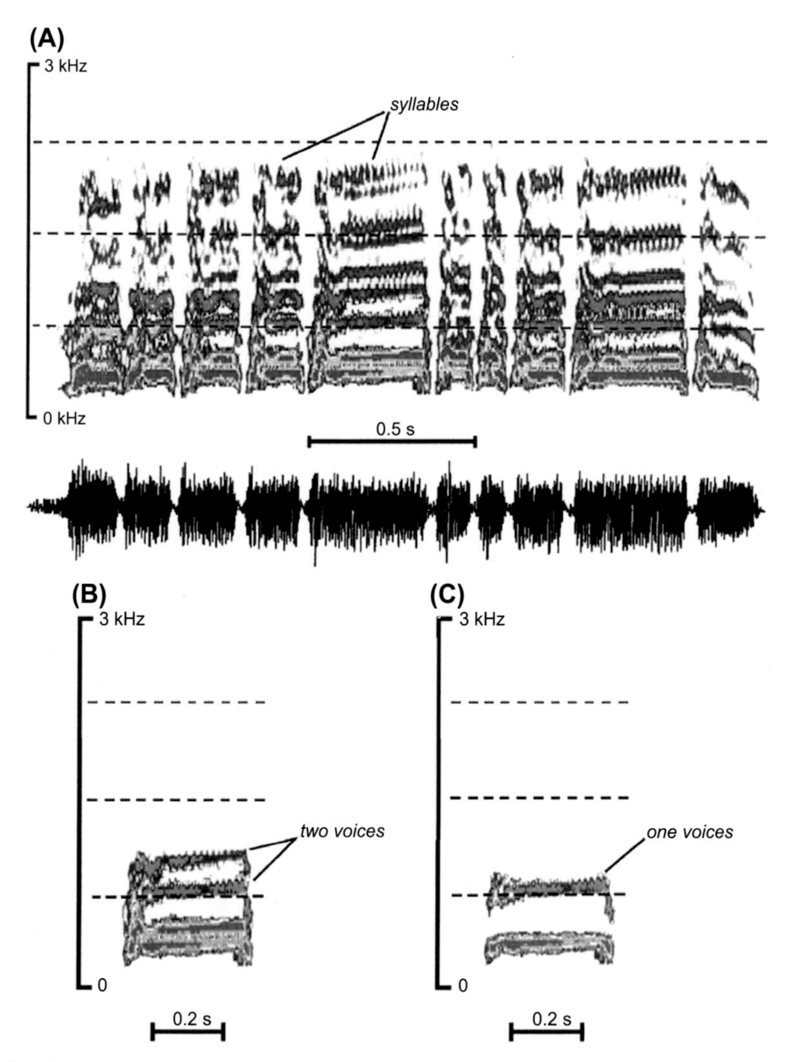


**FIGURE 5.22** Sound spectrograms of Emperor penguin songs. (A) A display song used in a two-voices playback (control) with oscillogram below, (B) One syllable of the low-pass altered experimental song with only the fundamentals and first harmonics retained, and (C) one syllable of the low-pass altered experimental signal with one voice removed. *From Aubin, T., Jouventin, P., Hildebrand, C., 2000. Penguins use the two-voice system to recognize each other. Proc. R. Soc. B Biol. Sci. 267, 1081–1087.*

The loudness of a King penguin breeding colony, measured at its edge, is 74.1 ± 3.4 dB (Aubin and Jouventin, 1998). Parental songs are given at 95.4 ± 0.4 dB, but the sound degrades rapidly over distance. Naturally, intervening penguin bodies attenuate the song, and at 14 m from a singing parent there are usually 15 adults between the parent and its chick. During moments of relative silence, the sound pressure level of the colony drops to below 44.1 dB. King penguin chicks have the remarkable ability to respond to the parental song, and thus to hear it, when the sound pressure level of the played back song is 6 dB below the background noise and given about 7 m away from the chick (Fig. 5.23). This is an even more extreme "cocktail party effect" that we described for Adélie penguins (Fig. 5.11). Six dB is about the loss of sound pressure that a song would suffer at a distance of about 7 m between parent and chick, if there were no intervening adult penguin bodies.

Emperor penguins breed on sea ice in the dead of the Antarctic winter, and commence courtship when temperatures can reach −40°C. They also have repeated short syllables of the song, though the length of the syllables may vary considerably between individuals. Each individual is characterized by the same number and length of syllables in a repeatable order. Once again, the song is consistent within individuals and varies greatly among individuals (Fig. 4.19; Table 4.4). Due to the consistent variation in each subsequent syllable length, the song looks a lot like a bar code on a sonogram, and it operates much in the same way in terms of specifying identity. Using computer sound equipment and software, the bar code songs of an adult Emperor penguin can be changed and then played back to the partner

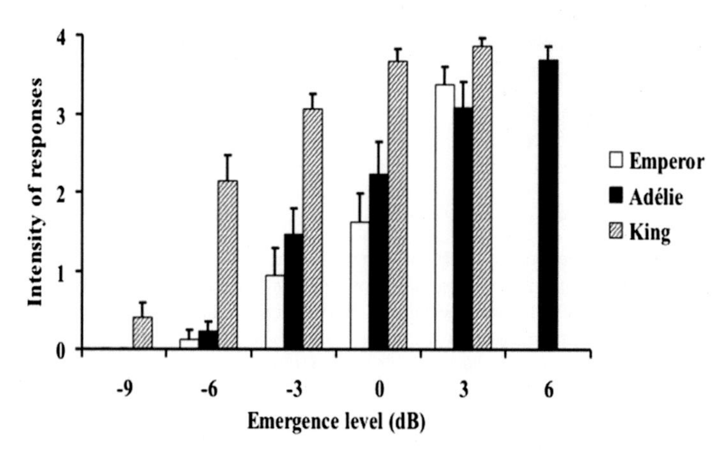


**FIGURE 5.23** Test of detection of the parental song from the colony noise by chicks of three penguin species (Emperor, Adélie, King penguins) in a jamming situation. The parental display song is mixed with five extraneous display songs to yield different emergence levels of the background noise. Nonnesting penguins are better at the cocktail party effect than the nesting ones (here Adélie penguins), and King penguins are the best performing of all penguins at recognizing a song among others that are louder. *From Aubin, T., Jouventin, P., 1998. Cocktail-party effect in King penguin colonies. Proc. R. Soc. B Biol. Sci. 265, 1665–1673.*

or chick to see whether they exhibit recognition via the presence or absence of a behavioral response. AM provides part of the song recognition, but it is less important than in the nesting penguin species. With shifts up or down in song frequency of up to 75 Hz, Emperor penguin chicks could still identify the song of their parents (Aubin and Jouventin, 2002a). This is greater than the shifts that still allowed recognition in the nesting penguins of only 25 Hz. As in the Macaroni penguin, the high or low frequency ranges could be masked, and chicks still responded to playbacks of parental songs. Unlike previous species, however, and probably because more information is needed to insure individual recognition, a single syllable was not sufficient to produce recognition (Table 5.2). Rather, at least three syllables were needed for individual recognition. There is a strong sexual dimorphism in Emperor penguin songs (Fig. 4.19), with the mean number of syllables/song about double for females (the syllable count is about 10 syllables/song for males and 20 syllables/song for females).

The accuracy of individual recognition in Emperor penguins is greatly increased by the use of "two voices." The difference in frequency between the two voices of the song varies from individual to individual (an average difference of about 60–96 Hz for Emperor penguins; Robisson, 1992). The range of difference between the two voices was much greater than in the King penguin, at about 25 Hz, increasing the range and so the accuracy of the vocal signature of this species. Removal of the difference between the two voices in Emperor penguins caused individual recognition to fail, a mirror result to those for the King penguin. And once again, the addition of the two-voice system multiplied the information content of the Emperor penguin songs. Of course, Emperor penguins also show the "cocktail party effect" as described previously for Adélie and King penguins (Fig. 5.23). Though not as extreme as in the King penguin, the perception of an individual voice at the same noise level as the background noise of the colony is an important aid to perception in a noisy environment.

## THE ECOLOGY OF VOCAL SIGNATURES

Summarizing the ecological context of the problem of individual recognition in penguins, the vocal signature of any of the species is of special importance because it is required for successful breeding. Together with the location of a partner or chick, the vocal signature provides the key to their identity. The different species of penguins, however, differ greatly in both the way that information about identity is provided by the voice and the precision required for vocal recognition. The simplest situation, which is the one with the least chance of confusion over the identity of a partner, parent, or chick, is that of Little penguins or banded penguins that use burrows. Next are sub-Antarctic Gentoo penguins, nesting in open areas on bluffs or plateaus above beaches; they usually have a relatively open and sparse colony. Since they have a nest, a parent has a rendezvous location at the nest that should give a good prediction of partner identity. In addition, since there are relatively few other penguins about, confusion is low and a complex vocal recognition would not provide an advantage over a simple one. In fact, the theoretical information content of these songs is relatively low. Nesting on rocky shores of the Antarctic, however, Adélie penguins have much denser colonies, but they still have a nest location that provides a good key to recognition and is on fairly flat ground. They have to breed quickly during the short summer in this extreme environment, and the displays are consequently numerous to prevent confusion. Accordingly, they have a repeated song that facilitates more precise recognition than sub-Antarctic Gentoo penguins, but the information content of the song is still relatively limited compared with nonnesting penguins.

Rockhopper and Macaroni penguins are also nesting species, but the nests are on rocky slopes and cliff faces that are difficult terrain. The density of their colonies and difficulty of the physical terrain add to their opportunities for confusion (Fig. 5.24). And for these two closely related species, information content of the songs is much higher than for sub-Antarctic Gentoo and even Adélie penguins, though Rockhopper penguins may be able to recognize slightly more of the song than Macaroni penguins (according to playback experiments, see above). The breeding colonies of Rockhopper penguins are not as dense as those of Macaroni penguins, and their song is not as informative (Searby and Jouventin, 2005). Due to the difficulty of finding their partner or chick in a jumbled and confusing colony, crested penguins have a more complex coding system than penguins that nest in burrows or on flat

**FIGURE 5.24** A colony of many thousands of breeding Macaroni penguins in the Kerguelen archipelago, showing dense packing of nests in rocky upland habitat where penguins returning from the sea have to find a partner or chick.

areas at lower densities. So the crested penguins have more sophisticated songs than Little and *Pygoscelis* penguins but not as sophisticated as the nonnesting and mobile large penguins.

Finally, we can see that the greatest sources of confusion occur in the noisy and very crowded King and Emperor penguin colonies. Due to frequency modulation and aural bar codes, respectively, and the recognition of the two-voice system of birds, these two species have the greatest information content of their songs. These two species are in the same genus and thus relatively closely related to each other. They also both lay a single egg and care for the egg and young chick on their feet. So they are unique among birds in being able to move with their offspring while breeding at the egg and young chick stages. The potential for confusion seems even higher in the Emperor penguin than one might expect from colony density alone, because blizzards cause extremely high density (up to 10 birds/m$^2$, according to Prévost, 1961) and movements both within the colony, and also of the colony itself. Indeed, King and Emperor penguins have songs that contain an astounding amount of information. They have two characteristics of vocal signature that seem different, frequency modulation and the bar code. But they are actually modifications of the same basic type of song.

The King penguin produces a song that raises and lowers in frequency over time (Fig 5.16B), in which only the initial changes in frequency are needed for correct recognition. As the frequency changes, amplitude also changes over time. This produces a melody of raising and lowering sound, but it is the initial frequency change that is most important to individual recognition. The Emperor penguin actually has the same system of song but with the sound strongly hatched (Fig. 5.16A). To hatch the song, the low amplitude sounds are dropped to virtual silence. This produces bars of sound as time goes by on a sonogram, with relative silent spaces in between. The more hatched song of the Emperor penguin compared to the King penguin is thus a new evolutionary novelty that results in the apparent "bar code" on the sonogram. The song represented is one of quickly alternating sound and silence, and thus an extreme AM over time. The hatched code of the Emperor penguin is more accurate due to the timing of sound in the bar code. It is also easier to locate because a listener's ears hear the sound differently and thus can turn to the direction of the sound by equalizing the amplitude of sound to both ears, and thus estimate the direction from which it comes. In the Antarctic cold, it is especially important for parents to find each other and especially find their chick quickly in a dense and moving crowd, to provide warmth.

The bar code is an extraordinarily efficient vocal signature, and hearing it is necessary for individual recognition when trying to differentiate among several thousand individuals. The confusion in a colony of Emperor penguins is exceptionally high because they are continuously moving. To identify the sophisticated bar code, an Emperor penguin needs to hear more of the song for individual recognition than a King penguin does. This helps to explain why the King penguin can recognize a song when hearing only one syllable (actually, even less at 200 ms), and the Emperor penguin cannot (Table 5.2). The basic structure of the calls of these two species, however, is very similar.

In addition, both species use the difference between the two voices to increase the information content of the vocal signature, producing more unique songs. The presence of two voices does not improve the accuracy of individual recognition as much as either frequency modulation of the King penguin or the bar code of the Emperor penguin. But, even if the two-voice system is less efficient than the impressive bar code, using the difference between the two voices considerably increases the number of recognizable calls by a multiplicative process. If 100 songs could be recognized from changes in frequency modulation, for example, use of the two-voice system for only another four-song possibilities would raise the total to 400 recognizable songs. The bar code of the Emperor penguin (or the frequency modulation of the King penguin) may have evolved separately from the two-voice system, but they are in fact linked. These two-coding systems may be different in origin and efficiency, but they are complementary because they constitute a sophisticated twin acoustic code. The weak code of the two voices increases the accuracy of the more efficient bar code. The Emperor penguin has a broader range of mean difference between the two voices (60.4 Hz for males, 95.9 Hz for females), than the King penguin (23.6 Hz for males, 27.1 Hz for females). Once again, the greater range for the Emperor penguin should produce greater accuracy, and probably results in more individually recognizable songs.

The evolution of songs of penguins has predictably followed the need for increased information in the different species faced with increasing difficulties in differentiating their partners, parents, and offspring from others. Song complexity has coevolved with the complexity of breeding biology and environments of the species. Differences in breeding biology and song complexity can be most easily seen in the differences between the nesting and nonnesting species, with colony density much higher for the nonnesting species. The penguin species occur in the southern hemisphere, where land, especially predator-free land to breed on, is scarce for most seabirds. Perhaps the most remarkable breeding feat of any bird occurs in the Emperor penguin, breeding on sea ice in the dead of the Antarctic winter. But it is clear that this could not occur only by metabolic adaptations and without an advanced and complex communication system for individual recognition.

To conclude, penguins have remarkable acoustical gifts. All six species of penguins that were studied experimentally can identify their partner with only a part of the song. In the first experiments with the King penguin, even the first half of only one syllable of the song was enough to elicit individual recognition (Derenne et al., 1979). Further experiments (Jouventin et al., 1999a; Lengagne et al., 2000) showed that only a small part of the song is was necessary to distinguish it among the songs of several hundred penguins. In fact, 200 ms were enough for recognition in the King penguins, although the whole song lasted between 3 and 6 s. This reflected a surprising capacity to obtain vocal individual recognition with only a small amount of information. Recognition by song is adaptive, as shown by the high redundancy in repeated songs in all of the species of penguins. This is particularly true for the King penguin, where recognition of the vocal signature of the partner, chick, or parent occurs in an extremely noisy background aural environment of the colony that produces a high and permanent level of jamming (about 74 dB for a King penguin colony, about the same level as traffic on the beltline around Paris).

The display song is transmitted in a context involving the background noise generated by the colony plus the screening effect of the birds' bodies, both reducing the signal-to-noise ratio. In addition, the signal is masked by background noise that has similar spectral and temporal characteristics to an individual's song (i.e., similar songs are being given by hundreds of neighboring penguins). To estimate the minimal discrimination threshold of the display song in a jamming situation, a series of mixed signals was broadcast to penguin chicks. The parental song was combined with five extraneous adult songs, each with different emergence levels, and the tested signal was increased in energy ratio compared to the extraneous noise. The superimposition tested a mixed signal with a total lack of silences, and with numerous overlapping frequencies. This test of jamming mimiced a situation frequently observed in a penguin colony. Chicks of three species were tested with signals of different emergence levels (i.e., the difference between the energy level of the tested song and that of the five extraneous songs) at a distance of 7 m. The emergence level was defined as $E = 20 \; log \; Ap/Ae$, where E represents the emergence level in dB of the parental song of the chick tested, *Ap* is the absolute amplitude of the parental song, and *Ae* is the absolute amplitude of the mixed extraneous songs.

In a comparison of several species of both nesting and nonnesting penguins, our experiments indicated that the chick could detect its parental song in an extreme jamming situation. With King penguin chicks, detection is possible even when the parental song intensity is well below (−6 dB) that of the noise of simultaneous songs produced by other adults (Aubin and Jouventin, 1998). At the same distance, Adélie and Emperor penguin chicks were not as good at detecting songs as the King penguin chick (Fig. 5.23). Nevertheless, Adélie and Emperor penguin chicks have an excellent ability to recognize the parental song even embedded in the noise of the colony (at a difference of 0 dB of emergence, Jouventin and Aubin, 2001). This capacity to perceive and extract the information from an ambient noise with similar acoustic properties to that of the signal, is termed the "cocktail party effect" or "intelligent hearing" in human speech intelligibility tests. Thus, the chick's ability to find its parents is enhanced. The process of perception is much better in these penguins than in humans, the only other species previously known to have a cocktail party effect.

It is obviously linked to an acoustic coding system that is adapted to the constraints of colonial life and nesting ecology in the different species.

According to the theory, to extract a signal from background songs, animals analyse either frequency bands or frequency/time modulations of the partner's song. The first coding–decoding system, used by nesting penguins, is easy to produce but costly in terms of analysis time. The second one, used by nonnesting penguins, is a vocal signature, which is quickly analysed but costly to produce. This acoustic signal is particularly efficient for rapid location of a partner that is in a noisy crowd and on the move. Briefly, the frequency analysis is enough to solve the relatively easy problem of individual recognition in nesting birds, while the complex temporal analysis of amplitude of the two nonnesting penguins is an adaptation to extreme acoustic and breeding conditions (Figs. 5.25 and 5.26).

## HOW TO MEASURE INFORMATION CONTENT

Two aspects of penguin songs might produce information useful to individual recognition. One is the rhythm of the song, measured over time and termed the

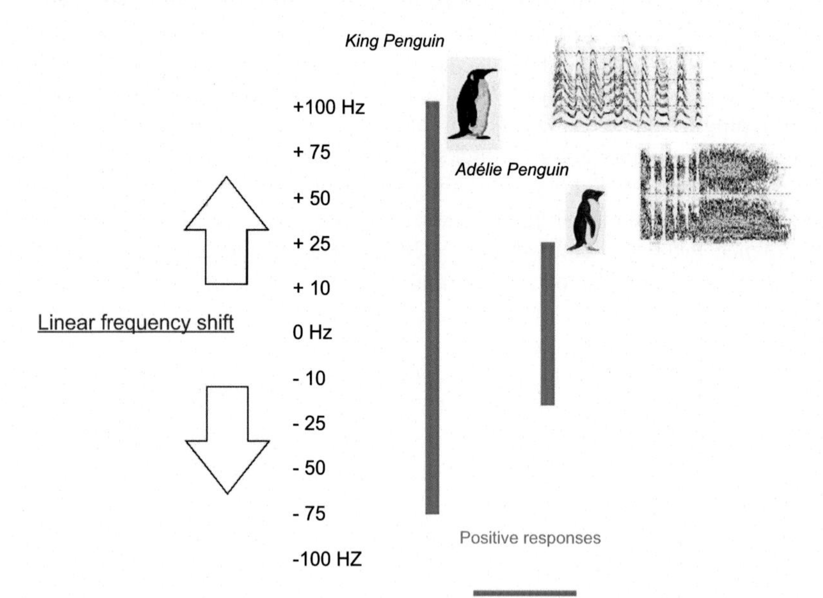


**FIGURE 5.25** According to our experiments with modified signals, King penguins (similar to Emperor and Macaroni penguins) tolerate more change in frequency than do Adélie or Gentoo penguins. The latter species' vocal signatures are simple because they code identity only by the analysis of the spectral profile and the precise frequency values of the harmonics (i.e., mainly a timbre analysis). *From Aubin, T., Jouventin, P., 2002a. How to vocally identify kin in a crowd: the penguin model. Adv. Study Behav. 31 (31), 243–277.*

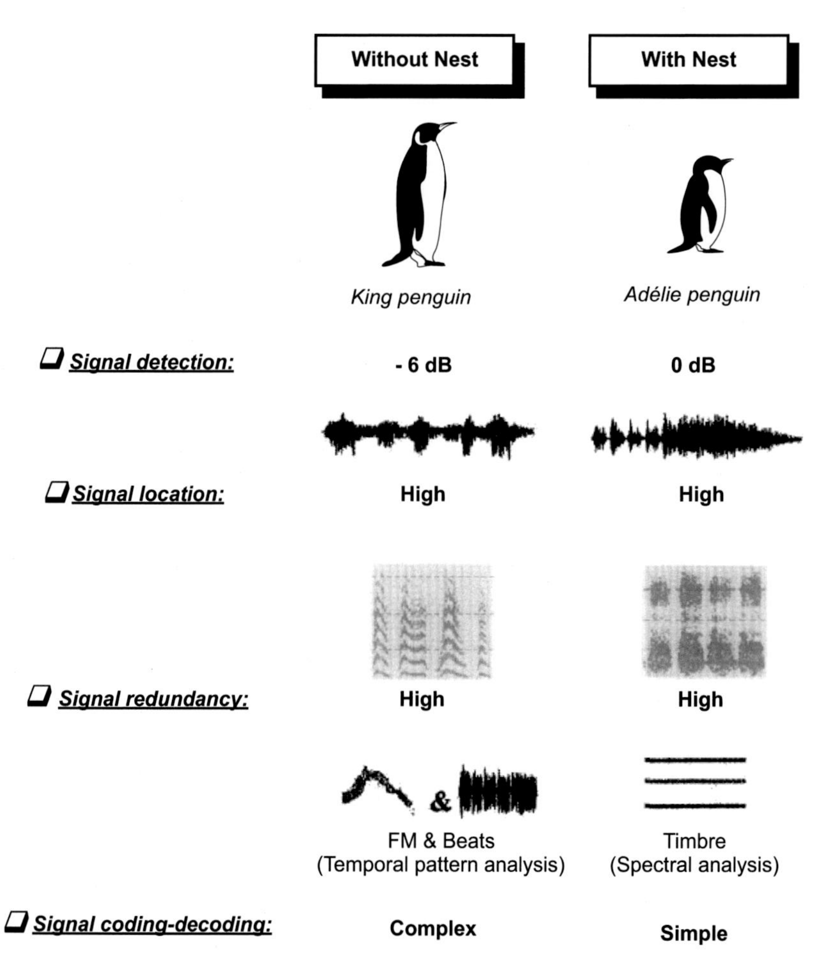


**FIGURE 5.26** Diagram comparing song characteristic of nesting and nonnesting penguins. Signal recognition with background noise (>70 dB) in two categories of penguins: a complex code in species without nest, like the King penguin, and a simple code in all seabirds with a nest, like the Adélie penguin. *From Aubin, T., Jouventin, P., 2002a. How to vocally identify kin in a crowd: the penguin model. Adv. Study Behav. 31 (31), 243–277.*

"time domain" or "temporal domain" of the song. Songs are made up of syllables that last fractions of a second, with silent fractions between the syllables. The alternation of syllable/silence, that is the amplitude modulation (AM), gives the sense of changes in the amplitude or intensity of the sound of the song. Most penguin songs have AM and it forms part of the individual recognition system of an individual's song; for example, in the Emperor penguin (Jouventin et al., 1979; Jouventin, 1982). But the frequency of the sound within a syllable may also change over time, as it would for a raising voice with a constant amplitude. This frequency modulation (FM) is a recognized main characteristic of the adult song of only one species, the King penguin, and of chick songs in all of the penguin species. The second aspect of penguin songs is their frequency content (as a low or high voice differs in sound frequency), termed the "frequency domain" or "spectral domain" of the song. Naturally, the temporal and frequency domains may interact in additive or even synergistic ways. If a song is to produce an individual signature, then the number of signatures will depend on how the time and spectral domains are used to produce distinctive sounds and how well these sounds are perceived. A more efficient system will allow the discrimination of more individuals.

The information potential for individual coding of songs depends on the songs exhibiting little variation within individuals and much variation among individuals (see Chapter 4; Beecher, 1982, 1989; Searby and Jouventin, 2004). The efficacy of the signature system, in this case the song, is thus enhanced by the stereotypy of individuals, defined as the ratio between intraindividual and interindividual variation. Mathematically, the information content of a song is:

$$H_S = log_2\sqrt{\frac{\sigma_A^2 + \sigma_W^2}{\sigma_W^2}}$$

where $\sigma_A^2$ is the among-individual variance and $\sigma_A^2$ is the within-individual variance.

In statistics, the analysis of within versus among group variation is called analysis of variance (ANOVA), and the significance of the among group variance (in our case, the variation among individuals in song measures) is given by an $F$ statistic. $H_S$ is related to $F$, but it does not change with changes in sample size (as $F$ does). Thus:

$$H_S = log_2\sqrt{\frac{\sigma_A^2 + \sigma_W^2}{\sigma_W^2}} = log_2\sqrt{\frac{F + n + 1}{n}}$$

Here $n$ is the number of songs sampled from each individual. So, $H_S$ can be estimated from an ANOVA, where several individuals contribute $n$ songs each to the comparison. As the value of $H_S$ goes up, the greater potential there is for encoding individual identity. An estimate of the number of potential signatures is $2^{H_S}$.

It would be easy if we could calculate the information content of songs in this way. But there are two problems. First, songs can be simple or complex, and the more complex songs have additional elements for analyses compared to the simple ones. For example, some species have an extremely "hatched" song (crested and Emperor penguins, explained in detail above) that makes the song easier to locate. Second, we need to know what parts of the song the penguins are listening too, and this differs among species. For example, all the penguin species have two voices (i.e., two voice boxes that produce slightly different song frequencies). But only the two large *Aptenodytes* penguins use the difference between the two voices for individual recognition in their vocal signatures.

The formula, then, only gives the potential information content of the song, it does not specify which aspects of the song are sensed or which are relatively most important for individual recognition. As for any signal, songs have a potential for carrying information and a realized or actual effect on a listener. In the case of penguins, partner and chick identification is so important that we might expect that all aspects of a song are used for identification. The problem with this simplistic viewpoint is that the environmental contexts in which songs are given are very different for different species. Thus, individual recognition is more difficult in some species than others. For example, the location of a nest provides a strong clue as to the partner or chick in the nest. On the other hand, intervening penguin bodies can "screen" songs so that they do not travel as far. We have discussed both of these topics above, and others that influence the efficacy of songs for individual recognition. But they serve to demonstrate that variations in environmental context may tell us a lot about the complexity and characteristics of the songs. That is, the context should provide clues to where more or less acute discrimination should occur.

What is needed in all cases is a vocal signature, so that individuals can be recognized. The more unique elements of the song that are recognized, the easier it will be to tell several individuals apart. Yet, some penguin species nest in relatively benign social and ecological environments where only a few individuals have to be discriminated from the partner and chick, while other species face a noisy environment where scores or even hundreds of individuals are nearby and have to be discriminated. So we might break the song down into some basic elements (Searby and Jouventin, 2004). The first is the way that the amplitude, or loudness of the song, changes over time, from loud to silent. The second is the way that the frequency distribution of the song changes over time, from no change to highly modulated. And the third is the way that amplitude and frequency interact, an important aspect of any song. Changes over time in amplitude and frequency map these three variables into two vectors. These vectors are simplifications of curves on a graph of amplitude over time (the output of an oscilloscope) and frequency over time (the output of a sonogram).

Now we have the elements needed to estimate the stereotypy, and thus information content, of a penguin song. The distance between two vectors is $d_{A_{ij}}$, where

$$d_{A_{ij}} = \frac{1}{N_{min}} \sum_{t=1}^{N_{min}} (A_j(t) - A_i(t))^2$$

and $N_{min}$ is the shortest length of the two vectors. When comparing a collection of many song recordings, the sum of the distance indices between all pairs of observations (*dist*) is

$$dist = \sum_{i,j} d_{A_{ij}}$$

So it is possible to estimate an $F$ statistic for the stereotypy of an individual penguin's songs by comparing the distances among all pairs of recorded songs ($dist_T$) and to the distances between songs for each individual ($dist_W$). For $g$ individuals and $n$ songs from each individual, this would be

$$F_{est} = \frac{n-1}{g-1}\left(\frac{dist_T - g \cdot dist_W}{dist_W}\right)$$

And the information content of the song would therefore be

$$H_S \approx H_{est} = log_2\sqrt{\frac{F_{est} + n - 1}{n}}$$

We used this procedure to provide a preliminary estimation of the information encoded in the songs of several penguin species. The estimates were then used to test our expectation that the information content of the songs should increase with the difficulty of individual recognition.

Among different penguin species, the problem of individual recognition might be considered a problem of confusion. Suppose a penguin needs to identify its mate. Then it needs to be able to distinguish the partner from all the other nearby breeding penguins. In a large densely packed colony this will be more difficult, or more confusing than in a smaller colony in which nests are widely spaced. An important question is how often penguins make mistakes, or become "confused" about who their partner or chick is. Under different conditions, the opportunity for confusion and mistakes in individual recognition change. To simplify, we assumed that confusion or recognition depends on the distance at which a penguin can hear another bird, and how many individuals are nearby within this distance, perhaps blocking the sound by intervening. In other words, the information accuracy of a song will depend on the density of breeders in the colony, and the distance that the sound of the song carries, to prevent confusion (Fig. 5.27). Then we can estimate the probability of confusion from the density of birds and the area in which the song can be heard (Fig. 5.28 top).

For a given spot in the breeding colony, the chance of a penguin being confused is one minus the chance that it can recognize its partner (recognition is a lack of confusion). The basic principle is that the chance of confusion and correct identification are directly related to each other:

$$P(\text{confusion}) = 1 - P(\text{recognition})$$

**FIGURE 5.27** An Adélie penguin is confused by the playback of its chick song emitted by the researcher in the background. The parent in the foreground comes singing, not to the loudspeaker 3 m right from the researcher, but to the previous nest where two chicks are waiting (and they used as a meeting site).

In turn, the chance of correct recognition depends on being close enough to the partner or chick so that they can hear the singer. If we model an area in which a song can be heard as a circle, then:

$$P(\text{confusion}) = 1 - \left(\frac{1}{\text{Density} \times \text{Area}}\right)$$

or, using the radius of a circle about the singer as the distance that the song can carry (Fig. 5.28 top):

$$P(\text{confusion}) = 1 - \left(\frac{1}{\text{Density} \times \pi \times \text{Distance}^2}\right)$$

Finally, we can predict that as the information capacity of the songs should depend on the probability of confusion; and that the more that confusion is possible, the more information the song should carry, so we can predict:

$$H_S \propto \text{Density} \times \text{Distance}^2$$

If a parent is trying to find their chick in a crèche, the problem of bringing the individuals together with the voice alone becomes more difficult because crèches are generally even more spread out than either nests of the previous breeding location (for the two large *Aptenodytes* species without a nest) (Fig. 5.28 bottom).

Our goal was to make an initial preliminary comparison the information content of the songs of different penguin species with the amount of confusion in their environments. So we used the simpler situation of partner recognition

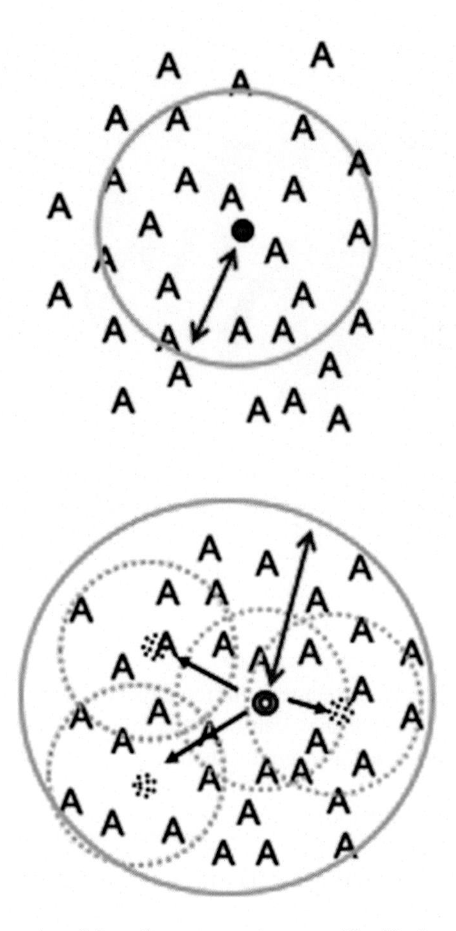

**FIGURE 5.28** In the noise of the colony, a penguin cannot identify the song of its mate more than about 10 m far. In this radius, it has to discriminate between a dozen other parents on their nest. Below, it is more difficult for a parent to find its chick among several in a crèche, unless the nest is used as a meeting point.

(Fig. 5.28 top). With estimates of colony density and the distance over which songs can be heard, it is possible to estimate the probability of confusion and compare it to the information capacity. Information context of the songs of the different penguin species appeared to increase exponentially with the chance of confusion in the penguin colonies (Fig. 5.29). In fact, we see a fairly tight fit between the information capacity of the song of the penguin species and the likelihood of confusion, given the breeding colony where the species reproduces and our very rough assumptions and estimates. The analysis suggests that the information content of the song is sufficient to produce recognition in

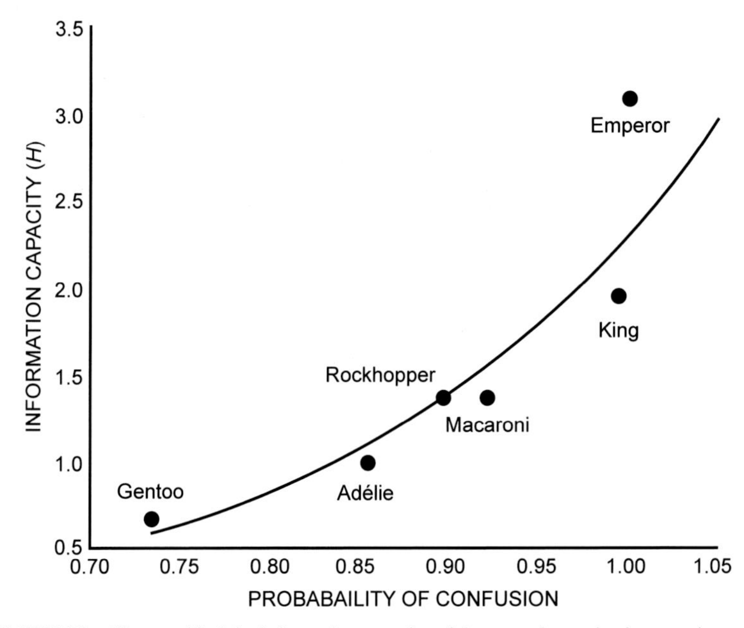


**FIGURE 5.29** We quantified the information capacity of the mutual song in six penguin species. As the estimated chance of confusion increases in the penguin colonies for the species shown, the information content of the species' song increases, confirming our prediction. The trend-line is an exponential curve fit ($y = 0.014e^{5.12x}$; $R^2 = 0.926$, $P = .001$). *From Searby and Jouventin, unpublished.*

environments with increasing sources of confusion. Since the sources of confusion are unique to each species, the information content of their songs should be unique as well. This suffices for a preliminary model of the efficacy of songs of penguin species for locating the partner of a breeding bird given their environmental constraints.

With some modifications, this model might be extended to finding the chick in a crèche (Fig. 5.28 bottom). But for the latter task, some walking around and singing may be necessary for the song to reach the right chick, especially if it has strayed from the nest site, as most chicks do. A study comparing Macaroni and King penguins showed that in the former species, with a nest, chicks were found quickly by parents (within half a minute, on average; and after giving 2–3 songs; Dobson and Jouventin, 2003a). The nonnesting King penguins not only took longer to find their chicks (about 3 min; and after giving 4–5 songs), but some parents had to search through the colony for their chicks, taking about three times as long and giving twice as many songs. Finding the chick is at least six times faster in a species using the nest location as a meeting point compared to a nonnesting penguin (Fig. 5.30). So, it is impressive that a King penguin can find its chick in such a large crèche as several hundred chicks.

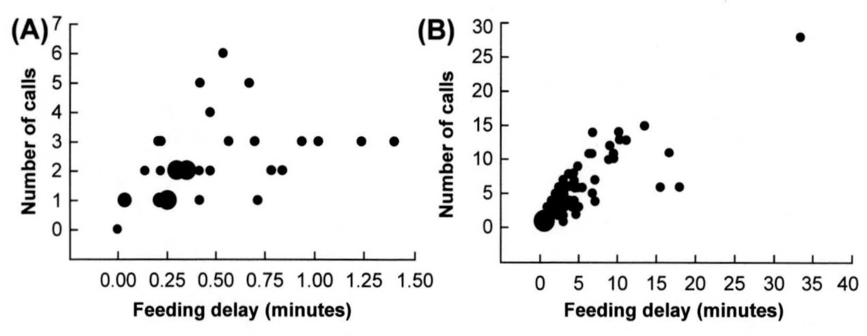


**FIGURE 5.30** The association of feeding delay (in minutes), the time between the first song given by a searching adult and the first feed, and the number of songs given during this period. (A) The nesting Macaroni Penguin ($r = 0.42$, $n = 32$, $P < .01$). (B) The nonnesting King Penguin ($r = 0.84$, $n = 80$, $P < .001$; without the outlier, $r = 0.74$, $n = 79$, $P < .001$). *Small dots* indicate one observation, medium dots show two observations, and *large dots* show three observations. Notice the differences in scales along both axes. On average, King Penguins sang more songs and took longer to find their chicks. *From Dobson, F.S., Jouventin, P., 2003a. Use of the nest site as a rendezvous in penguins. Waterbirds 26, 409–415.*

In terms of estimated information content, the Adélie penguin song had about 50% more information than the sub-Antarctic Gentoo penguin (Fig. 5.29). Of course, both these are nesting species, but with higher colony density in the Adélie penguin. Rockhopper and Macaroni penguin songs had more encoded information, at just over twice the information of the Gentoo penguins, and of course the crested penguins live in rock piles and cliff faces, with very dense colonies. The nonnesting King penguin had a song with just over three times the information of a Gentoo penguin song, was we would expect for a species without a nest and living in a very dense colony. Finally, the song of the Emperor penguin had about 50% more information than the King penguin song, as we would expect for a nonnesting penguin with packed breeders during blizzards and exceptionally dense aggregations at other times for a penguin of its large size.

Chapter 6

# The Evolution of Communication

## EVOLUTION OF AGGRESSIVENESS AND SHYNESS

Aggression is easy to identify in penguins. Beak-shaking, beak pecking, and flipper-flapping are common in all of the species of penguins (Table 2.1). These behaviors are not parts of ritualized displays but practical expressions of aggression and often accompanied by an attack posture. In species that show a twisting movement of the head, this seemingly aggressive behavior is part of a ritualized display. Twisting of the head may actually aid keenness of vision by refining a bird's perspective.

A horizontal head-circling motion is both ritualized and common in the two Antarctic species, the distantly related Adélie and Emperor penguins. Ritualization of aggressive postures confers a survival advantage by avoiding combat but at the same time defending the breeding territory or a personal space. Carried to an extreme, ritualized postures may even be adopted in the absence of congeneric individuals, that is, completely out of context. The most complex agonistic behavioral patterns, as well as clear-cut ritualization, appear in only three genera: *Aptenodytes*, *Pygoscelis*, and *Spheniscus*. The Emperor penguin is the only *Aptenodytes* species with ritualized aggression. This species is the least aggressive penguin species and requires a form of passive cooperation to survive winter blizzards that requires huddling with a minimum of personal space. On the other hand, the Adélie penguin is the most aggressive of all penguins, as they nest at high densities and are extremely territorial around the nest site. It is interesting that both conditions can lead to ritualization of agonistic behavior.

Within each of the penguin genera, aggressiveness varies considerably from species to species. For example, in *Aptenodytes*, Emperor penguins are much more pacific than the closely related King penguins. The low level of aggressiveness in the Emperor penguin is a consequence of the tolerance of close neighbors, required during the frequent huddles during blizzard conditions. Social thermoregulation, a means for surviving the Antarctic winter, would be unthinkable in a nesting and territorial species. Even the King

**Why Penguins Communicate. http://dx.doi.org/10.1016/B978-0-12-811178-9.00006-8**

penguin, though without a nest, defends a small personal space or territory for "standing" in the breeding colony, and the size of the territory is defined by the reach of frequent beak strikes at neighbors (Coté, 2000). In another example from the genus *Eudyptes*, the Macaroni penguin is much more aggressive than the Rockhopper penguin. These two species often inhabit separate habitats in the same locality. The smaller Rockhopper penguin nests under rocks, where both the need and opportunity for aggression is more limited. The open-nesting Macaroni penguin, usually on top of rocks in more open aspect, has far more opportunities for aggression near the nest.

The genus *Pygoscelis* provides the most extreme examples of aggression (Jouventin, 1982). Adélie penguins exhibit the highest level of aggression in penguins and they have the most varied repertoire of agonistic behaviors. On the other hand, and in the same genus, Gentoo penguins almost never quarrel seriously. Their shyness may largely explain their fear of man. But when humans approach, colonies differ considerably in their flight distances. On Possession Island in the Crozet Archipelago, Gentoo penguins along Marin Bay cannot be approached closer than 50 m without stimulating flight behaviors. In Marin Bay, there is a good harbor that has long been used by sealers and whaling vessels, so the penguins have a long experience of being approached by this potentially aggressive land predator. On the other side of this small island where there is not a sheltered harbor, however, Gentoo penguins can be approached to within 10 m of their nests. On both sub-Antarctic islands and on islands of the Antarctic Peninsula, visitors are often surprised to see similar differences in shyness. Until recently, sealers often overwintered on these islands, where Gentoo penguins breed in generally accessible colonies. These humans persecuted Gentoo penguins, collecting their eggs for food or egg collections. During sub-Antarctic winters, this species breeds and thus was available on land to be caught and eaten by early sailors. However, these practices ended nearly a century ago. Cultural transmission of fear and human-induced natural selection are two hypotheses that might explain the continued shyness of some Gentoo penguin populations, with the latter seeming more likely. Early sailors and sealers considerably reduced populations of Gentoo penguins, and only individuals who fled from these men were able to survive and reproduce.

Another good example of shyness is provided by the Yellow-eyed penguin, the least colonial of the penguin species. On the New Zealand mainland, Yellow-eyed penguin pairs nest separately and avoid terrestrial predators using specific behavioral techniques. This is unusual in penguins. Jouventin's (1982) observations revealed the absence of gregariousness, distrust of humans, and nesting habitats in the interior of bushes. Yellow-eyed penguins take care to enter the nest unobserved. When disturbed by humans, they flee toward the sea or into bushes, where they sometimes hide silently for an extended period. Nests may be hidden by tree trunks, and noisy duets between partners, and parents and chicks, are absent. Rather, they have short and quiet displays, especially during reliefs of the partner during the incubation and chick brooding periods. The number of Yellow-eyed penguins on the New Zealand mainland has decreased over the years, due to both visiting humans and dogs, and the destruction of their forested habitats as they were turned to farmlands. Where Richdale (1951) studied several dozen couples, only two or three remained 30 years later (Jouventin, 1982). Now, after a further 30 years, they have disappeared. However, the species still inhabits islands that are isolated from the mainland. Yellow-eyed penguins on a small island just off the shore, Campbell Island, are much less afraid of humans than those of the mainland. The distrust of humans may have been learned relatively early, since Maoris established themselves in New Zealand in AD 1200–1300. Before then, there may have been no terrestrial predators of Yellow-eyed penguins. Were it not for small populations like the one on Campbell Island, this species would likely be extinct.

## THE FUNCTION OF HEAD ORNAMENTS

Many of the visual characteristics that allow differentiation of the different penguin species occur in the head region (Wilson, 1907). Two explanations have been advanced to account for this phenomenon. According to Murphy (1936), the head is easy to recognize at sea, thus providing an accurate and quick way that the species can tell each other apart. Usually the head is visible when birds break through the surface of the water, so head appearance together with the contact song would aid the species in traveling and diving as a group. Subsequently, Roberts (1940) suggested the hypothesis that displays during mate choice and after tend to reveal neck and head ornamentation. He gave several examples: bowing in *Pygoscelis papua*, and mutual display in *Pygoscelis adeliae*, *Pygoscelis antarctica*, and *Eudyptes chrysocome*. An agonistic display of *Spheniscus magellanicus* involves twisting the head. Richdale (1951) agreed with Murphy, and Roberts' hypothesis was abandoned. These two explanations are not necessarily contradictory. The head markings might be helpful or useful characteristics both on land and at sea. In other words, distinctive markings or colors of the head could have complementary benefits in different ecological contexts.

The ever-popular tuxedo of penguins has a functional use when they are swimming in the sea. As Roberts (1940) suggested for penguins and Craig (1944) for seabirds in general, the "countercolored" white belly and contrasting dark back may serve as camouflage for catching prey (e.g., squid and fish), as well as in avoiding aquatic predators (the leopard seal, *Hydrurga leptonyx,* for example). Divers and swimming fish seen from above blend with the dark depths and from below cannot be distinguished from the light surface of the sea. Diving seabirds all exhibit countercolored body appearances to some degree. Penguins and Alcids, although belonging to unrelated orders, are often mistaken due to their similarity of form and marine way of life. These two groups offer remarkable evolutionary convergence in body coloration as well as head patterns in some of their species, such as head tufts, the white zone of the neck, sides or top of head, reddish beaks, red, or yellow eyes (Fig. 6.1). In auks as in penguins, there is a limited range of body appearance according to the genus (Fig. 6.2). In the two largest penguin species, coloration of the beaks (i.e., the conspicuous UV beak spot on the lower mandible), of the

**FIGURE 6.1** Ornaments in auks and penguins are similar (black and white, color, crest) in these two unrelated groups. Similarities are due to evolutionary convergence caused by the same way of life at sea. As for any diving bird, the throat is white and the back is black, the ornaments most visible only out of the sea and not developed in immature individuals. *From Jouventin, P., 1982. Visual and vocal signals in penguins, their evolution and adaptive characters. Advances in Ethology* ***24****, 1–149.*

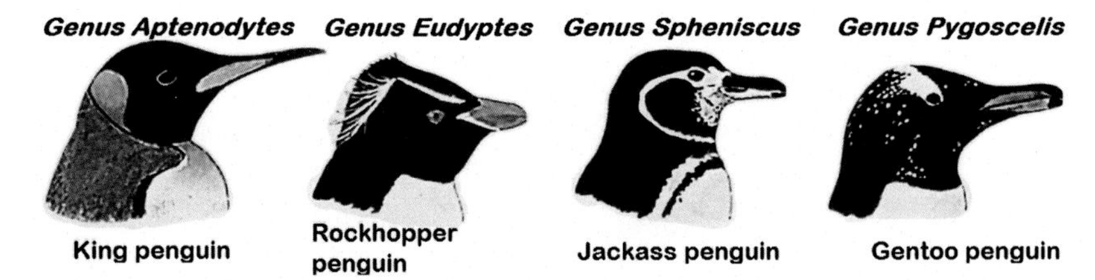


**FIGURE 6.2** Each genus has its pattern of ornaments based on orange feathers and beak in large penguins (*Aptenodytes*) and crested penguins (*Eudyptes*), and black and white head markings in banded penguins (*Spheniscus*) and Gentoo penguins (*Pygoscelis*).

auricular feathers, and of the breast are roughly similar. All crested penguin species have yellow tufts on the head, or "eyebrows." All banded penguins have body patterns of black and white stripes. Although their external appearance may have grouped these penguins together, they are now defined on the basis of DNA sequences (Baker et al., 2006; Thomas et al., 2013). So the similarity likely has been inherited from a common ancestor of the species. But the similar head patterns of penguins and auks are evolutionary convergences, due to the same marine life in which it is primarily the head that is most frequently visible, either in mating displays or at sea.

As in any bird species, colors and ornamental patterns of markings in penguins, as in the convergent Alcids of the Northern Hemisphere, have several biological functions. These external body markings serve to differentiate the species, both on land and at sea. Sexual maturity and body condition are also revealed to potential mating partners during mate choice early in the breeding season. Markings on the head may fulfill specific identification requirements at sea, as Murphy believed, and this use seems highly likely (Fig. 6.3). Other examples of distinctive head patterns can be found in groups of diving birds such as grebes. Such observations do not suffice to refute Roberts' hypothesis, however, because such characteristics may play also a role on land. In addition, ornamentation such as the colored feathers of crested penguins may not be attractive for sea predators because they are less visible underwater. When wet, the colorful yellow crest feathers are plastered against the head and more difficult to see. On the other hand, color patterns do not appear in young individuals and take several years to appear, and this is the case for at least some distinctive markings of nearly all penguin species. In penguins, as in Alcids (Fig. 6.1), immature individuals do not have colored crests; they have dull coloration (*Aptenodytes, Spheniscus, Eudyptes, Megadyptes*) or else their black and white patterning is different from the adult's (*Pygoscelis*). Nevertheless, there is an exception in Little penguins (*Eudyptula*). There are no patterns of colors or marking in the species and immatures resemble adults. The lack of markings for adults may reflect less reliance on vision for mate choice, since they forage during the day and display vocally to attract mates and to defend their burrow territories at night.

**FIGURE 6.3** Top: penguin species in the sea, with the head out of water, showing how the ornaments on the head are visible: (A) King penguin, (B) Emperor penguin, (C) Gentoo penguin, and (D) Jackass penguin. Middle: immature penguins lack ornament, (1) immature Macaroni penguin without crest, (2) adult Macaroni penguins, (3) immature Rockhopper penguin without crest, (4) adult Rockhopper penguin, and (5) Macaroni penguin chick. Bottom: (F) an albino Adelie penguin persecuted by its neighbors. *From Jouventin, P., 1982. Visual and vocal signals in penguins, their evolution and adaptive characters. Advances in Ethology* ***24****, 1–149.*

The head coloration and markings in penguin species that Roberts (1940) did not observe reinforces his hypothesis about their use in mutual display and mate choice. In fact, mating displays in the two large penguins may explain a very curious posture and support a role in mate choice. During the "face to face" display in Emperor and King penguins, partners stare at the head patterns. In these large penguins, the pair seems to keep its eyes closed during this display, but in fact they squint at the colored neck of the mate to judge first if it is the right species, head patterns being different between these closely related species. Second, by inspection they can discern if a potential partner is a mature bird that is able to reproduce. Moreover, displaying pairs may mutually evaluate their body condition because pigments and UV-reflecting microstructures are concentrated in head feathers and beak spots. When approaching an unmated singer, a Jackass penguin circles around it, staring at the head region. When Adélie penguin pairs meet on the nest for the first time, they approach each other from the side, while staring at their partner's neck. The Gentoo penguin has a distinctive spot on the top of its head, and bows especially low, probably to show it. Crested penguins shake their heads, flipping their crests about. Emperor and King penguins progressively inflate and extend their necks upward during the face to face display, obviously to increase the visibility of their auricular (ear) yellow–orange patch of feathers. Adélie penguins show the side of their head, as Penney (1968, p. 114) says in his description of the "oblique stare bow;" other behaviors like the "salute," the "gawky attitude," and the "shoulders-hunched posture" all reveal the cephalic region. In short, most of the postures adopted during pair formation of penguin species could be considered in terms of how various ornaments are presented.

Roberts' (1940) idea that the head markings of penguins are used in mate choice has been tested with experiments over a fairly long period of time. Early in his studies of King penguins and narrated in his chapter that described visual signals, Stonehouse (1960, p. 28) managed a small experiment on this subject by painting the colored feathers of two penguins that were searching for mates and found that the birds could not successfully pair. To enlarge the sample size of the experiment, Jouventin (1982) expanded the treatment to 200 couples of three sub-Antarctic species. While the birds might still have chosen mates via voice recognition (of mates from previous years), modified bachelor King penguins with shaved auricular and breast patches could seldom find mates, and they took longer to do so if they paired at all. He also cut the yellow crest feathers in Macaroni and Rockhopper penguins, and they encountered great difficulties in pairing as well. Sophisticated experiments on mating penguins (summarized in Chapter 3) confirmed these preliminary experiments. The results suggested that an important role is played by colored head feathers. Clearly, the head pattern is essential for pair formation, while later individual recognition of the partner is purely vocal. Observations under natural conditions indirectly support these experiments. For example, albino penguins or those with large unpigmented zones of feathers, particularly in the head region, never pair, and are often pecked by congeners (Fig. 6.3F). Our long-term and numerous experiments with nonnesting King penguins focused specifically on which colored ornaments were recognized and preferred during mate choice, revealing complex patterns of communication by aspects of different ornaments.

Such behaviors suggest that the lack of appropriate visual recognition of the head markings produces reproductive isolation. Hybridizations are very rare in penguins, except in captivity, and occur only within genera. In addition, the head patterns of sympatric species, particularly the three species of *Pygoscelis* (Adélie, Gentoo, and chinstrap penguins), and two species of banded penguins (*S. magellanicus* and *Spheniscus humboldti*) that occur in the same place, are more dissimilar than for the allopatric species (e.g., *S. magellanicus* and *Spheniscus mendiculus*; and King and Emperor penguins). To conclude, head markings and color ornaments in penguins might fulfill several functions. At sea, they aid (along with the "contact call") in keeping monospecific groups together while foraging, and at this time the head is the visible part of the body. On land, head markings and color ornaments reinforce reproductive isolation of the species (along with the "ecstatic display"). But ornaments also give information about sexual maturity and body condition of mating penguins. From a Darwinian viewpoint, penguins that are looking for a mate have to choose not only a bird of their own species and of the other sex but also an experienced and strong partner that will help rear their chick successfully.

## ADAPTATIONS OF VOCAL SIGNALS

Songs help listeners locate a penguin singer, a common attribute of seabirds that need to be recognized by their breeding partner in a crowded colony. In order to project vocal sounds that are easy to locate, many birds and mammals living in a variety of diverse environments produce the same type of vocalization as the hatched song of Emperor penguins. Examples include include Gannets (*Sula bassana*) that nest in dense colonies (Fig. 6.4A–F), White Pélicans (*Pelecanus onocrotalus*; Fig. 6.4G), Demoiselle cranes (*Anthropoides virgo*, Fig. 6.4H), snow petrels (*Pagodroma nivea*, Fig. 6.4I) (Guillotin and Jouventin, 1980), and even male Elephant seals (*Mirounga leonina* Fig. 6.4J–M) during combat, as they establish harems on seashores. Numerous other examples can also be found in amphibians, fish, and insects. What biological characteristics do these species have in common? A chopped signal is ideal for the rapid location of a singer (Marler, 1955, 1957). There are three ways that comparisons of sound between the two ears can be used to identify the location of a sound. These are comparisons of the intensities, time differences, and phase shifts of the sound. For example, temporal resolution of sound by Emperor penguins is at least three times as precise as our own. In the ecstatic display songs of penguins, especially Adélie, Little, Gentoo and

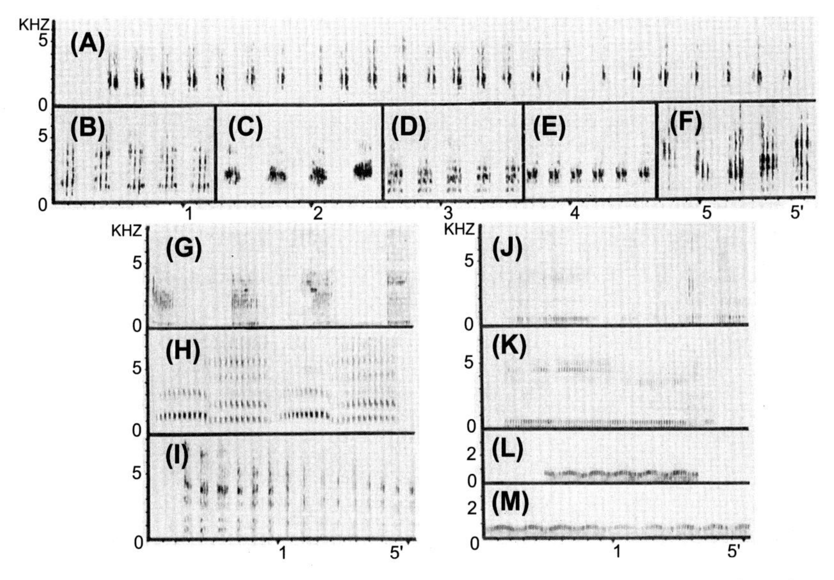


**FIGURE 6.4** Vocal convergence between hatched songs that is easy to locate in various species of birds and mammals. (A) Several songs of the same Gannet (*Sula bassana*) showing signal constancy of vocal signatures. (B–F) Songs of five Gannets showing interindividual variability. (G) Songs of a White pelican (*Pelecanus onocrotalus*). (H) Duet of Demoiselle cranes (*Anthropoïdes virgo*). (I) Songs of a Snow petrel (*Pagodroma nivea*). (J, K) Two songs of an elephant seal (*Mirounga leonina*). (L, M) Two songs of another elephant seal. *From Jouventin, P., 1982. Visual and vocal signals in penguins, their evolution and adaptive characters. Advances in Ethology* **24**, *1–149.*

Emperor penguins, aural comparisons between the two ears are facilitated by repetitions of identical short syllables that have a large frequency band. In Little penguins, interindividual variation is based on differences in the regular rhythm of syllable emission, and the physical form of the signal favors location of the singer at night. The redundancy of individualized songs is particularly useful for facilitating communication in a noisy environment.

The repertoire, that is, the form and variability of each type of penguin song, also has an adaptive aspect. The contact song of each species is stereotyped and easily distinguishable from those of other species, as well from the sounds made by ocean waves. There may be slight or no individual differences in the contact call, depending on the motivational state of the animal. A contact call has merely to indicate the species, and this may explain its stereotypy. Contact calls should also be audible over the greatest possible distance. These two functions, species identification and ability to carry over distance may explain why this monosyllabic song is so brief and simple for all of the penguin species that have been studied. Pitch and amplitude increase at the beginning of the song and decrease at the end, which may be related to an anatomical and physiological constraint. Agonistic songs in penguins are only slightly stereotyped, perhaps because they need to convey no more precise information than the motivational state of the individual. Agonistic songs are mainly for intraspecific interactions, since most penguin species have no real terrestrial predators. In the animal kingdom, the most stereotyped agonistic calls appear in species that are constantly threatened by predators. Such vocalizations can even state the nature of the danger and should then be considered alarm signals.

In penguins, the agonistic song simply aims at attracting the attention of a congeneric adversary (generally an intruder in the territory) and warning it of an imminent attack. It thereby indicates the mood of the singer and is not useful information for a broader audience. Its physical form is generally limited to a distorted abbreviation of the display song. On the other hand, display songs are clearly highly stereotyped, as they convey specific and complex information about identity. Consequently, display songs are long. Inside this rigid context, display songs must also express variation among individuals, an essential element for individual recognition. Since identification among thousands of possibilities demands more information than discriminating among a few species, these songs have acquired highly complex patterns. Moreover, the form of the display song varies over the phylogeny, but must also adapt to physical or biological environmental constraints inherent to each species. As in the case of head patterns, the structure of the songs is more dissimilar in sympatric species; the songs of the three species of *Pygoscelis*, which sometimes live together, are totally different in all respects. The songs of the sub-Antarctic sympatric Macaroni and Rockhopper penguins are not only different in structure, but also in pitch. It should be noted that Northern rockhoppers and Little penguins live at the edge of the sea and have chopped, high-pitched songs that are easily distinguishable from the dull and monotonous sound of waves breaking on the shore. Then again, the nocturnal Little penguin has no attractive color pattern, and its displays are not elaborate, whereas its vocal activity is particularly loud considering its size.

Adélie penguins reproduce during the brief Antarctic summer. The physical form of the male's "ecstatic display" song is perfectly adapted to facilitate location of the singer, being particularly hatched and associated with a head up position and the flapping of flippers. The "ecstatic display" song of the Chinstrap penguins inhabiting the milder Antarctic Peninsula is common to both sexes and is closer to the mutual display song, although much less chopped and thereby less easily located. As for the sub-Antarctic Gentoo penguin, its "ecstatic display" song can barely be distinguished from its "mutual display" song. The Gentoo penguin's reproductive cycle extends over a period of 7 months, and the interval between the first and last eggs laid is 2 months in the sub-Antarctic zone (slightly less in the Antarctic Peninsula). This can be compared to the only a 2-week interval between first and last eggs for Adélie penguins, whose entire reproductive cycle covers four and a half months due to its more Antarctic breeding. The optical and vocal signals of Adélie penguins seem to have been subjected to a more intense selective pressure than those of sub-Antarctic Gentoo penguins, the latter living in a milder climate and able to lay a second egg if the first egg fails to hatch.

The display song may also convey information relevant to the singer's sex. In those species studied with respect to sex, vocal dimorphism is clearest when physical dimorphism is least pronounced. No clear-cut vocal dimorphism was revealed in the songs of Adélie and Macaroni penguins, but the sex of partners can easily be identified in these species by size, aggressiveness, and width of the beak. In territorial penguins, males and females come back from foraging at sea and proceed to give ecstatic displays to whomever is at the nest site, and there is no confusion about sex because only the breeding partner takes care of an egg or chick at the nest site. In King and Emperor penguins that have no fixed site to return to from the sea, on the other hand, it is almost impossible to identify the sexes visually, but there is substantial vocal sexual dimorphism in their songs.

The form of chick songs is also adaptive, although they greatly resemble one another in different species. The Emperor penguin chick has a relatively long, stable, and complex song. The interval between maximum and minimum of the fundamental frequencies is more or less 2 kHz. King penguin chicks have a frequency range of only 1.5 kHz. This difference may be related to the ease with which this type of signal can be located. As discussed previously, in the more territorial King penguin, the song becomes longer (i.e., changes from the short song to the long song, the "short call" and "long call" of Stonehouse 1960), more complex, and less variable as the chick approaches emancipation. On the other hand, in the nonterritorial Emperor penguin where chicks may get lost, in the Antarctic winter, the chick song is almost as complex at the time of hatching as it is 5 months later. The type of ontogenesis of chick songs seems consequently related to environmental constraints particular to each species.

During the prelaying silence, Emperor penguins are mute from pair formation to egg laying (Isenmann and Jouventin, 1970; Jouventin, 1971b). Very occasionally, the male may sing during this period. The female then pecks at the male, but before he is silenced one or more females may have joined the pair. Trios may last several hours, during which females quarrel incessantly. Male songs played back near silent Emperor pairs attracted bachelorettes, which displayed to the presumed singer, generally resulting in the disruption of the original pair. Such a natural experiment shows the value of the prelaying silence in a species for which there is no territory to provide a site that facilitates the defense of the partner (Jouventin et al., 1979). The King penguin sings only part of the song when seeking a partner, the short song. King penguin pairs have a breeding site (or, rather, a "zone" in the breeding colony and a small defended "personal space"), so for them the prelaying silence is less respected. During this period, vocalization of one of the partners

**FIGURE 6.5** (A) King penguins incubate and brood on the seashore. (B) A passing bird (left) is attracted by the singer (right) as the partner (center) pecks the passing bird. (C) Now the passing bird (left) is singing as the partner (center) pecks at it, and the other partner (right) looks on. (D) Emperor penguins nest side by side and move about with their egg or young chick balanced on their feet, so individual recognition is most difficult. (E) The two parents brooding their chicks are flirting; they can do this because, in this nonnesting penguin, there are no territorial boundaries. *From Jouventin, P., 1982. Visual and vocal signals in penguins, their evolution and adaptive characters. Advances in Ethology* ***24****, 1–149.*

in a King penguin pair may attract a bachelor (Fig. 6.5B) and the unmated male may sing near the pair and is pecked at by the partner of the same sex (Fig. 6.5C). Trios form (usually, two males and a female) when the male partner pecks the female, then follows her on parade, while battling with the intruding male (Keddar et al., 2013; and see below for details).

## HOMOSEXUALITY, TRIOS, KIDNAPPINGS, ADOPTIONS, AND DIVORCES

Penguins have a long history of unusual sexual practices (Gillespie, 1932; Stonehouse, 1960; Prévost, 1961). Homosexual pairs and trios have been often observed, especially among displaying pairs, in captivity and in the field. The early observations of same-sex courtship display in penguins prompted a debate over a "trial-and-error" theory of sex recognition. This idea was also promoted by the difficulty for human observers in identifying individuals as males or females. For Roberts (1940, pp. 212–213), a typical penguin was "unaware of sex differences and does not differentiate between males and females even in mating." This theory was abandoned by modern authors that described sexual identification, particularly by vocal signals (for example, the acoustic studies and experiments on vocal signatures that we described in the previous chapters). Already, Richdale (1951, p. 86) concluded that "penguins can recognize the sex of another bird of its species on sight," thus disagreeing with the hypothesis of "trial-and-error" sex recognition. Whether they do this using vision, however, is questionable since the sexes often show little differentiation in external appearance and we saw that penguins are unable to find their mate when mute, experimentally or not. This historical controversy over sex recognition in penguins may seem a bit bizarre, but the occurrences of homosexual pairs have been confirmed in several species of penguins, and some penguin pairs of males in captivity have raised chicks (Davis et al., 1998; Zuk, 2006; MacFarlane et al., 2007).

A difficulty when studying this curious behavior is that the phenomenon is variable, depending on species, colonies, and years. Another problem is that in these monomorphic birds it is often difficult to discriminate the sexes in pairs, visual sexing being problematic. The similar appearance of males and females makes it even more difficult to estimate the presence or frequency of same-sex courtship display. Several articles suggest that homosexual pairs occur, but only one study on this topic sexed individuals using a DNA sex-specific marker (Pincemy et al., 2010). We studied 53 displaying pairs of King penguins from Kerguelen Island, and 28.3% were of the same sex. Of the homosexual displaying pairs, most (93.3%) were males. After displaying, many King penguins become bonded together, having learned the song of their partner, and they stand together until entering the breeding colony for copulation and egg laying. At the pair-bonded stage, among 75 pairs, only 2 were homosexual, one male and one female pair. Thus, homosexual behavior in penguins is confirmed and has to be explained. First, we concluded from zoo reports that this phenomenon is rarer in the field than in the zoo (Gillespie, 1932). Second, homosexual pairs appear much more infrequently in nesting penguins than in the two large *Aptenodytes* species. This likely occurs because nesting penguins have a meeting place at the nest that they defend against intruders (Jouventin, 1982; Bried et al., 1999; Dobson and Jouventin, 2003a). Occurences are most frequent in the King penguin, although they may be variable depending on location and year. Third, the homosexual pairs, common in displaying birds, are rare when pairs are bonded; that is, after they have learned the vocal signature of their partner. At least some information about sex is likely present in song characteristics in most of the penguin species, and this is certainly so for the Emperor and King penguins. Fourth, we found that displaying King penguins initially displayed significantly nonrandomly with respect to sex, despite the high frequency of homosexual pairs at this stage. This result is counter to the Robert's hypothesis of trial-and-error sex recognition.

For King penguins at South Georgia, Olsson (1995) found a slight surplus of adult males. Bried et al. (1999) found a sex ratio of 1.53 males per female among displaying adults in the Crozet Archipelago. Pincemy et al. (2010) found 1.65 males/females on Kerguelen Island, and for the same colony in another year, 1.96 males/female (Pincemy et al., 2009). Thus, a biased sex ratio could also explain the relative excess of same-sex display pairings by males. Also, a high level of circulating sex hormones could influence male King penguins to display in pairings with other males (Mauget et al., 1994, 1995; Garcia et al., 1996), especially since there are no nesting territories during pairing that in other nesting species keep males apart. Jouventin and Mauget (1996) found that when King penguins came back from the sea, males had high concentrations of testosterone and luteinizing hormone. These levels of sex hormones might render individuals extremely motivated to display, even to any individual regardless of sex. In addition, most brooding of unattended chicks is by failed breeders. Jouventin and Mauget (1996) also found that these failed breeders retain high levels of circulating prolactin in their blood for some time after losing their egg or chick. In this condition, they might be likely to feed chicks that are not their own, something that is frequently observed in breeding colonies, once chicks are semiindependent of their parents and in crèches.

Endocrinological mechanisms might also explain both chicknappings and alloparental care in Emperor penguins. Chicks beg from failed breeders to obtain care, such as extra-parental feedings or adoption. Thus, adults may be induced to become alloparents through a process of hormonal and associated social stimulation. This might explain why chicknapping and associated adoption occurs frequently and only in Emperor penguin colonies. Some breeders come back to the colony and do not find their mate (i.e., they have failed), but are still hormonally stimulated for parental care. Failed breeders are socially stimulated by begging chicks that are being fed by their own parents. The failed breeders feed the chicks and then attempt to brood them. This produces a chicknapping, and it is facilitated by the lack of territorial boundaries. The hormonal evidence suggests that chicknapping primarily results from an endocrinological motivation (Jouventin et al., 1995; Jouventin and Mauget, 1996; Garcia et al., 1996; Lormée et al., 1999; Angelier et al., 2006).

Finally, we can ask why trios occur only in the two large species of penguins. Systems of communication have to be adapted to the various ecological conditions found in this family. In small penguins, there are almost no trios because the territory prevents other birds, male or female, from trying to mate with one of the pair on the nest. The nest is aggressively defended, usually more so by the male. But in King and Emperor penguins, having no nest and no territory during the period of mate choice provides opportunities for competitors to approach from any direction and at any time. Thus, the two large penguins have to stay mute during this period to prevent or at least minimize trios, particularly the Emperor penguin where any male singing immediately attracts one or two birds of the opposite sex. The ensuing trio interaction can prevent pair formation and so reproduction, at least with the current partner. Penguins can have more than one song and one nuptial display to be more efficient at mating. The King penguin has two mutual songs. The short song ("short call" of Stonehouse 1960) promotes attractiveness during the search for a mate. The main information of this song is the sex of the singer. The long and very accurate song ("long call" of Stonehouse 1960), on the other hand, is given during incubation and brooding. At this time partners have to be positively identified, so that the exchange the egg or young chick can follow. In Adélie penguins, there are not only different songs according to the specific point in the breeding period but also different displays. This specificity probably facilitates the shortened breeding season and presence on land during the short Antarctic summer. Consequently, natural selection has favored an increase in the number of ritualized behaviors that prevent confusion in communication between breeding partners.

Curiously, but confirming our explanation, the sex ratio of trios is exactly the opposite in the two large species of penguins, likely due to patterns of mortality and associated sex-biased parental investment in offspring. Male Emperor penguins suffer about 10% greater mortality than females during the period of their heavy investment in incubation, which males accomplish on their own (Isenmann and Jouventin, 1970; Isenmann, 1971). At the beginning of the breeding season, females are consequently more numerous and strongly compete for mates during pairings. Thus, trios (or quartets) commonly have more females than males. In King penguins, mortality of females is higher, with males living about 1 year longer than females, on average (Weimerskirch et al., 1992). In other localities and years female mortality may be especially high when food resources are in short supply (Olsson and van der Jeugd, 2002). Thus, sex ratio of displaying prebreeding King penguins is biased toward males, and trios are usually composed of two males competing for a female (Fig. 6.6; Keddar et al., 2013). The sex biases in Emperor penguins (more females than males) is opposite to that of King penguins.

**FIGURE 6.6** In the King penguin, males are more numerous than females and trios are composed of a female and two males.

Another curious breeding fact was noticed by early observers of the Emperor penguin (Stonehouse, 1953; Prévost, 1961). Parents coming back to the colony to feed their young can cozy up close to a nonpartner parent of the opposite sex. Consequently, we can see two parents with their own young "flirting" with each other (Fig. 6.2E). This was first named by Richdale (1951) "Keeping company" in the Yellow-eyed penguin, but there it occurred between bachelors. Similarly, we can see in the two large penguin species, extra-feeding of chicks by adults that are not one of its parents (though they may be parents of another chick), or even feeding a foreign chick while they are brooding their own chick (Fig. 6.7).

These several surprising and odd breeding behaviors occur only in large penguins that have no nest. The behaviors include the prelaying silence,

**FIGURE 6.7** An Emperor penguin with its chick on its feet is giving an extra-feeding to another chick. *From Jouventin, P., 1971b. Comportement et structure sociale chez le Manchot Empereur. Revue d'Ecologie Terre et Vie 25, 510–586.*

advertisement walk, trios, kidnapping, adoptions, flirts, and extra-chick feedings. As well, Emperor and King penguins have a very high divorce rate (Bried et al., 1999) compared to other penguin species. In long-lived birds, monogamy enables breeders to retain the same breeding partner from year to year (e.g., the "divorce rate" of Wandering albatrosses was 0.3%; Jouventin et al., 1999b). But the two large penguin species have extremely low partner fidelity for a seabird, despite their high rate of survival (Table 1.3). In fact, some returning breeders retain their partner of the previous year because a few pairs were observed together for several successive years (up to five successive years in the Emperor penguin; Jouventin, 1971b). For large penguin colonies, this is nearly impossible statistically, without some previous partner recognition and pairing preference. But lacking the meeting point of the nest, these two large penguins cannot easily find their previous breeding partner, and to breed in time, they have usually to mate with a new partner (Bried et al., 1999). Nevertheless, even though they appear to remember their previous partner, their mate fidelity is much lower than for the species of nesting penguins.

## ADAPTIVE ASPECTS OF VOCAL SIGNALS AND VISUAL DISPLAYS

All vocal signals are associated with characteristic postures, whether ritualized or not. But not all postures are associated with vocalizations. Indeed, the "face to face," "waddling gait," "gawky attitude," "shoulders-hunched posture," "turn around stare," and "bowing" (except in Adélie and chinstrap penguins) are, as a general rule, performed silently. Also performed silently are the appeasement postures "slender walk" and "sheepish look." The only optical signals associated with vocal signals are all the ecstatic, mutual, and most of the agonistic displays. Recognition of sexual and later individual postures in ecstatic and mutual displays is facilitated by or perhaps completely dependent on song. Agonistic postures are much more efficient for rebuffing fights if songs attract the adversary's attention from a distance. Conversely, appeasement gestures, which are not meant to stimulate the partner, are silent.

There is a special association of vocal and visual signals in the chick of the Emperor penguin (Fig. 6.8). During the Antarctic winter, the greatest danger for the small chick is in getting lost, even if only for a few minutes. This seems to be the selective constraint responsible for the "bullseye" head pattern (see below), unique among penguin chicks, and the physical form of the chick song. This song consists of repetitive chopped signals with a large frequency band, similar to the vocal signals of dolphins or bats. Highly modulated, this song is easy to locate by a binaural process that compares the sound that arrives at each of the two ears. This provides information about direction, so that a parent can quickly find the source, such as a lost chick that is exposed to the cold. In addition, a singing chick stands with the beak raised and body extended (Fig. 6.8) and flaps its flippers rhythmically.

**FIGURE 6.8** For Emperor penguin chicks, associations of optical and vocal signals that are adapted for rapid recognition. Singing, the head is swung up and down to make it even more obvious because the winter can freeze parentless young chicks in a few minutes. *From Jouventin, P., 1982. Visual and vocal signals in penguins, their evolution and adaptive characters. Advances in Ethology* ***24****, 1–149.*

The "ecstatic display" of the Adélie penguin (Fig. 5.5) is a strongly communicative display of both vocal (ecstatic display song with clappings), and visual aspects within two vertical zones (one white, one black) repeatedly crossed by the flipper, and alternatively white or dark depending on which side is exposed ("maximum contrast," reinforced by a repetitive movement). This association of characteristics produces a remarkable signal that can be seen at a distance. This posture plays a role in finding a previous or new partner and synchronizes the physiological cycles of the members of a pair. The posture also minimizes delay in pair formation and thus confers an undeniable adaptive advantage to this summer breeder. A bird can find a mate as soon as possible, and thus have a better chance of rearing a chick before the expansion of the sea ice. As the sea ice freezes with the approach of the Antarctic winter, chicks can be isolated from the sea by the growing shelf of sea ice and exposed to extremely cold temperatures.

The head pattern of Emperor penguin chick (Fig. 6.9) consists of a dark eye, surrounded by a white and then dark zone, while the rest of the body is gray. Seen from the side, the pattern resembles a bullseye and is clearly visible at a great distance. This contrasts with all other penguins where chicks are

**FIGURE 6.9** In the Emperor penguin, the chick head looks like a target and is easy to see when lost and singing.

either uniformly dark (King, Adélie, and Chinstrap penguins) or else resemble their parents (dark back and light belly). Moreover, when singing, Emperor penguin chicks move their long-necked heads in a circular up and down motion (Fig. 6.6), which renders them even more conspicuous. Similar to the ecstatic display of the Adélie penguin, the appearance of Emperor penguin chicks render them as easy to find as possible.

According to the theory of information, a signal must be distinguishable from background noise. This is equally true for both optical and acoustical signals of penguins. Hatched songs used in the "song" of Emperor penguins or the "ecstatic song" of Adélie penguins were favored by natural selection because they are easy to locate. Agonistic displays (Fig. 6.10C), such as "sideways stare" and "beak-to-axilla" (Fig. 6.10D) or sexual displays (Fig. 6.10E) like "ecstatic display" and "mutual display" (Fig. 6.10F) must be distinct from common postures such as walking (Fig. 6.10A) or incubation (Fig. 6.10B). The origin of such ritualized postures becomes more and more enigmatic when considered from their functional and adaptive aspects, revealing convergences of postures (Fig. 6.10G and H) in species that are not closely related. Because behaviors are often extremely plastic, it is often difficult to discern their evolutionary development over phylogenetic history.

The ecstatic displays of Gentoo, King, and Banded penguins are identical, and merely consist in raising the head and sometimes slightly lifting the flippers (Fig. 6.11). Ecstatic displays are clearly visible at a distance, and a bird in this position is easily distinguished from lying or even standing conspecifics. In display, the bird holds the beak skyward and the neck is extended upward. In standing birds, the beak is held horizontally, usually in association with a compact stump-like body that is scrunched rather than stretched. The

**FIGURE 6.10** Ritualized postures as an agonistic (C, D) or sexual (E, F) posture must be differentiated from common postures such as walking (A) or incubating (B), for example in Adélie penguins. The ecstatic displays are similar in penguins (E, F) and albatrosses (G, H), not because they are closely related, but by evolutionary convergence because "head up" postures are easy to see in a colony. *From Jouventin, P., 1982. Visual and vocal signals in penguins, their evolution and adaptive characters. Advances in Ethology* ***24****, 1–149.*

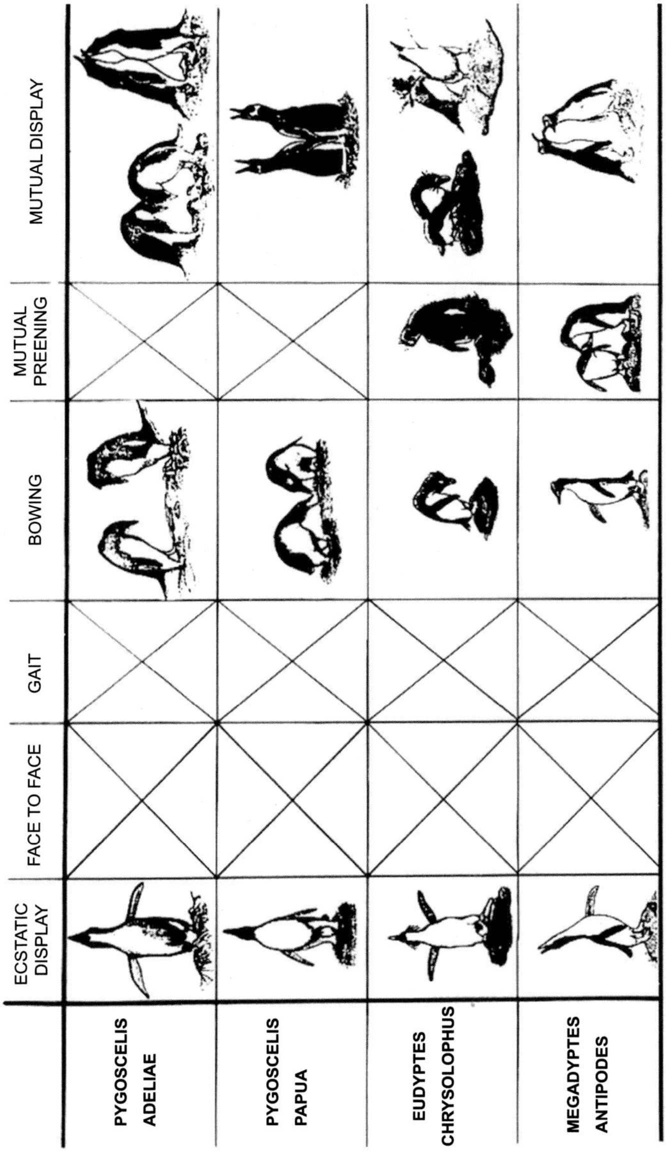

ECSTATIC DISPLAY
FACE TO FACE
GAIT
BOWING
MUTUAL PREENING
MUTUAL DISPLAY
PYGOSCELIS ADELIAE
PYGOSCELIS PAPUA
EUDYPTES CHRYSOLOPHUS
MEGADYPTES ANTIPODES

**FIGURE 6.11** Comparison of courtship displays in penguins.

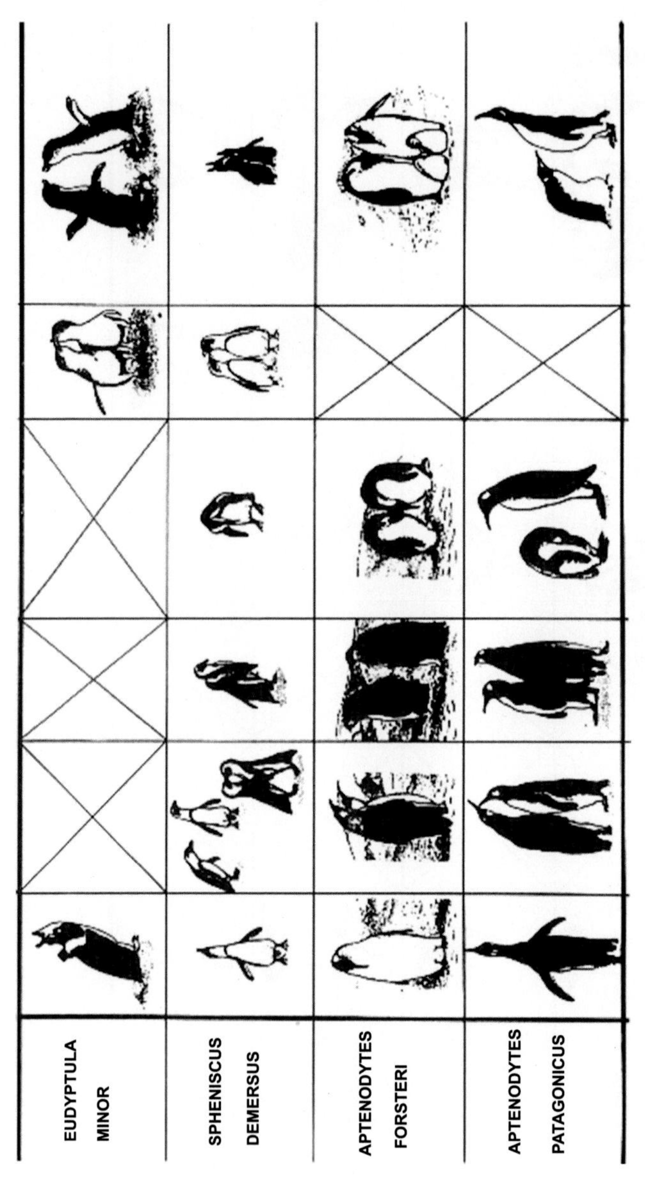


**FIGURE 6.11** cont'd

Emperor penguin is the only species to drop its head in this posture, and this stereotyped movement is clearly an evolutionary innovation. From a functional point of view, such a posture would seem to hinder rather to facilitate location of the display. In this case, however, there is a conflict between the song and the posture, the vocal signal predominating the visual signal. In fact, this quite uncommon song is extremely and precisely "hatched" from a control of expiration that can only be achieved with a lowered head. An opposing example are the ecstatic displays of other species, the Adélie, Chinstrap, Little, and crested penguins. In these species, songs are not so hatched and moreover not so precisely given as an accurate code of vocal identification, and they are produced with the head up. At the same time, the flippers are raised and flapped rhythmically and the head is sometimes shaken. These displays are much more spectacular and add new elements to the penguins' basic posture.

Special mention must be made of the nocturnal Little penguin, which lifts and shakes it flippers in the semidarkness as a very efficient visual signal. If the Emperor and the Little penguins are considered special cases, the other species fall into two groups. Some species exhibit a simple ecstatic display (King, Gentoo, and banded penguins), while in other species it is spectacular (Adélie, Chinstrap, and crested penguins). In the first group, reproductive cycles are poorly synchronized. There is a 2-month interval between the first and last laying of eggs in the Gentoo penguin, and a near absence of synchronization of egg laying in the King and banded penguins. In the second group, the breeding cycle is very short, about 2 weeks for egg laying in Adélie, Chinstrap, Macaroni, and Rockhopper penguins. In the Adélie penguin, the ecstatic display is the most exceptional in appearance, and the breeding cycle is the most contracted. A more complex ecstatic display in this species confers an evident adaptive advantage. Adélie penguins are the only species in which the pair formation period can be hastened.

It is also instructive to compare the "face to face" and the "waddling gait" displays of penguin species. The two large *Aptenodytes* penguins are the only species in which the "face to face" display is unambiguously present. In this case, it can be attributed to the reduced dependence on a fixed nest and associated territory. There is a need in these species to maintain close and permanent contact between partners during pair formation but without resorting to constant singing. Without the help of a nest, Emperor and King penguins need to carefully choose a partner by closely examining the colored feathers and beak in the head region of potential partners. During mate choice, these birds need to keep track of potential partners, and at the same time display information about their own breeding status and body condition. In the banded penguins of the genus *Spheniscus*, the "face to face" display is only a short element in a sequence of postures comprising the "turn around stare" display and "beak-sharpening" display (Fig. 6.11), followed by the "automaton" posture. This last posture clearly corresponds to the "gait" category.

Exaggerated gaits seem to fulfill the same function in all species. They are used to attract a partner to the nest (in banded penguins), the egg laying and incubating site (King penguin) or an isolated site (Emperor penguin) where the pair will remain alone until the egg is laid.

The "waddling gait" is exaggerated walking, and an easy motion that distinguishes a partner from common walkers (Fig. 6.11, "Gait"). In Emperor penguins, the waddling gait enables partners to remain in visual contact at a time when all vocalizations stop after pair formation. An association exists between the inability of partners to locate one another by singing and this characteristic gait, which newly formed pairs adopt when moving about. That Emperor penguins have developed a ritualized gait in the absence of a territory that is quite understandable. But one wonders why the King penguin or species of Banded penguins do not settle directly on their previous standing or nesting sites, rather than risk being separated. For King penguins, the answer to this might also explain the existence of such original postures as the "face to face," the "turn around stare," and others. The breeding cycle of King penguins is very different from that Emperor penguins. In King penguins, the previous standing site is often already occupied by many breeders standing with an egg already, so that partners must search for an incubation site in the shifting colony. If they are dominant enough, they may "force" a space between the standing breeders, thus causing a gentle wave of respacing among birds in the local area. But at the same time, partners must stay in close contact and remain mute to prevent the attention of other competing adults that are still looking for good mates. With no nest, one cannot tell if the specific location of a previous breeding spot is taken, and even with an egg or young chick on their feet, breeding King penguins shift about the breeding colony (Lengagne et al., 1999a).

"Bowing" (Fig. 6.11) is a display that occurs in all species but may play different roles and has different contexts in different species. In both species of large *Aptenodytes* penguins, bowing is involved in partner synchronization. First the most motivated partner executes a bow, and then the mate is progressively stimulated to respond. Thus, a sequence of mutual bows is produced, ending in copulation. Later, this same stimulation of mutual bowing occurs when the egg or young chick is exchanged between partners during relief exchanges. At this time, the rhythm of bowing is accelerated (cf. Fig. 2.11). In all the other penguin species, the main function of bowing is to inhibit the partner's aggressiveness, particularly when a bird approaches their partner on the nest. These two types of bowing in two separated branches of the penguin family show a convergence of form but not of function.

"Mutual preening" (Fig. 6.11), although uncommon, is sometimes observed in King penguins but never in Emperor penguins. It is not present in any of the *Pygoscelis* species but exists in the other genera of penguins. Thus, mutual preening appeared—or disappeared—haphazardly in the evolutionary lineage of penguins. The hypothesis of Lack (1956), Goodwin (1960), and

Sparks (1962, 1965) is that "mutual preening" is related to the level of aggressiveness, but this idea is not strongly supported for penguins. It would agree with the higher level of aggressiveness in King penguins compared to Emperor penguins, but mutual preening is not an element of the particularly aggressive Adélie penguin's behavioral repertoire. It is possible, however, that the rate of mutual preening is related to the presence of ectoparasites. Although mutual preening may have a social value, it might also occur due to its primitive antiparasitic function. Indeed, mutual preening is absent in Antarctic penguins (Emperor and Adélie penguins) that have almost no parasites, owing probably to their cold environment. And mutual preening is frequent in temperate and equatorial species. Banded, crested and Little penguins all exhibit mutual preening. Little penguins, which live in burrows and spend most of their time preening, are densely covered with ticks and other parasites. Furthermore, the parts of the body that are most commonly mutually preened involve those that penguins cannot reach by themselves, usually the head and neck.

"Mutual displays" (Fig. 6.11) accomplish the basic function of partner or chick recognition and are present in all species. They are similar to the "ecstatic display" and are also involved in territorial assertion, though with an exception. In Adélie penguins, mutual and ecstatic displays are different and highly specialized to accelerate the pace of breeding, and the "mutual display" can be displayed by a single adult used by mates or between parents and chick(s). The main function of the ecstatic display is the location of the partner. The sequence of the mutual display consists of two successive phases in Adélie penguins, as well as in the also aggressive crested penguins. First, bows are completed to reduce aggressiveness. Second comes a duet with the head raised, which accomplishes territorial assertion and identification. The role of the mutual display in territorial assertion in the genera *Eudyptula, Spheniscus, Eudyptes, and Pygoscelis* produces a contagious effect (also evident for the ecstatic display). If these displays were classed according to the intensity with which they are performed, the species follow the same order as when the level of aggressiveness and thus their degree of territoriality. Penguins living in dense colonies fiercely defend their nests (such as crested penguins and *Pygoscelis* species) and show the most complex and agitated displays. Next comes the animatedly displaying species that have more space and are less engaged in territory defense. Finally, the nonterritorial Emperor penguin shows an almost motionless mutual display. The form of the mutual display is related to the type and strength of the penguins' nesting habits.

Behaviors are notoriously subject to evolutionary convergence, but nonetheless it can be interesting to interpret them in a phylogenetic context. In particular, stereotyped behaviors, such as displays, are less flexible than other behaviors, and thus might in some cases follow a phylogenetic pattern (e.g., Lorenz, 1951). For example, the large species of the genus *Aptenodytes* have long been considered a highly evolved group and their behavior patterns seem highly derived as well. *Aptenodytes* is a sister taxon to all the other living penguin species and probably split from the penguin line more than 10 million years ago (Baker et al., 2006; Thomas et al., 2013). This particular systematic position in the order and family of penguins is consistent with the highly original optical signals, as well as life-history characteristics (e.g., large size, lays a single egg, thin beak and beak spot, and incubation of the egg on the feet). The uniqueness of these two species seems to go beyond the generic level, as all the other penguin genera are much more closely related to each other than to the large penguins.

The questionably monospecific genus of Little penguins (*Eudyptula*) has few ritualized attitudes, probably a consequence of its nocturnal and also more primitive way of life, which is more similar to the burrowing petrels than to other penguins. Indeed, certain characters pertaining to nesting habits, territoriality, size, morphology, and the absence of head patterns seem more similar to the Procellariiformes. Nonetheless, like all penguins, Little penguins cannot fly. Such arguments agree with the thesis of Furbringer (1888), now widely accepted, that penguins are related to petrels and derived from winged ancestors. Comparatively, the genus *Pygoscelis* is much larger, both in body size and in the number of currently recognized species. The Chinstrap penguin, an intermediate species, reveals the relationships regarding courtship display and head patterns between the other species of the genus, Adélie and Gentoo penguins, which otherwise would have been hard to establish. The three species are undeniably related, according to optical signal characteristics, but the relationships are more tenuous than between species of other genera. It may be possible to find several species within Gentoo penguins, as suggested by DNA results of long-term population differentiation among some islands (de Dinechin et al., 2012; Vianna et al., 2017).

The species and subspecies of crested penguins (*Eudyptes*) will be detailed in the following section on evolution of vocal signals. This genus is quite homogeneous and only slight variations in courtship displays differentiate the species. For example, the Rockhopper penguin shakes its beak more rapidly than the Macaroni penguin. However, careful observation shows that the colored plumes that ornament the head differ greatly in pattern. The most important characteristics for differentiating these species are the color of the eye and beak, the naked skin at the base of the beak, and the disposition, color and length of the plumes of their crests. Speciation, although very active in this group, superficially appears to involve only a few relatively inconspicuous characteristics that insure reproductive isolation of the species. When considering behavior and morphology, it is surprising that, before DNA analyses, a close phylogenetic relationship was not suggested sooner between the genera *Megadyptes* (Yellow-eyed penguin) and *Eudyptes* (crested penguins). Optical signals such as the red color of the eye, the hint of a crest, and similarities of courtship displays indicate the close evolutionary association of these genera (Jouventin, 1982).

## COMPARATIVE STUDY OF VOCAL SIGNATURES

Penguins recognize their partner and the chick, but only vocally. This unimodal identification provides a good model for a comparative study of individual acoustical recognition in animals. Identification in almost all penguin species is apparently based on two main parameters of their songs: temporal sequence and sound frequency. These characteristics generally interact, and thus can be difficult to dissociate experimentally. They often do not seem to be perceived separately by the penguins. Recognition systems based on temporal sequence are used mainly to indicate the location of a singer. On the other hand, song characteristics based on frequency variation with time seem to contain less information. Frequency variation is more difficult to study and is mainly used when partner identification is already facilitated by the presence of a nest or when vegetation or other sound screens block syllable transmission. Two other recognition systems have been described, one based on the imitation of the partner's pattern (Gwinner and Kneutgen, 1962). The other system is based on imitations as well but with recognition of the partner's song followed by development of a contrasting song (Grimes, 1965; Thorpe and North, 1965, 1966). However, these systems seem more specialized, rare, and without evidence of applications among the penguin species.

The family of penguins is much more diversified than it might at first appear. In particular, it is instructive to contrast nesting and nonnesting species. This contrast provides a good opportunity to compare vocal signatures according to breeding ecology. The two large penguin species that breed without the advantage of a rendezvous site of the nest provide "natural experiments" that reveal the importance and biological function of the nesting territory for the other penguin species and for nesting birds in general. The nest location greatly facilitates identification of partner or chick by providing a familiar meeting place. In a sense, this is demonstrated by the difficulty of individual recognition in the two penguin species that lack a nest. Penguins have adopted different solutions to the problem of individual recognition that depend on the difficulties presented by both the ecological and social environments of the species. First, there are many highly territorial species of penguins, as aggressive in defense of territory as any of the seabird species. Second, in the King penguin, we have a species that has no nest and is able to move about with an egg on its feet, but having a laying and brooding site, if a somewhat mobile one. Finally, the Emperor penguin is a completely non-territorial species that breeds on a surface (sea ice) that seasonally disappears and may change somewhat from year to year.

This gradation is associated with changes in sonogram patterns that are easily quantified and compared statistically among the penguin species (Table 6.1). For example, in Emperor penguins, the mean coefficient of variation for all song characteristics yields a value of 60.4% for the population

**TABLE 6.1** Relationship Between Variability of Physical Song Characteristics in Penguins and the Nesting Ecology for Each Species

| Species Analyzed | Possibility of Topographic Localization of Partner | Songs of Different Individuals (Mean of Variation Coefficients for all Characteristics) (%) | Songs of Same Individuals (Mean of Variation Coefficients for all Characteristics) (%) | Ratio of Song Variability of Population Sample Over Individuals |
|---|---|---|---|---|
| *Aptenodytes forsteri* | Species without a nest and mobile during reproductive period (no topographic landmark) | 60.43 | 5.5 | 10.98 |
| *Aptenodytes patagonicus* | Species without a nest but having a relatively fixed site (i.e., a zone) on the seashore | 33 | 5.10 | 6.47 |
| *Eudyptes chrysolophus chrysolophus* | Highly territorial species having a nest, which aids partner and parent–chick rejoining | 67.2 | 16.33 | 4.11 |
| *Eudyptula minor* | Highly territorial species having a nest, which aids partner and parent–chick rejoining | 46.40 | 14.99 | 3.09 |
| *Eudyptes chrysocome moseleyi* | Highly territorial species having a nest, which aids partner and parent–chick rejoining | 20.15 | 8.49 | 2.37 |
| *Pygoscelis adeliae* | Highly territorial species having a nest, which aids partner and parent–chick rejoining | 19.28 | 10.44 | 1.81 |

From Jouventin, P., 1982. Visual and vocal signals in penguins, their evolution and adaptive characters. Advances in Ethology **24**, 1–149.

sample, while only 5.5% for repeated songs of one individual. When individual songs are fairly constant and variability within the population is large, confusion among individual songs is minimal. In the example given above, the ratio of population variation to that of individuals is 60.4%/5.5% = 10.9. So different songs of an individual Emperor penguin are about 11 times less variable than the range of songs found in the whole population. This value agrees with expectation that in the uniqueness of the song, the Emperor penguin is the most specialized of all penguins studied. Individual variability is about the same (5.1%) for King penguins. However, since variability of the population (33.0%) is about half that of Emperor penguins, the population/individual ratio (6.47) is much lower.

In four other species of penguins (*Eudyptes chrysolophus chrysolophus*, *Eudyptes chrysocome moseleyi*, *Eudyptula minor*, and *P. adeliae*), all highly territorial, these values range between 1.8 and 4.1, representing the smallest differences to be found in penguins between individual and population songs. Thus, the relationship between variability of physical song characteristics and the problem of individual recognition is evident. To avoid bias, all the measured characteristics of songs were used to calculate the mean of coefficients of variation. However, several parameters play no role whatsoever in individual recognition, while one measure, the length of the syllables, is extremely important. If the mean of coefficients of variation for syllable length is noted in the ordinate and the order number of the syllable in the abscissa (Fig. 6.12), two sets of points appear, one corresponding to the individual variation coefficient and the other to the population's. The more the sets are separated, the more the song is individualized. In the four species for which syllable length was measured, results again coincide with the social and ecological characteristics of the species. These results accord with the more sophisticated results from information theory, described in Chapter 5 and obtained by a completely different method from coefficients of variation (Fig. 5.29).

For Macaroni and Adélie penguins, values corresponding to population and individual coefficients of variation are very close, rendering confusion likely. Both these species are highly territorial, and singing serves only to identify the bird beside the nest as the expected partner. On the other hand, the two sets of points are widely separated in King penguins, and even more so in Emperor penguins, both only slightly or not at all territorial. In these latter two species, this parameter reflects the easier recognition of a singing individual. In Macaroni and Adélie penguins, the first syllable is particularly variable and is immediately followed by a decrease in variation in components of the song, which increases again at the end of the song. Thus, the message is most precise when attention is at its peak, that is, just after the first syllable. Since these two species are not closely related, this coincidence seems to be an adaptation to a similar communication problem for territorial species.

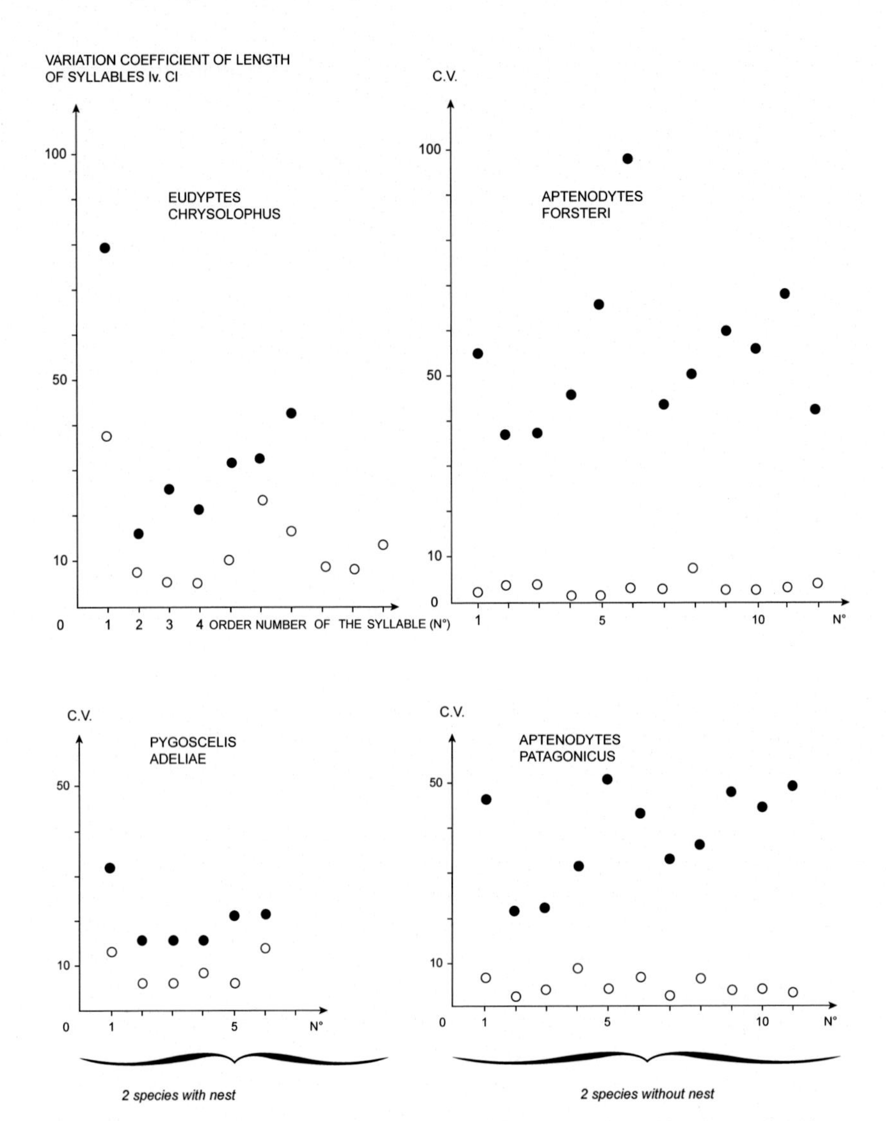


**FIGURE 6.12** The variability in the length of different song syllables in relationship to the manner of nesting in four species of penguins (●: for different individuals, and ○: for one individual). Nesting species, not so informative because they have a meeting point, are on the left, and nonnesting species, with a precise timing of the vocal signature, on the right. *From Jouventin, P., 1982. Visual and vocal signals in penguins, their evolution and adaptive characters. Advances in Ethology* ***24****, 1–149.*

Frequencies (song pitch) and temporal patterns were analyzed by the same method. In Gentoo penguins, the sequence is only slightly more variable than the pitch, and both ratios of population to individual coefficients of variation are relatively low compared to, for example, those in the King penguin.

In Southern rockhopper penguins, temporal sequence (2.7) is about twice as variable as pitch (1.2). In Little penguins, frequencies are more variable (6.0) than the sequence of the song (4.3) and both parameters are more variable than in the banded penguins. In Macaroni penguins, frequencies are not variable (1.3), although the sequence (4.7) is a bit more so. In King penguins, frequencies (10.6) as well as sequence (8.0) are remarkably variable. In Emperor penguins, frequencies (1.9) are not variable, but the temporal pattern (14.5) is the most variable of all. From the data provided (Table 6.2), the ratio of variability of temporal sequence to that of pitch can be computed. In Emperor penguins, the ratio is 7.5 (14.5/1.9) whereas it ranges from 0.7 to 3.7 in the other species. The Emperor penguin, then, uses a system almost entirely based on temporal sequence.

We summarized the difference between the vocal signatures of nesting and nonnesting penguins, and combined them with information on visual signals, to show the extreme dissonance of these two groups of penguins (Table 6.3). The complexity and sophistication of visual and vocal signals increases as we move from the nesting to nonnesting penguins. While these differences follow a phylogenetic pattern, they also reflect ecological differences in the challenges of giving up a fixed nest site. In losing the nest site, Emperor and King penguins require frequent changes in partners and thus frequent mate choice, this encouraging sexual selection and facilitating more complex visual ornaments that reveal mate quality. At the same time, the increased density of colonies of mobile individuals makes the accurate identification of faithful partners more difficult, this facilitating natural selection of a highly complex and accurate vocal signature.

Jouventin (1972a) suggested that individual recognition based on temporal sequence occurs in many other species of birds (Fig. 6.4), although not to the same extent as in Emperor penguins. In Fig. 6.4A, the sonogram of a northern gannet (*Sula bassana*) shows the characteristic regularity and stability of this species' songs, consisting of a series of identical phrases formed by several equally spaced syllables. On the other hand, intraspecific variability is evident when comparing songs of five different Northern gannets (Fig. 6.4B–F). The similarity with the Emperor penguin's song is obvious, although the main frequencies are more stable in Northern gannets. This, again, is a convergent adaptation. Northern gannets nest about 1 m from one another in very large colonies, generally on rocky plateaus that are devoid of visual landmarks. Before landing on the nest, they identify themselves vocally. Their partners, that otherwise would have attacked them, then respond. The White pelican (*Pelecanus onocrotalus*) also nests in large dense colonies, where the chicks are gathered as in a penguin crèche, and it has the same type of song (Fig. 6.4G) as the preceding groups of birds. Similar interindividual variation was also evident in recordings (not shown).

In seabirds, the hatched pattern of the songs is used to locate the singer. It is a redundant song so that it can "get through" to the receiver in a noisy

**TABLE 6.2** Comparing Variation of Related Parameters (Temporal Sequence Song Pitch) in Six Species of Penguins

| | *Aptenodytes forsteri* | *Aptenodytes patagonicus* | *Eudyptes chrysolophus chrysolophus* | *Eudyptula minor* | *Eudyptes chrysocome moseleyi* | *Pygoscelis adeliae* |
|---|---|---|---|---|---|---|
| Mean of coefficients of variation for parameters related to frequencies (main and maximum frequencies) in one individual | 22.62% | 1.88% | 13.89% | 16.68% | 10.12% | 9.28% |
| Mean of coefficients of variation for parameters related to frequencies in several individuals | 43.34% | 19.87% | 17.6% | 69.61% | 12.10% | 15.17% |
| Mean of coefficients of variation for parameters related to temporal sequence (length of different syllables or phrase period) in one individual | 3.48% | 4.64% | 15.63% | 5.50% | 7.07% | 9.91% |
| Mean of coefficients of variation for parameters related to temporal sequence in several individuals | 50.77% | 37.1·2% | 73.73% | 23.45% | 19.14% | 19.38% |
| Population/individual ratio for parameters related to frequencies | 1.91 | 10.55 | 1.26 | 5.99 | 1.19 | 1.63 |
| Population/individual ratio for parameters related to sequence | 14.58 | 8.06 | 4.71 | 4.26 | 2.70 | 1.95 |

From Jouventin, P., 1982. Visual and vocal signals in penguins, their evolution and adaptive characters. Advances in Ethology **24**, 1–149.

**TABLE 6.3 Our Thesis is That Large Penguins That Lack a Nest Have to Communicate More Accurately Than Nesting Penguins, Adding Visual Signals (Beak Colors to Mate Better) and Vocal Signals (Double Voice to Better Identify the Partner)**

| | Visual Signals | Vocal Signals |
|---|---|---|
| Biological function | Mainly mate choice for pairing | Mainly vocal signature for brooding and rearing |
| Small penguins (common colonial nesting seabirds) | Simple visual ornaments = spheniscin pigmented feathers | Simple vocal signatures = Timbre Analysis (Spectral profile and pitch of the song) |
| | +Nest to know social status of the possible mate (a nest? a big nest? center or edge of colony?) | +Nest to confirm partner and chick identity |
| Large penguins (special colonial birds without nest) | Complex visual ornaments = pigmented feathers (after molt) *long-term body condition* (yearly) | Complex vocal signatures = temporal analysis (on amplitude time in Emperor penguin and on frequency time in King penguin) |
| | +UV (structural color) & beak carotenoids permanently produced *short-term body condition* (reflect recent body condition) | +Double voice (beats coding) for confirmation |

environment, during brief periods of relative silence. The hatched pattern is the component of the song is initially identified by the receiver. By acoustic convergence, we find the same pattern in colonial mammals, such as fur seals and elephant seals (Charrier et al 2002, 2003; Dobson and Jouventin 2003b; Aubin et al 2015). As has been shown for Northern Elephant seals (Mirounga angustirostris), males recognize other males by the rhythm of their calls (Casey et al 2015; Mathevon et al 2017).

Moreover, Jouventin (1972a,b) suggested that the songs of some bird families, such as Alcids, could help in understanding the general problem of vocal signatures. Tschanz (1968) published spectrographic analyses of three different Common guillemot (*Uria aalge*) songs, and eight songs of the same individual. These sonograms showed the similarity of sequence in one individual's songs, although this is much less precise than in Emperor penguins. He concluded that the length and succession of syllables were quite stable, whereas tone, pitch and strength of the signal varied to a greater extent. Guillemots have no formal nest and incubate their egg on bare rock. They breed in groups of about 10 on small cliff ledges. It was apparent from Tschanz's (1968) acoustic experiments that modifications of pitch and tone influenced the response of a chick to a certain degree, although the chick still recognized its parent's song. If syllable succession was modified, however, hardly any change in chick behavior was observed. Since Alcids are far phylogenetically from Sphenicids, convergence between systems of recognition is obvious and can be attributed to the similar way of life of the two groups of colonial seabirds. This lifestyle is responsible for convergence in morphology and head patterns, and on land, dense colonial nesting that increases the probability of confusion over the identity of breeding partners. Similar studies in the bird families of flamingos, crows, and certain ducks may show that they also have a system of recognition based on temporal sequence. This acoustical convergence is not limited to adults. In guillemots (Alcid seabirds) and flamingos, chicks have the same way of life and type of songs as young emperor penguins.

## ISLAND SPECIATION AND TAXONOMY

The family of penguins is not only a model for individual recognition via vocal signatures, it is also a model for studying the mechanisms of speciation. It would seem that there is little opportunity for speciation in this family, since all the species breed on islands and fish at sea. The southern oceans are their habitat, in a hemisphere where water circles the globe. These oceans might seem to have no distinct boundaries that would isolate subpopulations and begin the process of subpopulation divergence that initiates speciation. At the same time, however, some species by their nature may be more mobile than others, such as nearshore feeders versus broadly pelagic species. The biological definition of the species (Mayr, 1963) is based on an interfertility criterion for populations under natural conditions. Disparate subpopulations often belong to the same species specifically because of the broad movements of individuals among subpopulations in different geographic locations. This concept is expressed in a somewhat different manner by Dobzhansky (1937), who considered a species as a "closed gene pool." However, breeding on different islands separates subpopulations and may lead to speciation if dispersal among islands for breeding purposes is low. In addition, different islands are separate environments and may be surrounded by separate seas, both in terms of food resources and the physical conditions of the waters. Geographic isolation makes it difficult to determine whether a local variety is in fact a species. This explains the controversial and often contrite views of systematists on taxonomy.

The key to documenting speciation is the evolution of behavioral isolating mechanisms. When such mechanisms occur, for example, by incompatibilities in mating choices, genetic mixing in penguin breeding colonies on land can be curtailed. In such cases, the extensive abilities of penguins to migrate at sea may not homogenize Dobzhansky's closed gene pool. Thus, such behaviors as mating displays might provide a clue to the degree of speciation that has occurred among subpopulations, and where lines between such divergent subpopulations can be drawn. Courtship displays in penguins reveal differences at a generic level, and sound recordings of songs from different species provide confirming data. An objective study of vocal signals was attempted by Jouventin (1982), using statistically analyzed sonograms. The songs were considered a species-isolating mechanism, similar in concept to different patterns of head markings that appeared to function in mate choice. Similarity of songs between two island populations of penguins was interpreted as reflecting gene flow between them, whereas greater differences favored the thesis of a genetic barrier between populations, giving rise to new species.

We will first summarize the conclusions of Jouventin (1982), comparing behaviors and particularly songs in different penguin populations. Then we compare these results to more recent analyses using DNA. No difference was perceived between recordings of Emperor penguins at Caird Coast (British Institute of Recorded Sounds, BIRS) (recordings, $n = 10$) and those made at Adélie Land ($n > 50$). Nor were differences found between recordings of Adélie penguins in the South Orkneys (BIRS) ($n = 12$) and in Adélie Land ($n > 50$). Thus, at opposite points of the Antarctic Continent, Emperor and Adélie penguin populations seemed fairly homogeneous. The songs of Australian Little penguins ($n > 50$) seemed slightly lower-pitched in the field than those of Little penguins from New Zealand ($n = 10$), but individual variations were too large within the samples to confirm this impression.

The songs of King penguins from Macquarie Island provided by the Commonwealth Scientific and Industrial Research Organisation (CSIRO) of Australia ($n = 10$) seemed identical to those from Crozet and Kerguelen, while those from the Falkland Islands communicated by the BIRS ($n = 10$) were slightly different. The whole song from the Falkland Islands lasted up to 7 s (2 s longer than in the preceding localities) and consequently contained cyclical patterns (up to 5). This slight acoustic difference agreed with modern classification based on morphological characteristics, distinguishing two subspecies of King penguins. *Aptenodytes patagonicus patagonicus* occurs on the Falkland Islands and *Aptenodytes patagonicus halli* in the Indian Ocean.

Generally, two subspecies of Gentoo penguins are described: *Pygoscelis papua ellsworth* lives on the South Orkney Islands, South Shetland, and South Sandwiches, while *P. papua papua* occupies the rest of the distribution area. The analysis of song recordings also disclosed two quite distinct groups, not however matching the preceding distribution. Recordings from Macquarie Island (provided by the CSIRO) were close to those made at the Crozet and Kerguelen Archipelagos. The rhythm was slow, the tone low-pitched and the phrases long. But songs of Gentoo penguins from the Falkland, South Orkney, and South Georgia Islands (communicated by the BIRS and B. Despin) show a rapid rhythm, short phrases, and higher pitch. These differences can be visualized and compared in the sonogram of a Gentoo penguin song from South Georgia (Fig. 4.1) and from the Crozet Archipelago (Fig. 4.9A–C). The differences are evident and involve all characteristics. The differences between sonograms seem to pertain to at least a subspecific level because the period, length of phrase, the inspiration and number of syllables do not or only slightly overlap. Differences in main and maximum frequencies between the sonograms are less clear but are nevertheless noticeable.

The Rockhopper penguin was considered a complex "species," and heterogeneity among populations was recognized long ago and attributed by some authors to geographic variation (Hagen, 1952). However, most often two (Peters, 1931; Murphy, 1936; Prévost and Mougin, 1970) or three (Mathews, 1932; Falla, 1937; Jouanin, 1953) subspecies were distinguished, while several other authors considered them as two (Oustalet, 1875) or three species (Hutton, 1879). There are several reasons for this confusion. One has to do with sampling all of the available populations. Naturally, it was difficult to sample all localities within the widely distributed area of the species. In addition, classical methods of body or skeletal measurements were inadequate to resolve this particular problem. Individuals from these populations show almost no difference in size or weight, characteristics which turn out to be irrelevant as systematic characteristics in Rockhopper penguins. In penguin species, size varies considerably with the measuring method, and weight changes daily, depending on feeding trips and the stage in the reproductive cycle. Even measurements better adapted to this group of birds vary with method and measurer. Hagen (1952) attempted to group data from early publications and concluded that subspecific distinctions were not evident, as only head tufts showed clear differences in length. Segonzac (manuscript note) measured a large number of sexed animals from St. Paul and Amsterdam (1969-70-71) as well as from Crozet (1971). Length of flipper, foot, and above all beak differentiated the two populations, if same-sex birds were compared. If the sexes were not separated there was a strong overlap.

But the main difference among populations of Rockhopper penguins lies in the songs. Should the songs be considered differentiated local dialects, or as evidence of two species corresponding to the two populations from which data were collected, St. Paul–Amsterdam Islands and the Crozet and Kerguelen Archipelagos? Although the latter islands are quite distant from each other, the St. Paul and Amsterdam Island populations should be regarded as one, for these islands are less than 100 km apart. An answer to the problem of whether populations are disjoint was provided by recordings of the other populations of the same subspecies, i.e., the Gough and Tristan da Cunha Island groups. Song recordings from the Gough Islands ($n = 10$) were acquired from the BIRS. The songs vary only slightly from recordings of songs from Rockhopper penguins at St. Paul and Amsterdam Islands ($n > 50$, cf. Fig. 4.16). Moreover, additional recordings from the Rockhopper penguin population at Staten Island ($n = 10$) at the tip of South America, also obtained from the BIRS, and recordings from Macquarie Island (n = 8) were obtained from the CSIRO. The recordings both resemble those of Rockhopper penguins from the Crozet and Kerguelen Archipelagos, although recordings from Staten Island differed slightly, understandable as these populations are somewhat distant. Thus, the main separation of different Rockhopper penguin populations into two groups was confirmed, as each has a completely distinct type of song with only minor differences within each group.

Another character, length of head tufts, follows the same pattern. Although noted by several authors, few data were available as this parameter was not classically measured. However, we grouped tuft length values provided by different authors (Roberts, 1940; Hagen, 1952; Segonzac Ms. note) as individual errors were not important in this case. Three graphs are presented in Fig. 6.13. The first indicates the mean of maximum frequencies for Rockhopper penguin songs in relation to the latitude of the island where they were recorded. The St. Paul, Amsterdam, Gough Islands group is shown to be quite distinct from the rest (Crozet, Kerguelen, Staten, and Macquarie Islands). The second graph gives the mean of the maximum length of tufts according to latitude. In this case also, the St. Paul, Amsterdam, Tristan da Cunha, and Gough Islands group together (the last locality is not included as only one measure was available) were quite distinct from the other islands (Crozet, Kerguelen, Antipodes, Snares, Campbell, and Falkland Islands).

The third graph in Fig. 6.13 shows the date of maximum egg laying in relation to mean sea temperature, which more or less matches latitudinal effects. This graph indicates that the Rockhopper penguins of more northerly (Tristan da Cunha, St. Paul, Amsterdam, and Gough) islands are again, although slightly less evidently, different from those of the more southern (Antipodes, Campbell, Crozet, Ildefonso, Falkland, Macquarie, Marion, Kerguelen, and Heard) islands. Thus, a vocal character, a morphological character (tufts), and a physiological one (the reproductive cycle) all vary in the same way in the populations studied. What causes this nonrandom variation, or better, what have the populations within each group of islands in common? Rockhopper penguins (generally named as the subspecies *Eudyptes chrysocome moseleyi*), inhabits the northern group of islands, ranging from 37° to 41° latitude. Rockhopper penguins (*Eudyptes chrysocome chrysocome*) of the other subspecies occupy the southern islands, between 44° and 54° latitude. There is then a different latitudinal extension of each "subspecies." Alternatively, could their formation be explained by geographic isolation? The answer is no, as the St. Paul and Amsterdam Islands are inhabited by *Eudyptes chrysocome moseleyi* and are 1850 km distant from Kerguelen, where *E. chrysocome chrysocome* lives; Kerguelen is some 6000 km from Macquarie, in which the latter subspecies is also present, while the St. Paul—Amsterdam group is still further than this from Gough—Tristan da Cunha (with *E. chrysocome moseleyi*). However, within each group, geographic isolation has certainly played a role, although not a very important one, since some of the far-isolated populations show original characteristics.

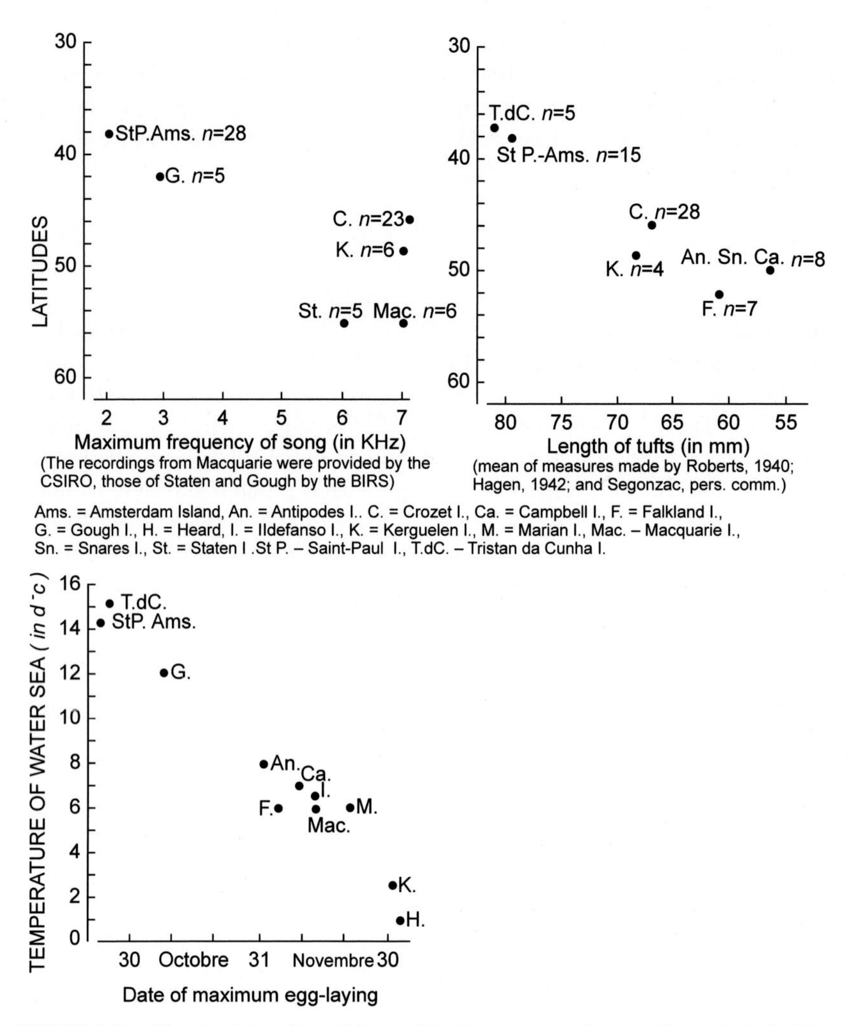


**FIGURE 6.13** Characteristics of populations of Rockhopper penguins according to latitude. Top left, maximum frequency of song. Top right, length of crest. Bottom, date of egg laying at islands with different sea-surface temperatures. *From Jouventin, P., 1982. Visual and vocal signals in penguins, their evolution and adaptive characters. Advances in Ethology **24**, 1–149.*

Speciation in crested penguins is more related to latitude than to interisland distance. What specific influence has latitude on the two groups of populations? Obviously, external temperature decreases toward the pole. Even though courtship displays are mainly performed on land, the ecology of penguins is marine. The average difference in sea temperature at the northern and southern groups of islands where Rockhopper penguins live and breed is 10°C, which must result in different thermoregulatory adaptations (Fig. 6.14). Following Mayr (1963, 1970), we can conclude that each population is a finely

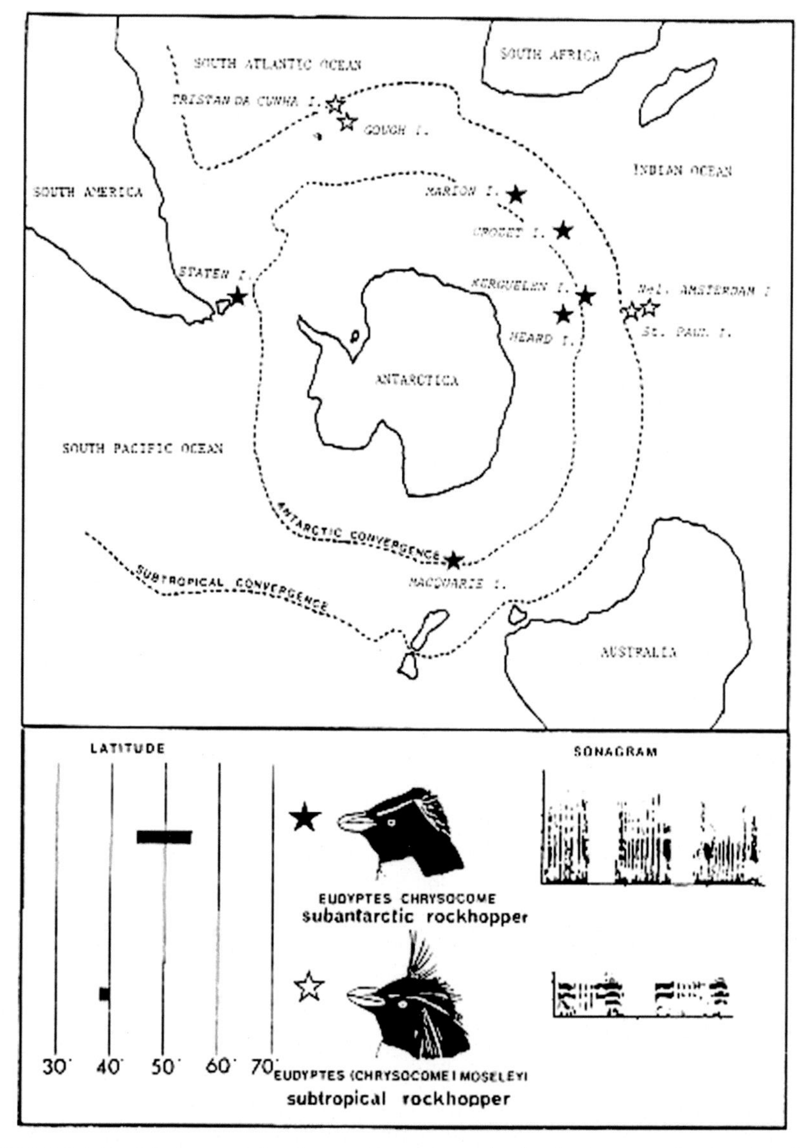


**FIGURE 6.14** Northern and Southern rockhopper penguin distributions are related to latitude and not to the distance between islands. Both species have same body measurement but really different mitochondrial DNA haplotypes and according to different characteristics used in nuptial displays, such as crests and songs, as noticed in 1982. *From Jouventin, P., Cuthbert, R.J., Ottvall, R., 2006. Genetic isolation and divergence in sexual traits: evidence for the northern rockhopper penguin* Eudyptes moseleyi *being a sibling species. Mol. Ecol. 15, 3413–3423.*

integrated genetic system adapted to a distinct ecological niche. Isolating mechanisms should prevent hybridization between the two groups of Rockhopper penguins, and such characters as song and crest length evolved much faster than other morphological characters such as length of flippers, beak, and feet. These are traits that evolved differences to a lesser degree than those related to the breeding cycle (the date of laying and crest length are important to mate choice).

In effect, preservation of the genetic integrity of populations in each habitat is fundamental to producing the two gene pools of Rockhopper penguins, as required by Dobzhansky (1937) for defining species. Ecological differentiation is maintained by reproductive isolation. The existence of two groups of populations that live in different habitats and have reproductive isolating mechanisms give a clue to the direction of evolution. The disproportionate behavioral barrier, mainly between the nearest populations (Crozet and Kerguelen Archipelagos vs. St. Paul and Amsterdam Islands), indicates that genetic drift (i.e., random chance) is not as involved in the process of divergence of the two groups of Rockhopper penguins as is sexual selection and mate choice. Each group appears to constitute a genetically closed pool. In other words, there are at least two species. Given the biological definition of a species, it cannot be argued through artificial crosses in captivity that the two groups are the same species. Indeed, hybrids exist in zoological parks between the very distant Adelie and Gentoo penguins, as well as between Jackass and Humboldt penguins. But such crosses do not occur in nature. Even if first-generation hybrids were found to be fertile, this would not constitute proof of subspecificity (Mayr, 1970, p. 164).

The taxonomic status of Royal or Schlegel penguins is uncertain and presents an even more complex problem. This species cooccurs with Macaroni penguins at Macquarie Island, and only rarely occurs at other localities. Depending on various systematists, the Schlegel penguin constitutes a morph, a geographical race or a subspecies (*Eudyptes chrysolophus schlegeli*). Other authors considered it a species. Known morphological data show the two species, the Schlegel and Macaroni penguins, to be quite similar, so they need to be reexamined by other methods (Barré et al., 1976). The Schlegel penguin has a white throat (Fig. 6.15) whereas the Macaroni penguin (*E. chrysolophus chrysolophus*) is black throated like most other members of *Eudyptes*. Knowing the importance of head patterns in penguin speciation, one is tempted at first sight to consider the Schlegel penguin as a species. The problem is even more complicated by the presence in each population of breeding birds with the other group's characteristics, as well as intermediate animals with a gray throat. Are these different morphs, or visitors that settled in the population? First of all, morphs showing a partly white plumage occur in all penguins, and particularly in Schlegel penguins. Also, recordings made of visiting individuals in Adélie Land and those provided by the CSIRO show

**FIGURE 6.15** A Macaroni penguin (right) breeding with a Schlegel-like penguin (white throat, left) in a colony on Kerguelen Island.

that Schlegel penguin songs are almost indistinguishable from those of Macaroni penguins (Fig. 4.15A–D).

The Macquarie population of Schlegel penguins was carefully studied by Shaughnessy (1975), who found coloration to be independent of age but not of sex (48% of males and 22% of females have a white throat). Pairs with the same throat pattern were too frequent to be random. At the Crozet Archipelago, the rare white-throated individuals occur in breeding colonies of Macaroni penguins (1%–4% of the colony are white throated), and intermediate individuals occur as well. These Schlegel-like individuals mate preferentially with each other, so pairing appeared to be nonrandom for the color of the throat feathers. This type of observation acts as a natural experiment and confirms the role of head patterns as a mechanism for reproductive isolation. The "white throats" are much less numerous at Macquarie Island than first believed, and their proportion differs considerably on different parts of the island. According to Shaughnessy's counts, the percentage can be estimated at 56% on the east coast of Macquarie Island (1641 individuals, six colonies) and only 9% on the west coast (803 individuals, five colonies).

This interesting admixture of phenotypically Schlegel and Macaroni penguins raises several questions. Is the genotype of white-throated individuals advantageous on one side of Macquarie and disadvantageous on the other, as well as perhaps at other localities? Is the difference in phenotype frequency between sexes due to a gene on the sex chromosome, or to the differential selection of white-throated males by females with both phenotypes? Are the percentages stable or changing? The only certain fact is that current data are insufficient to provide adequate answers. On taxonomic grounds, one cannot argue for a separate species with certainty. The similarity between songs and the heterogeneity of the Macquarie population does not even allow consideration of Schlegel penguins as a geographical race or a subspecies, until further evidence is provided. Additional evidence suggests geographic variation of Macaroni penguins. Song recordings provided by the BIRS from South Georgia ($n = 10$) were different from those of Crozet and Kerguelen ($n > 50$). In the South Georgia population, the rhythm of the song is more rapid and the voice slightly lower pitched. Therefore, isolated populations seem to exist in these localities, and their distributions need to be defined.

Finally, the monospecific genus *Megadyptes antipodes*, the Yellow-eyed penguin, can be compared to the different species of *Eudyptes*. The different song characteristics of the Yellow-eyed penguin lie within the variability of the *Eudyptes* species. This is particularly evident for the main and maximum frequencies. However, the phrase periods, their length and above all the number of syllables set it apart from *Eudyptes* songs. Song differences were greater between the Rockhopper penguins of St. Paul (*E. chrysocome moseleyi*) and Crozet (*E. chrysocome chrysocome*) than between the latter and Yellow-eyed penguins. However, we must be much more cautious than Chappuis (1969) when applying bioacoustics to systematics. Leaving the population level for the species, genus, and family, selective pressures increase and convergences render homologies more and more difficult to establish. Knowledge of results acquired in other domains are imperative, as well as of the detailed biology of species. The comparative study of courtship displays provides indications of affinities mainly at the species and within the generic levels, and should augment acoustic studies, to produce insights about the genetic isolation of populations.

Conclusions of Jouventin (1982) on the subject of speciation in penguins are summarized in Fig. 6.16 and they clarify the mechanisms of speciation. The principle is that of Mackintosh (1960). Continuous lines join reproductive areas. Species and subspecies are differentiated according to our conclusions. When speciation seems to be due only to geographic isolation, it is suggested by dotted lines. Cold currents are indicated by arrows and are all used by penguins. The 10° and 20°C sea isotherms explain, firstly, the colonization of only the western coast of America and Africa by *Spheniscus* penguins, and secondly the distribution of penguins in concentric circles. As shown for Rockhopper penguins, speciation follows latitudinal gradients rather than interisland proximity and stems from adaptation to different sea temperatures. A cross section of the diagram is presented in Fig. 6.17, showing the link between mean latitude (i.e., sea temperature) and phylogenetic relationships. Differentiation between species occurred mainly between 40° and 50° South latitude, centering on the zone surrounding New Zealand. The latter area contains the greatest number of original species, particularly the Little penguins, seemingly the most primitive. On the other hand, the genera *Spheniscus* and *Aptenodytes*, inhabiting opposite extremes of the distribution range of the family, appear to have the most differentiated behavior patterns. So penguins

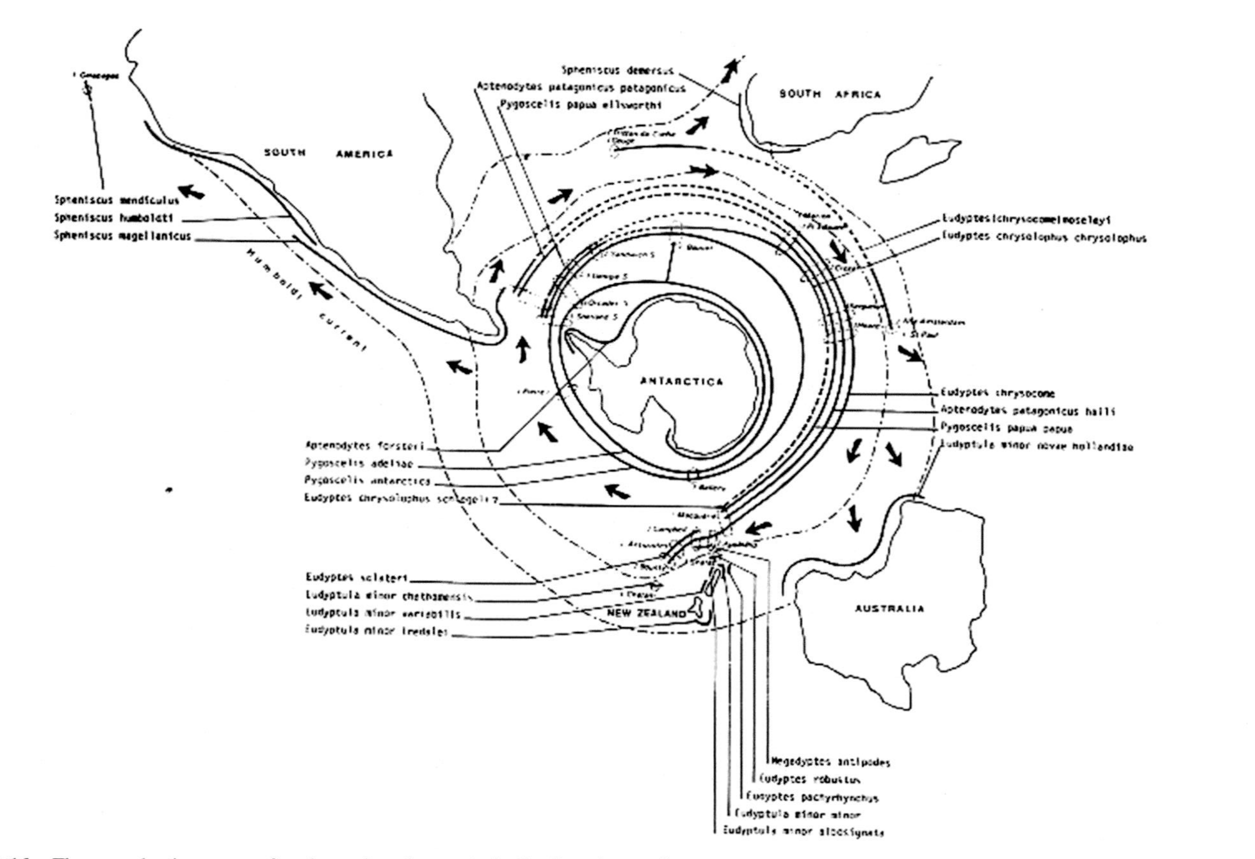


**FIGURE 6.16** The reproductive range of each species of penguin is distributed according to water temperature (as in Rockhopper penguins) and in some cases distances between breeding islands for species with limited vagility (as in Gentoo penguins). Cold currents and 10 and 20°C sea isotherms are shown. *From Jouventin, P., 1982. Visual and vocal signals in penguins, their evolution and adaptive characters. Advances in Ethology* ***24****, 1–149.*

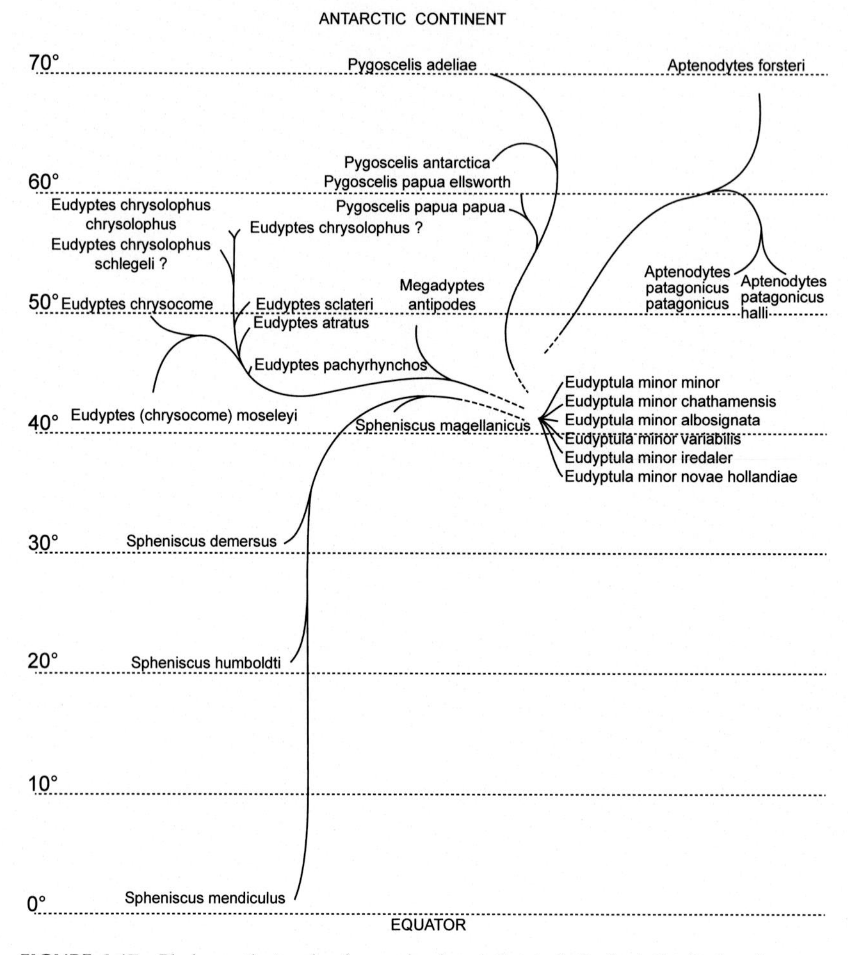


**FIGURE 6.17** Phylogenetic trends of penguins in relation to latitudinal distribution (i.e., sea-surface temperatures in the Southern hemisphere). *From Jouventin, P., 1982. Visual and vocal signals in penguins, their evolution and adaptive characters. Advances in Ethology* ***24****, 1–149.*

would seem to originate from the New Zealand zone, contrary to the historical hypothesis of an Antarctic origin.

In conclusion, the genus *Aptenodytes*, as shown previously, clearly stands apart from the five others. The two *Aptenodytes* species are closely related phylogenetically but differ greatly in their way of life. The study of their behavior confirms the likelihood of only one subspecies of King penguin in the Falkland Islands, *A. patagonicus patagonicus*. The genus *Spheniscus* is more derived than previously thought and is not closely related to the highly homogeneous and primitive genus *Eudyptes*. The genus *Pygoscelis* is a variegated group as its three species, especially *P. adeliae* and *P. papua*, are very different, the latter species being very differentiated from the other two species. No indication of a subspecies was found in *P. adeliae* and *P. antarctica*, while *P. papua* seems to exhibit at least two subspecies or species. The monospecific genus *Megadyptes* is closely related to the genus *Eudyptes* with six species: *E. pachyrhynchus*, *E. sclateri*, *Eudyptes atratus* (=*robustus*), *Eudyptes chrysolophus*, *E. chrysocome*, and *Eudyptes moseleyi*. A geographical race of *E. chrysocome* (subspecies *chrysocome*, which lacks a fleshy gape) inhabits Staten Island and should also be found on the Falkland Islands. Furthermore, a geographical race of *E. moseleyi* is present at Gough and perhaps Tristan da Cunha Islands (and is perhaps a subspecies). A morph of *E. chrysolophus* is perhaps found mainly at Macquarie (corresponding to the species or subspecies *E. chrysolophus schlegeli* of some systematists), and an undescribed geographical race in South Georgia (and perhaps neighboring islands), thus confirming the possible phylogeny of penguins (Figs. 6.16 and 6.17; from Jouventin, 1982).

## DNA RESULTS

New DNA studies of the phylogeny of penguins changed the scope of research on speciation of penguins (Fig. 6.18) giving more robust results on the link between behavior and systematics. This new form of data was used to test our hypotheses from 40 years previously, and based solely on behaviors (for example, for Rockhopper penguin songs; Fig. 6.19). So, after the acoustical and optical ways to study the behavioral ecology of penguins, we opened a third research track focused on evidence of speciation among populations of penguins, especially Rockhopper and Gentoo penguins (Jouventin et al., 2006; de Dinechin et al. 2007, 2009, 2012).

The most interesting of our predictions was to test whether there were several Rockhopper penguin species. For this, there were measurements of behavioral signals from several populations that were different for Northern and Southern groups of Islands where the penguins breed. The signals were both visual (the length of the crests) and vocal (very different nuptial songs). As it turned out, genetic distances and phylogenetic analyses showed a clear split between the two Rockhopper penguins with Northern and Southern breeding distributions (Fig. 6.20; Jouventin et al., 2006). These results added further support to the proposal of two Rockhopper penguin taxa, often considered subspecies (*Eudyptes chrysocome chrysocome* and *E. chrysocome moselyei*). The consistent split between these forms suggests that they should be recognized as two true species, *E. chrysocome* and *E. moseleyi*. A similar molecular result that separated Northern and Southern rockhopper penguins was obtained by Banks et al. (2006), but with the addition of Rockhopper penguin specimens from the Falkland Islands. The latter was a more moderate split with specimens of the Crozet and Kerguelen Archipelagos. This last split, perhaps not surprisingly, corresponds to the differences in song structure and

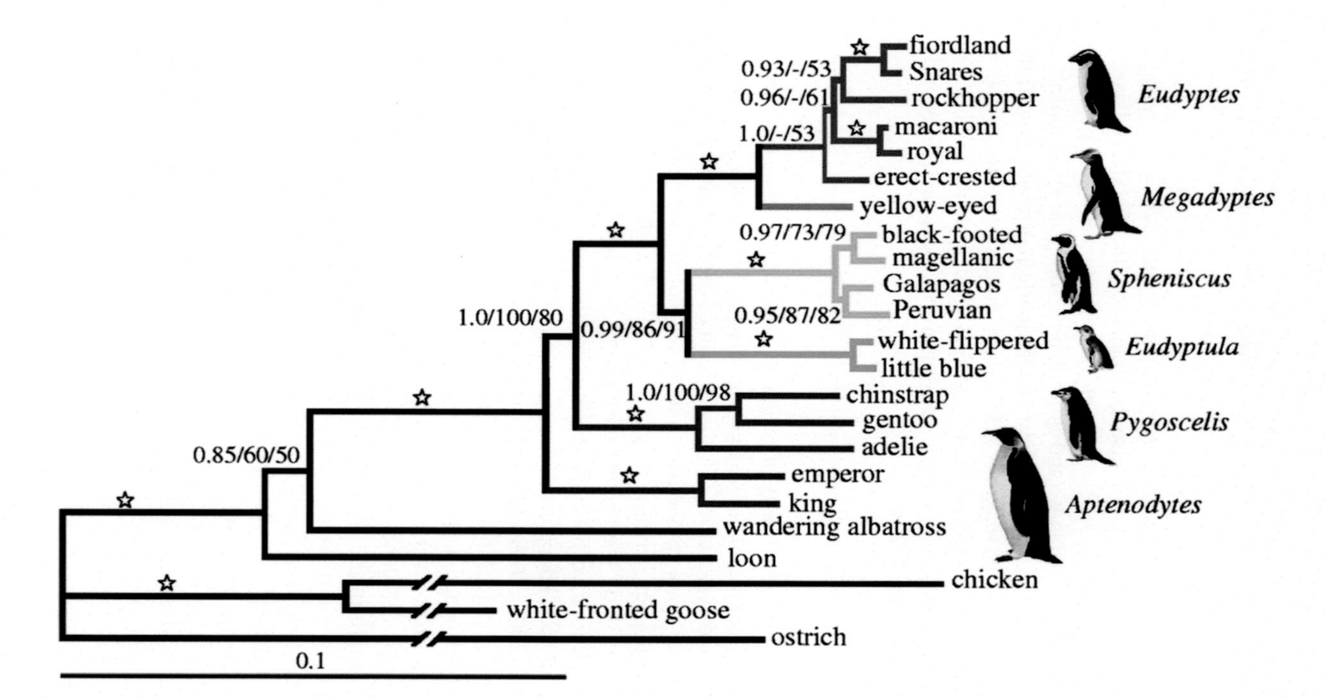


**FIGURE 6.18** DNA analyses clarify the global systematics of penguins. *From Baker, A.J., Pereira, S.L., Haddrath, O.P., Edge, K.A., 2006. Multiple gene evidence for expansion of extant penguins out of Antarctica due to global cooling. Proc. R. Soc. B Biol. Sci. 273, 11–17.*

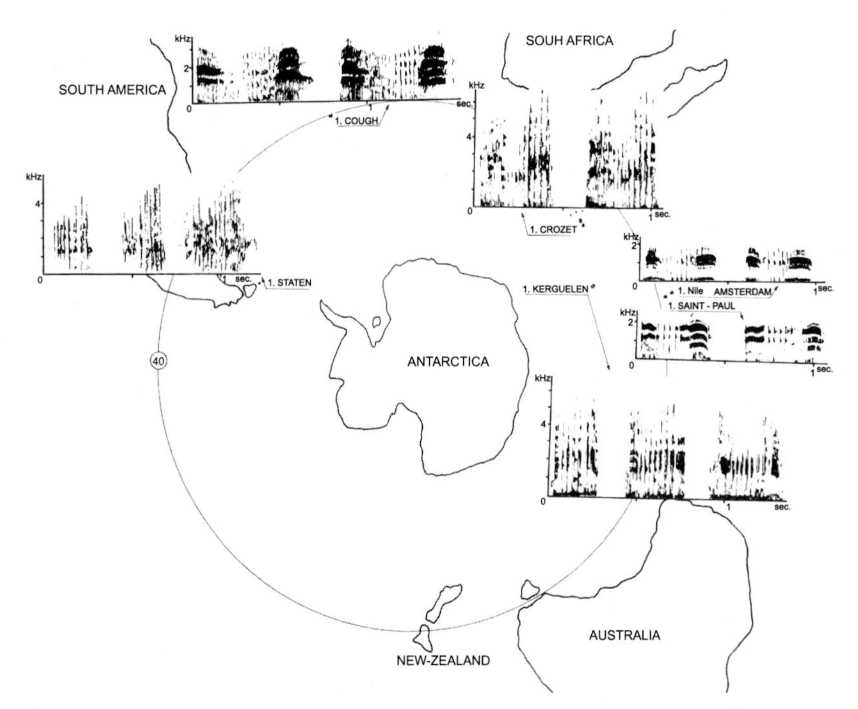


**FIGURE 6.19** Rockhopper penguin songs according to latitude fit into two categories, similar Gough–Amsterdam–Saint Paul Island populations, and Possession–Kerguelen–Staten Island populations. Speciation of Rockhopper penguins was more influenced by sea temperatures than by distance between islands. Behavioral mechanisms of sexual isolation, such as the song and the crest, maintain the isolation between Rockhopper penguins adapted to different water masses. *From Jouventin, P., 1982. Visual and vocal signals in penguins, their evolution and adaptive characters. Advances in Ethology* ***24****, 1–149.*

fleshy gape reported by Jouventin (1982) for Staten Island (just a few hundred km from the Falkland Islands) and the Crozet and Kerguelen Archipelagos. This difference possibly indicates a third Rockhopper penguin species, termed the Eastern rockhopper penguin (*Eudyptes filholi*) but closely related to the Southern Rockhopper penguin of the Falkland Islands (Fig. 6.21; de Dinechin et al., 2009).

Are the differences between the three possible Rockhopper penguin species a recent divergence due to broadly separated populations, or a functional difference associated by reproductive isolation and adaptations to specific niches associated with differences in water masses? First, the division of Macaroni and Rockhopper penguins occurred about 1.8 million years ago (de Dinechin et al., 2009). Northern and Southern Rockhopper penguins in different water masses then diverged about 0.9 million years ago. The Southern and Eastern Rockhopper penguins diverged about 0.5 million years ago, perhaps caused by a blockage of gene flow between these populations that

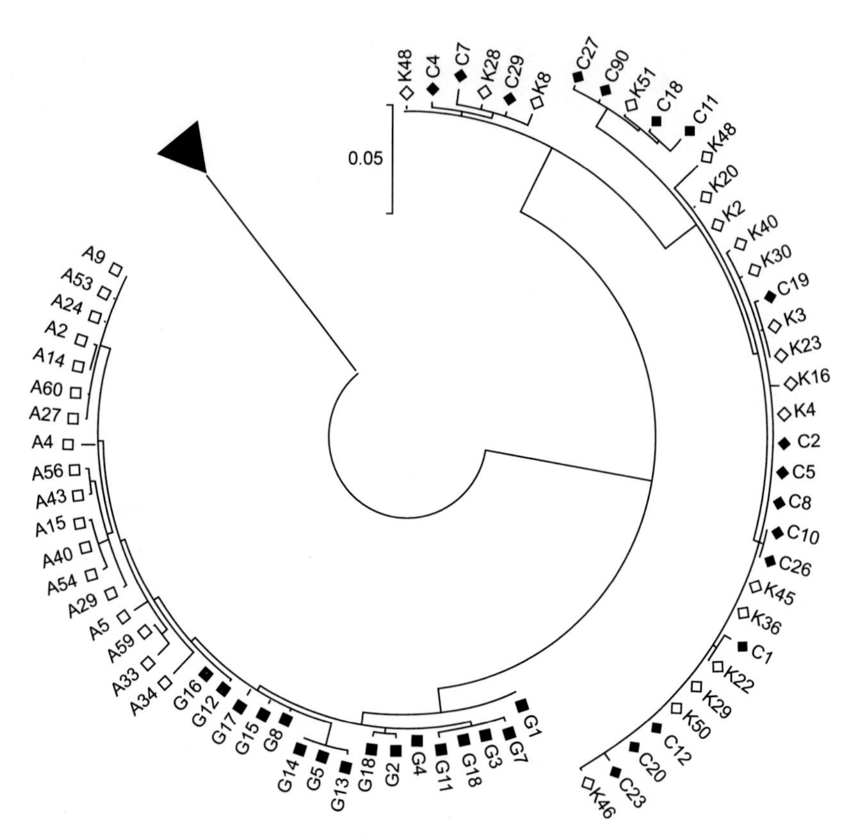


**FIGURE 6.20** Three clades and at least two species of Rockhopper penguins, from an analysis of the control region of mitochondrial DNA. Shown is a maximum likelihood phylogram with bootstrap values. Two species at least were predicted 35 years previously from differences in nuptial displays. *From Jouventin, P., Cuthbert, R.J., Ottvall, R., 2006. Genetic isolation and divergence in sexual traits: evidence for the northern rockhopper penguin* Eudyptes moseleyi *being a sibling species. Mol. Ecol. 15, 3413–3423.*

was due to icing up of the Antarctic Peninsula and southern Argentina and Chile during a glacial period. Finally, the St. Paul and Amsterdam Islands were colonized soon after their uplift from the sea floor about 0.25 million years ago. The last forms are not greatly divergent, nor perhaps are the Southern and Eastern forms sufficiently diverged to be recognized as species. Nonetheless, the divergence in mating signals found between the Northern and Southern Rockhopper penguins seems to have occurred relatively recently (according to the mitochondrial DNA analyses; de Dinechin et al., 2009).

Thus, the behavioral changes (crest and voice) may have been enough to isolate the Northern and Southern Rockhopper penguin species without substantial associated morphological differentiation. These systematic results may have important conservation consequences, since *E. moseleyi* is far less

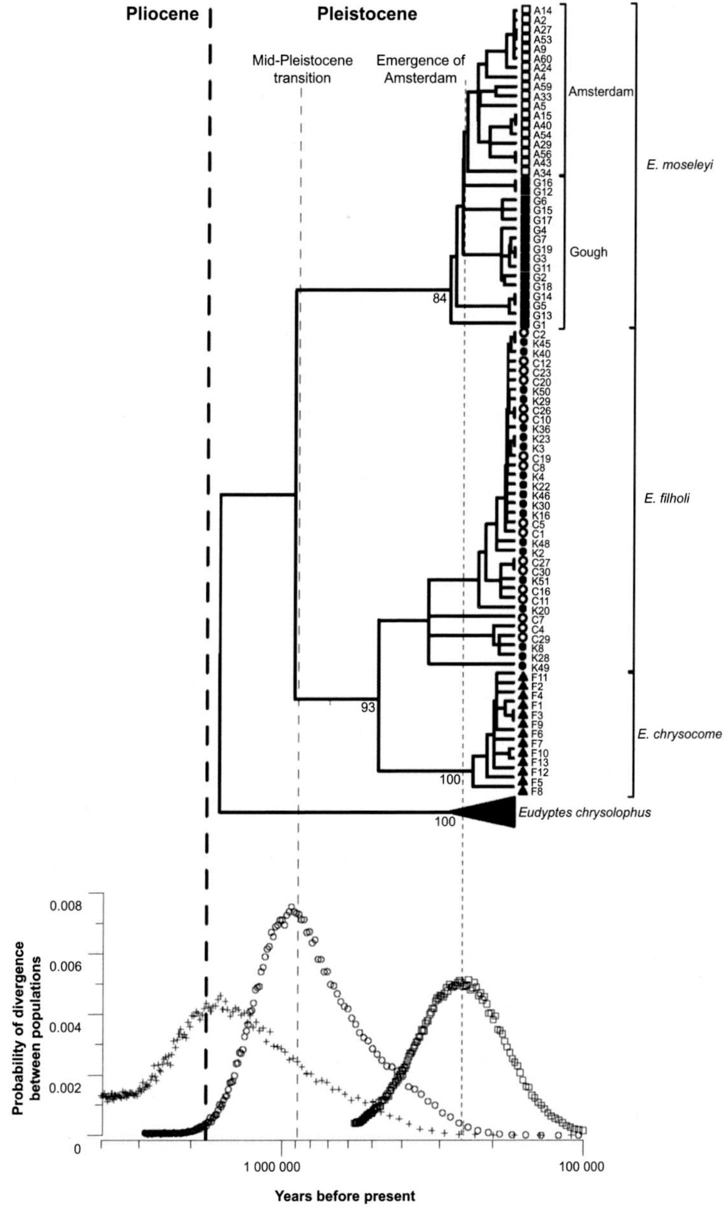


**FIGURE 6.21** Clades of Rockhopper penguins, produced from a different DNA analysis, confirming the previous mitochondrial DNA analysis. *From de Dinechin, M., Ottvall, R., Quillfeldt, P., Jouventin, P., 2009. Speciation chronology of rockhopper penguins inferred from molecular, geological and palaeoceanographic data. J. Biogeography 36, 693–702.*

abundant than the southern forms. The results also have important implications for understanding the mechanisms of divergence and speciation in penguins via the link between nuptial displays and marine ecology (Fig. 6.14). The different voice and head feathers in each group of Rockhopper penguins, acted as ethological mechanisms of sexual isolation, to separate the two groups of penguins. These two groups are narrowly adapted to two water masses of very different temperatures and salinities, implicating different adaptations for feeding, diving, and thermoregulating in different environmental conditions. Nevertheless, penguins can swim across the subtropical convergence, a major ecological boundary for marine organisms, but the vocal and visual signals used in the nuptial display erected a gene flow barrier between northern (subtropical) and southern (sub-Antarctic) populations.

Behavioral and ecological evidence supports a taxonomic division into Northern and Southern rockhopper penguins. At Kerguelen Island, field observations revealed a Northern rockhopper penguin from Gough Island (easy to distinguish by their song and crest, and confirmed by DNA profile) that migrated into a population of Eastern rockhopper penguins (de Dinechin et al., 2007). This migrant, however, was unable to mate with the Eastern rockhopper penguins that populate Kerguelen Island. Also, Thiebot et al. (2013) used satellite transmitters to follow the feeding paths of three species of crested penguins (Macaroni, Northern rockhopper, and Eastern rockhopper penguins). They found differences in foraging trips according to sea-surface temperature, with the Northern rockhopper penguin favoring warmer seas and Eastern rockhopper penguins favoring cooler seas. Future research should examine further the degree of population and niche differentiation between Southern and Eastern rockhopper penguins, and in particular the associations of the Rockhopper penguins from Macquarie Island.

Another case of speciation concerns Gentoo penguins. These penguins had historically been divided into two subspecies, the larger and northern *P. papua papua* and smaller *P. papua ellsworth* (Forster 1781; Murphy, 1936). But examining the nuptial songs of Gentoo penguins from different populations, Jouventin (1982) found evidence of differences between the southern Indian Ocean basin (Maquarie, Crozet, and Kerguelen Islands) versus the south Atlantic (Falkland, South Orkney, and South Georgia Islands). Gentoo penguins are coastal feeders, and not truly pelagic, so we might expect a pattern of "isolation by distance" to explain genetic divergence of populations. We tested this idea with molecular markers from several populations of Gentoo penguins. The polar front divides cold Antarctic waters from the warmer sub-Antarctic water mass (Fig. 6.22 from Jouventin, 1982), and Gentoo penguins breed on either side of the polar front. A phylogenetic tree was again constructed from a sample of 110 Gentoos penguins with 58 haplotypes from the control region of the mitochondrial DNA (Fig. 6.23; de Dinechin et al., 2012). The current systematics division into two subspecies based on differences in body size (morphological differences) was not supported. Rather, the division between

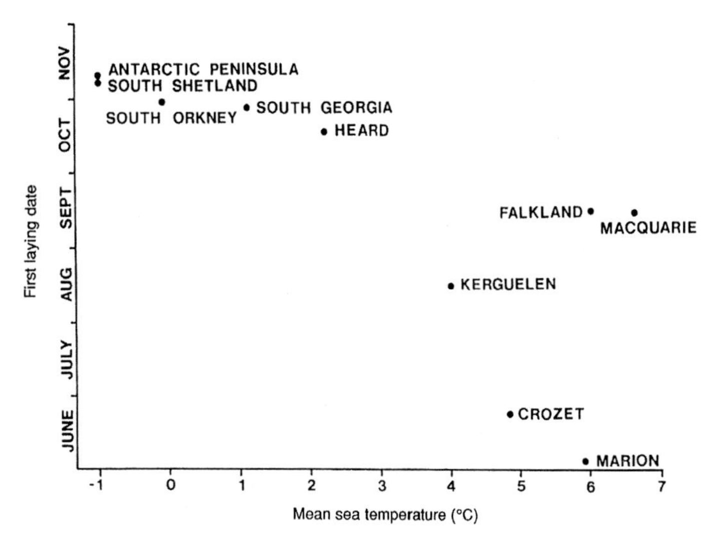


**FIGURE 6.22** For Gentoo penguin populations (similar to Rockhopper penguins), a big difference was found in the breeding biology, specifically laying dates as sea-surface temperatures surrounding islands change. These results support at least two distinct species following a north–south pattern (with a gap between Antarctic and sub-Antarctic populations). *From Jouventin, P., 1982. Visual and vocal signals in penguins, their evolution and adaptive characters. Advances in Ethology **24**, 1–149.*

populations in the Indian and Atlantic Oceans was great enough to justify taxonomic revision, with at least three distinct clades. Two of these groups were in the respective sub-Antarctic and Antarctic zones of the Atlantic Ocean, and a deeply divergent and unnamed third clade occurred in the sub-Antarctic Indian Ocean (Fig. 6.24; de Dinechin et al., 2012). A subsequent test by Vianna et al. (2017) found the same division of Gentoo penguins into several clades, but increased the number of islands sampled and found additional diversification into still more apparently isolated groups. In particular, populations of Gentoo penguins in the Crozet and Kerguelen Archipelagos were highly divergent. It is an unexpected result because these populations breed in the same southern Indian Ocean at a distance of only 1340 km (Fig. 6.25).

How can we explain the divergence of such close populations of breeding Gentoo penguins? Vianna et al. (2017) suggested an oceanographic explanation in relation to paleobiogeography, proposing a similar process of ecological speciation between Rockhoppers and Gentoo penguins. We suggest a different mode of speciation between Gentoo and crested penguins, due to their differences in marine ecology. Highly pelagic species like Rockhopper penguins are constrained mainly by marine boundaries and sea-surface temperatures. The restricted coastal foraging ranges of Gentoo penguins, however, induce a stronger geographical isolation, rather than an oceanographic one

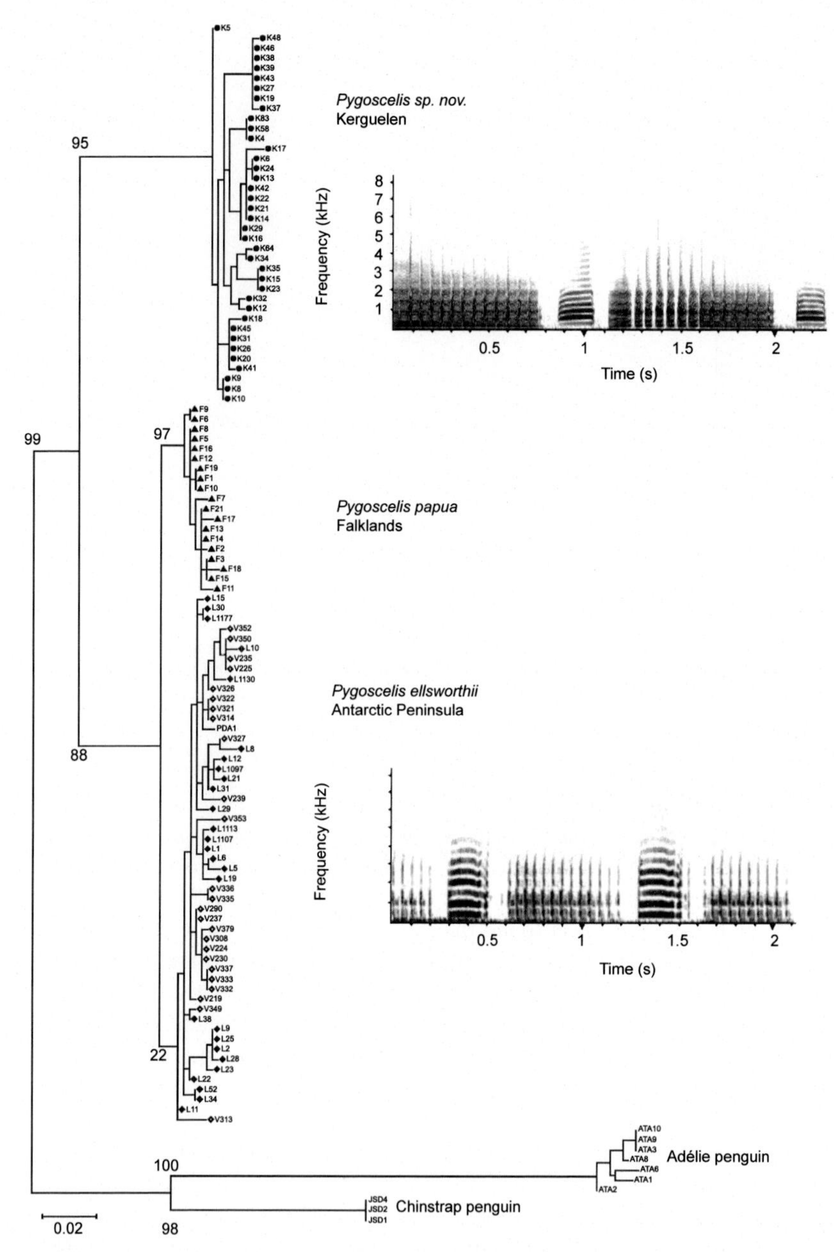


**FIGURE 6.23** Among Gentoo penguin populations, analysis of the control region of mitochondrial DNA maximum likelihood revealed two major clades in the Atlantic and southern Indian Oceans. Shown is a phylogram with bootstrap values. Differences in songs recorded by Jouventin (1982) predated the molecular pattern. *From de Dinechin, M., Dobson, F.S., Zehtindjiev, P., Metcheva, R., Couchoux, C., Martin, A., Quillfeldt, P., Jouventin, P., 2012. The biogeography of Gentoo Penguins (*Pygoscelis papua*). Can. J. Zool. 90, 352–360.*

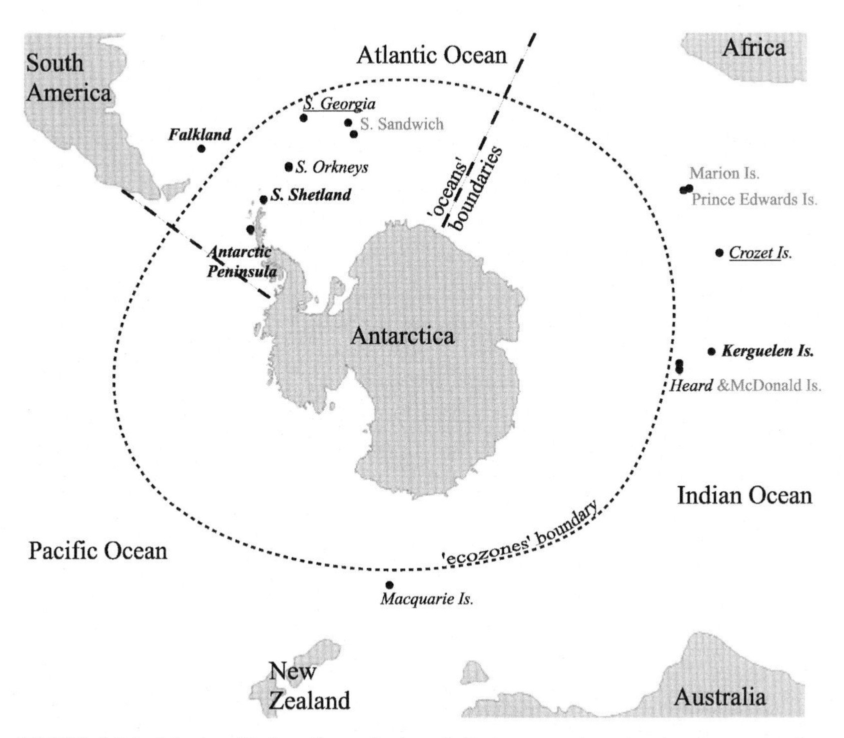


**FIGURE 6.24** Islands with breeding colonies of Gentoo penguins, showing two competing hypotheses of phylogenetic relationships, an ecozone hypothesis dividing populations into northern and southern forms and an ocean basin hypothesis dividing populations into forms bases on geographic proximity of islands. In fact, both hypotheses are confirmed and complementary. Macquarie Island is shown but was not surveyed for mitochondrial DNA. *From de Dinechin, M., Dobson, F.S., Zehtindjiev, P., Metcheva, R., Couchoux, C., Martin, A., Quillfeldt, P., Jouventin, P., 2012. The biogeography of Gentoo Penguins (*Pygoscelis papua*). Can. J. Zool. 90, 352–360.*

based on sea temperatures. Although both Gentoo and Rockhopper penguins are strongly genetically separated in the southern Indian Ocean, the former shows few differences in songs and head patterns between the two sibling species when, in the latter, the Northern and Eastern rockhopper penguins are very different in their nuptial displays. These ethological mechanisms of sexual isolation appear only in Rockhopper penguins because these pelagic penguins have adapted to different masses of water. However, individuals are able to reach colonies of the other sibling species to try to mate (as we saw at Kerguelen Island, recorded by de Dinechin et al., 2007).

Gentoo penguins in the same ocean seem too philopatric to disperse between the Crozet and Kerguelen Archipelagos. Consequently, no ethological mechanisms of sexual isolation have been selected to keep the two populations of Gentoo penguins from interbreeding. The distance among isolated oceanic archipelagos explains mainly the distribution of genetically more

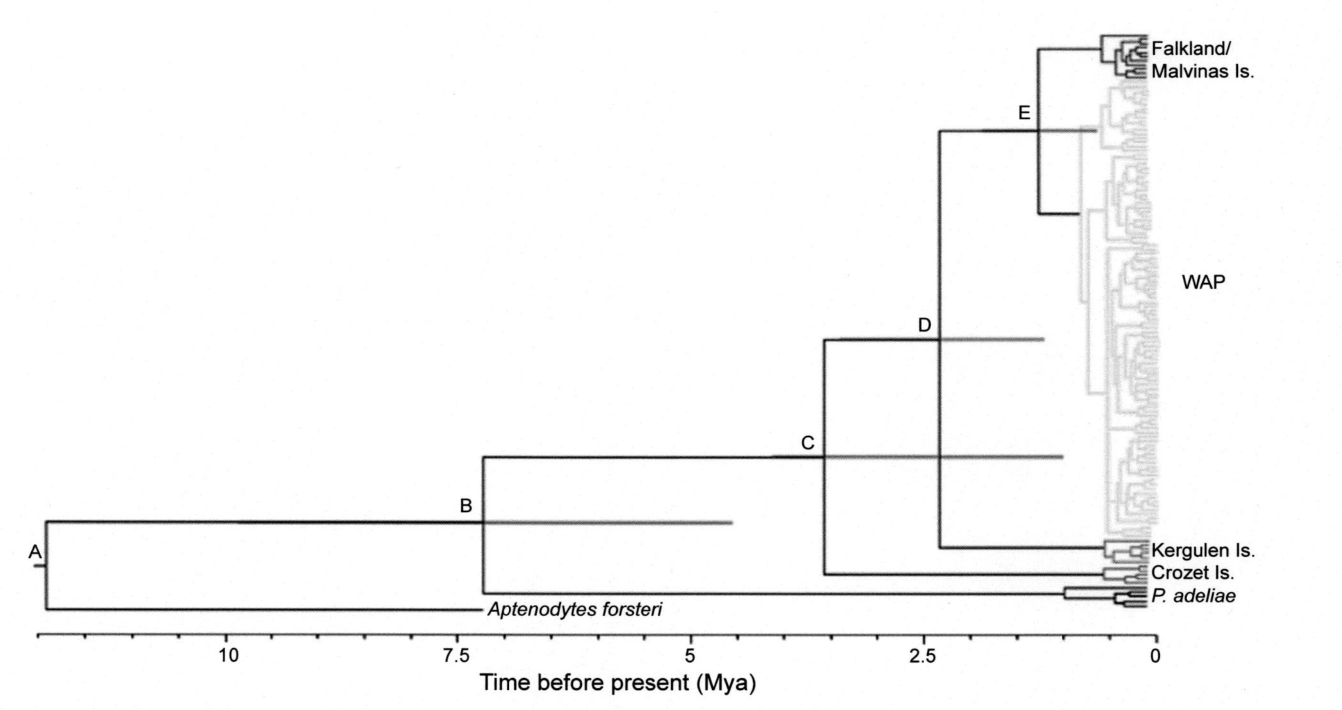


**FIGURE 6.25** Phylogenetic tree for Gentoo penguins, from mitochondrial DNA data. Falkland Islands are in red, western Antarctic Peninsula and nearby islands in yellow, Kerguelen Island in green, and the Crozet Archipelago in blue. Letters show major splits in the phylogeny, divergence time in millions of years ago. Purple lines indicate the Bayesian highest posterior density of the estimated time periods, a measure of variability. Crozet and Kerguelen populations from the same Indian Ocean are surprisingly very different. *Modified from Vianna, J.A., Noll, D., Dantas, G.P.M., Petry, M.V., Barbosa, A., Gonzalez-Acuna, D., Le Bohec, C., Bonadonna, F., Poulin, E., 2017. Marked phylogeographic structure of Gentoo penguin reveals an ongoing diversification process along the Southern Ocean. Mol. Phylogenet. Evol. 107, 486–498.*

differentiated populations of Gentoo penguins. The Falkland Island Gentoo penguins diverged from those of the Antarctic Peninsula and associate islands an estimated 1.27 million years ago (Vianna et al., 2017), longer than the divergence of Northern and Southern rockhopper penguins. The divergence of Gentoo penguins of the Crozet and Kerguelen Archipelagos occurred earlier, an estimated 3.57 million years ago. Yet, these populations have hardly diverged in their nuptial songs. The explanation for this is likely a lack of recontact between these distant populations of coastal penguins. It will be interesting to analyze the DNA of Gentoos of other islands such as Heard and Macquarie, to see if they will also prove to be highly divergent due to their distances from other islands.

These mitochondrial DNA results should change the current taxonomy and substantially increase the number of penguin species. But moreover, the genetic systematic analyses reveal the directing role of the behavior in evolution, pointing to the relationship between courtship display and song, and the divergence and niche specialization of populations over time. In the ecologically different Gentoo and Rockhopper penguin groups, we see the mechanisms of speciation in all penguins. Geography and oceanography are always two selective influences on populations of penguins but with different intensities according to the species.

In pelagic species, such as crested penguins, oceanography appears to be the main ecological parameter that influences population divergences. But in the coastal Gentoo penguins, geography (viz., the distances among islands and perhaps the direction of major ocean currents) appears to be the stronger influence on speciation. Consequently, all species of penguins are distributed around the Antarctic continent like a target due to the decreasing water temperatures around the South Pole, but ecological dispersion may result in an overlap of greater or lesser genetic differentiation among islands. Each species is narrowly adapted to its water mass thermally and ecologically through marine prey. When they move easily, the marking of the head as well as the nuptial song isolate these penguins, preventing interbreeding on land and the loss of their unique genetic adaptations (Fig. 6.16). When foraging is off shore, rather than pelagic, like those of Gentoo penguins on the Crozet and Kerguelen archipelagos, nuptial displays show no special divergence.

In the relationship of animal and environment (and here more than the morphological changes used classically in ornithology and systematics), behavior plays a major role in speciation. Alterations in behavior can induce the splitting of a population into two species and even sometimes facilitate the transfer of a subpopulation to a new ecological niche (Mayr, 1970). A modification of "wiring" in stereotypic displays seems more labile, in an evolutionary sense, than changes in morphological structures. In Gentoo and Yellow-eyed penguins, for example, the higher fleeing distance from man has probably preserved populations from extinction. In Rockhopper penguins, the modifications of song pitch and tuft length sufficed to isolate two groups of populations genetically. The rest of their morphology has remained practically identical, whereas eco-physiological adaptations have diverged. So modifications of behavioral patterns seem more economical, thereby reducing the "cost" of evolution.

## SCIENTIFIC AND ETHICAL QUESTIONS

What is the future of penguins? Historically, the number of penguin species was much greater than it is today. After the disappearance of dinosaurs, many marine reptiles became extinct, but the penguin family diversified. Their species numbers increased during the last several million years. But today their populations are threatened and reduced, particularly in places where humans are especially active. Curiously, the number of species needs to be increased due to the discovery of sister species that were detectable by their differences in nuptial displays, and later confirmed by differences in DNA patterns.

Due to the lack of terrestrial predators and the isolated parts of the globe where they live, penguins are usually easy to approach and easy to study, once transportation to their distant islands is achieved. Observing these nonflying birds is a good opportunity to combine field work in natural habitats with close observations. Most penguins act as though they do not care about the presence of researchers, as long as they are several meters distant. Some species of penguins behave as though curious about human observers, even gently poking at them with the beak. Consequently, observations of penguins are often easier than for shyer animals observed in the wild in Europe or America. A good example of the latter might be wolves, clearly fearful of humans, and thus significant behaviors like mutual assistance can be missed (Jouventin et al., 2016). The crowded colonies of penguins are also unique because the behavior of numerous adults allow multiple repeated observations. Such data allow quantitative generalizations about penguin behaviors.

Natural observations are, of course, only the first layer of behavioral study, and we can go much deeper in our investigations. Using working hypotheses, we can examine the ecological context of behaviors. Finally, our hypotheses about the interaction of behaviors and the environment can be tested by conducting carefully designed comparative and experimental studies in the field. In this way, we can go beyond describing how penguins behave to examine the question of why penguins exhibit particular behaviors. The series of experiments that we summarized in this book were fruitful for understanding much about the communication processes of penguins. For example, penguins are the best studied group of birds for acoustic individual recognition.

Such scientific progress, of course, has disturbing impacts on populations of penguins in the field. The behavioral impact, however, is actually rather minor and perhaps even benign when compared to commercial treatments of wildlife in general during historical times. Past commercial activities, such as

**FIGURE 6.26** Scientific studies now use temporary markings such as plastic numbered bands to identify individual penguins.

sealing and whaling, clearly had a greater detrimental impact on penguins than behavioral and ecological research. When Pierre Jouventin began studying Emperor and Adélie penguins in 1969, there was little concern about the impact of monitoring techniques, even those used to study penguin demography. But in the 1980s, the common marking technique of banding the flippers of Adélie and Emperor penguins with numbered metal bands fell out of favor, due to observations of injuries to molting birds. After 2000, an committee was created by the French Polar Institute to insure ethical treatment in all of their scientific programs. Since 1999, when Stephen Dobson joined the penguin research program, and strictly for all French research since 2010, individual penguins have been followed using either plastic bands that can be easily removed from the birds at the end of a study period (Fig. 6.26), or by painting colored numbers on the penguin's chests (with dye that lasts until the next feather molt).

The problem of increased mortality from using metal flipper bands for monitoring King penguins was clearly demonstrated by Saraux et al. (2011). An alternative technique is to implant penguins with passive radio-frequency identification (RFID) tags. An electronic scanner can then be used to read the tags. RFID tags, however, are only readable at a short distance of some tens of centimeters. One solution is to fence a colony in, and provide openings that are equipped with electronic RFID readers. This procedure, however, limits the free movement of birds to and from the colony. It remains an unsolved problem to identify and follow unconstrained and RFID marked penguins (for robot trials, see Le Maho et al., 2014). Critics have also argued against radio-tracking of diving birds and mammals at sea. No doubt some species may be equipped with satellite transmitters too frequently. The cultural concerns of scientists have thus changed considerably over the years, rendering studies more cautious over concerns about marking diving animals, such as penguins, seals, and whales.

When we played backed the nuptial song of the partner to an incubating penguin, or of the parent to a chick, we fooled birds about the presence of their partner when it was absent. When we masked a colored ornament of a King penguin to know if it was used in mating and was attractive to potential partners, we sometimes prevented the experimental penguins from reproducing for a year. Do behavioral experiments need to be stopped, even if they don't kill or injure penguins? These are serious questions that should be pondered by researchers, always searching for way to make research less invasive or harmful for the animals. Although we disturbed some individuals, we worked in colonies of several thousand pairs (in King penguins, about 100,000 pairs in the colony at Cape Ratmanoff, Kerguelen Island, where we conducted several experiments and observations). Several penguins were unable to breed because of our experimental manipulations. To learn more about penguin communication and to increase our knowledge of behavioral adaptations of the birds, we had to accept that we disturbed some scores or even a 100 individuals at a time. However, we limited our study activities not only to a small subset of the penguin population, but also to a small area in the huge breeding colonies where we conducted research.

Our experiments on penguin communication thus seem relatively harmless and justified on ethical and scientific grounds. Results obtained in the wild are better than those recorded in captivity, where rearing conditions are artificial. Nevertheless, we tested captive penguins when possible, such as our tests of whether the origin of the beak color in Gentoo penguins was from their diet (Jouventin et al., 2007b). It would have been impossible to conduct this study in the field, and the feeding experiment was conducted without disturbance to the captive animals. So our series of experiments on communication in penguins constituted a long-term commitment to the study of communication in the wild. Besides the small number of manipulated birds, experiments produce more powerful results than more natural comparative studies. Thus, the experiments that we used to answer many of our questions do not need to be repeated. One or a few replications are usually enough if the result was clear and the sample of tested penguins large. There is no question, however, that any scientific study on animals, even on the communication in penguins, has an impact that should be minimized.

Part of the reason that we studied the behavioral ecology of seabirds was to protect them and the islands where we worked for so many years. The need for protection is obvious from many years of historical exploitation. For example, during three years of a voyage in the 1860s, sailors from five ships killed about 450,000 penguins in two of the French archipelagos, harvesting their skin and their fat (Chansigaud 2007, page 177). The Kerguelen, Crozet, and Amsterdam Archipelagos are now legally protected by the Administration of Terres Australes et Antarctiques Françaises (Territory of the French Southern and Antarctic Lands), thanks to a meeting organized by Hubert Reeves between Pierre Jouventin and the French Minister of Ecology, Nelly Ollin. A natural reserve was officially created in October 2006 covering 7781 $km^2$ of terrestrial habitat and 1,570,000 ha of marine habitat surrounding these islands in the middle of the southern Indian Ocean. These remote islands have no inhabitants, except for the about 100 researchers and technicians that rotate through each year. Pierre Jouventin served as the expert biologist in the French delegation to the Madrid Protocol, an international agreement for a moratorium until 2048 on the exploitation of mineral resources in the Antarctic. J.Y. Cousteau led the international public discourse for this effort, and France was joined by Australia and many other countries with interests in the Antarctic and its important natural resources. The knowledge of marine birds and mammals was thus useful for protecting the Antarctic continent. It is our hope that the future of the millions of penguins will be ensured for many years to come. We still have much to learn about how natural selection has produced their numerous sophisticated adaptations for communication.

Due to their bipedal and human-like locomotion on land, it is easy to anthropomorphize about penguins, especially when observing their nuptial displays. But as we have seen, their behavioral adaptations are much more sophisticated than we might imagine from fanciful tales and cartoons. Penguins are not so difficult to understand if we observe them for a long time, conduct careful experiments, and try to "think" about their world not as humans, but as penguins.

# References


Ainley, D.G., 1975. Displays of Adelie penguins, a reinterpretation. In: Stonehouse, B. (Ed.), The Biology of Penguins. Macmillan, London, pp. 503–534.

Ainley, D.G., DeMaster, D.P., 1980. Survival and mortality in a population of Adélie penguins. Ecology 61, 522–530.

Ainley, D.G., LeResche, R.E., Sladen, W.J.L., 1983. Breeding Biology of the Adélie Penguin. University of California Press, Berkeley.

Andersson, M., 1986. Evolution of condition-dependent sex ornaments and mating preferences – sexual selection based on viability differences. Evolution 40, 804–816.

Andersson, M., 1994. Sexual Selection. Princeton University Press, Princeton, New Jersey.

Angelier, F., Barbraud, C., Lormee, H., Prud'homme, F., Chastel, O., 2006. Kidnapping of chicks in Emperor penguins: a hormonal by-product? J. Exp. Biol. 209, 1413–1420.

Aubin, T., 1994. Syntana: a software for the synthesis and analysis of animal sounds. Bioacoustics 6, 80–81.

Aubin, T., Jouventin, P., 1998. Cocktail-party effect in King penguin colonies. Proc. R. Soc. B Biol. Sci. 265, 1665–1673.

Aubin, T., Jouventin, P., 2002a. How to vocally identify kin in a crowd: the penguin model. Adv. Study Behav. 31, 243–277.

Aubin, T., Jouventin, P., 2002b. Localisation of an acoustic signal in a noisy environment: the display call of the King penguin *Aptenodytes patagonicus*. J. Exp. Biol. 205, 3793–3798.

Aubin, T., Jouventin, P., Hildebrand, C., 2000. Penguins use the two-voice system to recognize each other. Proc. R. Soc. B Biol. Sci. 267, 1081–1087.

Bagshawe, T.W., 1938. Note on the habits of the Gentoo and Ringed or Antarctic penguin. Trans. Zool. Soc. Lond. 24, 185–306.

Baker, A.J., Pereira, S.L., Haddrath, O.P., Edge, K.A., 2006. Multiple gene evidence for expansion of extant penguins out of Antarctica due to global cooling. Proc. R. Soc. B Biol. Sci. 273, 11–17.

Banks, J., Van Buren, A., Cherel, Y., Whitfield, J.B., 2006. Genetic evidence for three species of rockhopper penguins, *Eudyptes chrysocome*. Polar Biol. 30, 61–67.

Barrat, A., 1976. Quelques aspects de la biologie et de l'écologie du Manchot Royal (*Aptenodytes patagonicus*) des iles Crozet. Comité National Français des Recherches Antarctiques 40, 9–52.

Barré, H., Derenne, P., Mougin, J.-L., 1976. Les "Gorfous de Schlegel" des iles Crozet. Comité National Français des Recherches Antarctiques 40, 177–189.

Beaglehole J.C. (Ed.), 1963. The *Endeavour* Journal of Joseph Banks 1768–1771. Two volumes. Second Edition. Angus and Robertson, Sydney, Public Library of New South Whales, New South Wales, Australia: Angus and Robertson.

Beecher, M., 1982. Signature systems and kin recognition. Am. Zool. 22, 477–490.

Beecher, M., 1989. Signaling systems for individual recognition – an information-theory approach. Anim. Behav. 38, 248–261.

Beecher, M., Campbell, S., Burt, J., 1994. Song perception in the song sparrow – birds classify by song type but not by singer. Anim. Behav. 47, 1343–1351.

Beletsky, L.D., 1983. An investigation of individual recognition by voice in female red-winged blackbirds. Anim. Behav. 31, 355–362.

Bennett, P.M., Owens, I.P.F., 2002. Evolutionary Ecology of Birds: Life Histories, Mating Systems and Extinction. Oxford University Press, Oxford.

Black, J.M., 1996. Introduction: pair bonds and partnerships. In: Black, J.M. (Ed.), Partnerships in Birds: the Study of Monogamy. Oxford University Press, Oxford, pp. 3–20.

Boersma, P.D., 1976. An ecological and behavioral study of the Galapagos penguin. Living Bird 15, 43–93.

Bonadonna, F., Cunningham, G.B., Jouventin, P., Hesters, F., Nevitt, G.A., 2003a. Evidence for nest-odour recognition in two species of diving petrel. J. Exp. Biol. 206, 3719–3722.

Bonadonna, F., Hesters, F., Jouventin, P., 2003b. Scent of a nest: discrimination of own-nest odours in Antarctic prions, *Pachyptila desolata*. Behav. Ecol. Sociobiol. 54, 174–178.

Bost, C., Jouventin, P., 1990. Laying asynchrony in Gentoo penguins on Crozet Islands – causes and consequences. Ornis Scand. 21, 63–70.

Bost, C., Jouventin, P., 1991. The breeding performance of the Gentoo penguin pygoscelis-papua at the northern edge of its range. Ibis 133, 14–25.

Bowmaker, J., Martin, G., 1985. Visual pigments and oil droplets in the penguin, *Spheniscus-humboldti*. J. Comp. Physiol. A 156, 71–77.

Brémond, J., Aubin, T., 1992. The role of amplitude-modulation in distress-call recognition by the black-headed gull (*Larus-ridibundus*). Ethol. Ecol. Evol. 4, 187–191.

Bried, J., Jouventin, P., 2001. The King penguin *Aptenodytes patagonicus*, a non-nesting bird which selects its breeding habitat. Ibis 143, 670–673.

Bried, J., Jouventin, P., 2002. Site and mate-choice in seabirds: an evolutionary approach to the nesting ecology. In: Schreiber, E.A., Burger, J. (Eds.), Biology of Marine Birds. CRC Press, New York, pp. 263–305.

Bried, J., Jiguet, F., Jouventin, P., 1999. Why do *Aptenodytes* penguins have high divorce rates? Auk 116, 504–512.

Brodin, A., Olsson, O., Clark, C.W., 1998. Modeling the breeding cycle of long-lived birds: why do King penguins try to breed late? Auk 115, 767–771.

Buhrman-Deever, S.C., Hobson, E.A., Hobson, A.D., 2008. Individual recognition and selective response to contact calls in foraging brown-throated conures, *Aratinga pertinax*. Anim. Behav. 76, 1715–1725.

Capuska, G.E.M., Huynen, L., Lambert, D., Raubenheimer, D., 2011. UVS is rare in seabirds. Vis. Res. 51, 1333–1337.

Carrick, R., 1972. Population ecology of the Australian Black-backed magpie, royal penguin and silver gull. In: Population Ecology of Migratory Birds: a Symposium. U.S. Department of the Interior Wildlife Research Report No 2, pp. 41–99.

Casey, C., Charrier, I., Mathevon, N., Reichmuth, C., 2015. Rival assessment among northern elephant seals: evidence of associative learning during male–male contests. Royal Society Open Science. http://dx.doi.org/10.1098/rsos.150228.

Chansigaud, V., 2007. Histoire de l'Ornithologie. Delachaux et Niestlé, Paris.

Chappuis, C., 1969. Apport de la Bio-acoustique en systematique. Alauda 37, 206–218.

Charrier, I., Mathevon, N., Jouventin, P., 2003. Vocal signature recognition of mothers by fur seal pups. Anim. Behav. 65, 543–550.

Cherry-Garrard, A., 1922. The Worst Journey in the World. Carrloo & Graf, London.
Clark, J.A., Boersma, P.D., Olmsted, D.M., 2006. Name that tune: call discrimination and individual recognition in Magellanic penguins. Anim. Behav. 72, 1141–1148.
Cook, J., 1785. A Voyage to the Pacific Ocean. H. Hughs, London.
Coté, S.D., 2000. Aggressiveness in King penguins in relation to reproductive status and territory location. Anim. Behav. 59, 813–821.
Coulson, J., 1968. Differences in quality of birds nesting in centre and on edges of a colony. Nature 217, 478.
Craig, K.J.W., 1944. White plumage of seabirds. Nature 153, 288.
Croxall, J.P., Rothery, P., 1995. Population change in gentoo penguins *Pygoscelis papua* at Bird Island, South Georgia: potential roles of adult survival, recruitment and deferred breeding. In: Dann, P., Norman, I., Reilly, P. (Eds.), The Penguins: Ecology and Management. Surrey Beatty and Sons Chipping-Norton, Australia, pp. 26–38.
Cuthill, I.C., Bennett, A.T.D., Partridge, J.C., Maier, E.J., 1999. Plumage reflectance and the objective assessment of avian sexual dichromatism. Am. Nat. 153, 183–200.
Dampier, W., 1699. A New Voyage Round the World. James Knapton, London.
Darwin, C.R., 1859. On the Origin of Species. John Murray, London.
Darwin, C.R., 1871. The Descent of Man, and Selection in Relation to Sex. John Murray, London.
Darwin, C.R., 1872. The Expression of the Emotions in Man and Animals. John Murray, London.
Davis, L.S., 1988. Coordination of incubation routines and mate choice in Adélie penguins (*Pygoscelis adeliae*). Auk 105, 428–432.
Davis, L.S., Hunter, F.M., Harcourt, R.G., Heath, S.M., 1998. Reciprocal homosexual mounting in Adelie penguins *Pygoscelis adeliae*. Emu 98, 136–137.
de Dinechin, M., Pincemy, G., Jouventin, P., 2007. A northern rockhopper penguin unveils dispersion pathways in the Southern Ocean. Polar Biol. 31, 113–115.
de Dinechin, M., Ottvall, R., Quillfeldt, P., Jouventin, P., 2009. Speciation chronology of rockhopper penguins inferred from molecular, geological and palaeoceanographic data. J. Biogeogr. 36, 693–702.
de Dinechin, M., Dobson, F.S., Zehtindjiev, P., Metcheva, R., Couchoux, C., Martin, A., Quillfeldt, P., Jouventin, P., 2012. The biogeography of Gentoo penguins (*Pygoscelis papua*). Can. J. Zool. 90, 352–360.
Derenne, P., Jouventin, P., Mougin, J.-L., 1979. The King penguin call *Aptenodytes-patagonica* and its evolutionary significance. Le Gerfaut 69, 211–224.
Dhondt, A., Lambrechts, M., 1992. Individual voice recognition in birds. Trends Ecol. Evol. 7, 178–179.
Dobson, F.S., Jouventin, P., 2003. Use of the nest site as a rendezvous in penguins. Waterbirds 26, 409–415.
Dobson, F.S., Jouventin, P., 2007. How slow breeding can be selected in seabirds: testing Lack's hypothesis. Proc. R. Soc. B Biol. Sci. 274, 275–279.
Dobson, F.S., Jouventin, P., 2010. The trade-off of reproduction and survival in slow-breeding seabirds. Can. J. Zool. 88, 889–899.
Dobson, F.S., Jouventin, P., 2015. Does feeding zone influence egg size in slow-breeding seabirds? Can. J. Zool. 93, 589–592.
Dobson, F.S., Nolan, P.M., Nicolaus, M., Bajzak, C., Coquel, A.-S., Jouventin, P., 2008. Comparison of color and body condition between early and late breeding King penguins. Ethology 114, 925–933.
Dobson, F.S., Couchoux, C., Jouventin, P., 2011. Sexual selection on a coloured ornament in King penguins. Ethology 117, 872–879.

Dobzhansky, T., 1937. Genetics and the Origin of Species. Columbia University Press, New York.

Downes, M.C., Ealey, E.H.M., Gwynn, A.M., Young, P.S., 1959. The Birds of Heard Island. Australian National Antarctic Research Expedition. Report Series B 1, 1–135.

Dresp, B., Langley, K., 2006. Fine structural dependence of ultraviolet reflections in the King Penguin beak horn. Anat. Rec. - Part A Discov. Mol. Cell. Evol. Biol. 288, 213–222.

Dresp, B., Jouventin, P., Langley, K., 2005. Ultraviolet reflecting photonic microstructures in the King penguin beak. Biol. Lett. 1, 310–313.

Eggleton, P., Siegfried, W., 1979. Displays of the jackass penguin. Ostrich 50, 139–167.

Emlen, J.T., Penney, R.L., 1964. Distance navigation in the Adelie penguin. Ibis 106, 417–431.

Endler, J.A., 1986. Natural Selection in the Wild. Princeton University Press, Princeton, New Jersey.

Endler, J.A., Mielke, P.W., 2005. Comparing entire colour patterns as birds see them. Biol. J. Linn. Soc. 86, 405–431.

Falla, R.A., 1937. Birds. BANZARE Rep. Ser. B 2, 1–304.

Falls, J.B., 1982. Individual recognition by sounds in brids. In: Acoustic Communication in Birds, vol. 2. Academic Press, New York, pp. 237–278.

Favaro, L., Ozella, L., Pessani, D., 2014. The vocal repertoire of the African penguin (*Spheniscus demersus*): structure and function of calls. PLoS One 9, e103460.

Fisher, R.A., 1934. The Genetical Theory of Natural Selection. Clarendon Press, Oxford.

Fitch, W.T., 1997. Vocal tract length and formant frequency dispersion correlate with body size in rhesus macaques. J. Acoust. Soc. Am. 102, 1213–1222.

Fitch, W.T., 2000. The evolution of speech: a comparative review. Trends Cognit. Sci. 4, 258–267.

Forster, J.R., 1781. Historia *Atpenodytae* Generis, avium orbi australi proprii. Societatis Regiae Scientiarum Gottingenses 3, 121–148.

Fox, D., Smith, V., Wolfson, A., 1967. Carotenoid selectivity in blood and feathers of lesser (African) chilean and greater (European) flamingos. Comp. Biochem. Physiol. 23, 225–232.

Fridolfsson, A.K., Ellegren, H., 1999. A simple and universal method for molecular sexing of non-ratite birds. J. Avian Biol. 30, 116–121.

Furbringer, M., 1888. Untersuchungen zur Morphologie und Systematik der Vogel. van Halkema, Amsterdam.

Garcia, V., Jouventin, P., Mauget, R., 1996. Parental care and the prolactin secretion pattern in the King penguin: an endogenously timed mechanism? Horm. Behav. 30, 259–265.

Gauthier-Clerc, M., Le Maho, Y., Clerquin, Y., Drault, S., Handrich, Y., 2000. Ecophysiology – penguin fathers preserve food for their chicks. Nature 408, 928–929.

Gilbert, C., Robertson, G., Le Maho, Y., Naito, Y., Ancel, A., 2006. Huddling behavior in Emperor penguins: dynamics of huddling. Physiol. Behav. 88, 479–488.

Gillespie, T.H., 1932. A Book of King Penguins. H. Jenkins, London.

Goldsmith, T., 1990. Optimization, constraint, and history in the evolution of eyes. Q. Rev. Biol. 65, 281–322.

Goodwin, D., 1960. Observations on avadavats and golden-breasted waxbills. Avic. Mag. 66, 174–199.

Grafen, A., 1990. Sexual selection unhandicapped by the fisher process. J. Theor. Biol. 144, 473–516.

Grimes, L.G., 1965. Antiphonal singing in *Laniarius barbarus barbarus* and the auditory reaction time. Ibis 107, 101–104.

Guillotin, M., Jouventin, P., 1979. La parade nuptiale du manchot empereur *Aptenodytes forsteri* et sa signification biologique. Biologie du comportement 4, 249–267.

Guillotin, M., Jouventin, P., 1980. Le Petrel des Neiges a Pointe-Geologie. Le Gerfaut 70, 51–72.

Guinet, C., Koudil, M., Bost, C.A., Durbec, J.P., Georges, J.Y., Mouchot, M.C., Jouventin, P., 1997. Foraging behaviour of satellite-tracked King penguins in relation to sea-surface temperatures obtained by satellite telemetry at Crozet Archipelago, a study during three austral summers. Mar. Ecol. Prog. Ser. 150, 11–20.

Guinard, E., Weimerskirch, H., Jouventin, P., 1998. Population changes and demography of the northern rockhopper penguin on Amsterdam and Saint Paul Islands. Colon. Waterbirds 21, 222–228.

Gwinner, E., Kneutgen, J., 1962. Uber Die Biologische Bedeutung Der Zweckdienlichen Anwendung Erlernter Laute Bei Vogeln. Naturwissenschaften 49, 615.

Hackett, S.J., Kimball, R.T., Reddy, S., Bowie, R.C.K., Braun, E.L., Braun, M.J., Chojnowski, J.L., Cox, W.A., Han, K.-L., Harshman, J., et al., 2008. A phylogenomic study of birds reveals their evolutionary history. Science 320, 1763–1768.

Hagen, Y., 1952. Birds of Tristan da Cunha. Vicenskaps-Akademi, Oslo, Norway.

Hart, N.S., 2001. The visual ecology of avian photoreceptors. Prog. Retin. Eye Res. 20, 675–703.

Hawkins, G.L., Hill, G.E., Mercadante, A., 2012. Delayed plumage maturation and delayed reproductive investment in birds. Biol. Rev. 87, 257–274.

Higham, J.P., Hebets, E.A., 2013. An introduction to multimodal communication. Behav. Ecol. Sociobiol. 67, 1381–1388.

Hill, G.E., 2002. A Red Bird in a Brown Bag: The Function and Evolution of Colorful Plumage in the House Finch. Oxford University Press), Oxford.

Hill, G.E., McGraw, K.J., 2006. Bird Coloration, vol. 1. Harvard University Press, Cambridge, Massachusetts.

Hoogland, J., Sherman, P., 1976. Advantages and disadvantages of bank swallow (*Riparia-riparia*) coloniality. Ecol. Monogr. 46, 33–58.

Hutton, F.W., 1879. Notes on a collection from the Auckland Islands and Campbell Islands. Trans. N. Z. Inst. 10, 337–343.

Huxley, J.S., 1914. The courtship-habits of the great crested grebe (*Podiceps cristatus*); with an addition to the theory of sexual selection. Proc. Zool. Soc. Lond. 491–562.

Isenmann, P., 1971. Contribution, à l'éthologie et à l'écologie du Manchot Empereur à la colonie de Pointe Geologie. L'Oiseau et La RFO 41, 9–64.

Isenmann, P., Jouventin, P., 1970. Eco-ethologie du Manchot empereur (*Aptenodytes forsteri*) et comparaison avec le Manchot adelie (*Pygoscelis adeliae*) et le Manchot royal (*Aptenodytes patagonica*). L'Oiseau et La RFO 40, 136–159.

Jacot, A., Reers, H., Forstmeier, W., 2010. Individual recognition and potential recognition errors in parent-offspring communication. Behav. Ecol. Sociobiol. 64, 1515–1525.

Jiguet, F., Jouventin, P., 1999. Individual breeding decisions and long-term reproductive strategy in the King penguin *Aptenodytes patagonicus*. Ibis 141, 428–433.

Jouanin, C., 1953. Le materiel ornithologique de la mission "Passage de Venus sur le Soleil" (1874), station de l'Ile St. Paul. Bulletin Du Musée National d'Histoire Naturelle 2, 529–540.

Jouventin, P., 1971a. Incubation et elevage itinerants chez les manchot empereurs de Pointe Geologie (Terre Adelie). Revue du Comportement Animal 5, 189–206.

Jouventin, P., 1971b. Comportement et structure sociale chez le Manchot Empereur. Revue d'Ecologie Terre et Vie 25, 510–586.

Jouventin, P., 1972a. A new system of acoustic recognition in birds. Behaviour 43, 176–185.

Jouventin, P., 1972b. Note sur l'existence et la signification d'une rythmicite des parades mutuelles. Alauda 40, 56–62.

Jouventin, P., 1975. Mortality Parameters in Emperor Penguins *Aptenodytes forsteri*. In: Stonehouse, B. (Ed.), The Biology of Penguins. Macmillan, London, pp. 435–446.

Jouventin, P., 1977. Olfaction in snow petrels. Condor 79, 498–499.

Jouventin, P., 1978. L'evolution du comportement chez les spheniscides (manchots). Memoires et Travaux de l'Institut de Montpellier de l'Ecole Pratique Des Hautes Etudes 185–196.

Jouventin, P., 1982. Visual and vocal signals in penguins, their evolution and adaptive characters. Advances in Ethology 24, 1–149.

Jouventin, P., Aubin, T., 2000. Acoustic convergence between two nocturnal burrowing seabirds: experiments with a penguin *Eudyptula minor* and a shearwater *Puffinus tenuirostris*. Ibis 142, 645–656.

Jouventin, P., Aubin, T., 2002. Acoustic systems are adapted to breeding ecologies: individual recognition in nesting penguins. Anim. Behav. 64, 747–757.

Jouventin, P., Lagarde, F., 1995. Evolutionary ecology of the King penguin (*Aptenodytes patagonicus*): the self-regulation of the breeding cycle. In: Dann, P., Norman, I., Reilly, P. (Eds.), The Penguins: Ecology and Management. Surrey Beatty and Sons, Chipping Norton, Australia, pp. 80–95.

Jouventin, P., Mauget, R., 1996. The endocrine basis of the reproductive cycle in the King penguin (*Aptenodytes patagonicus*). J. Zool. 238, 665–678.

Jouventin, P., Mougin, J.-L., 1981. Les stratégies adaptatives des oiseaux de mer. Revue d'Ecologie Terre et Vie 35, 217–272.

Jouventin, P., Robin, J., 1984. Olfactory experiments on some Antarctic birds. Emu 84, 46–48.

Jouventin, P., Roux, P., 1979. Le chant du manchot adelie (*Pygoscelis adeliae*). Role dans la reconnaissance individuelle et comparaison avec le manchot empereur non territorial. Oiseau et La Revue Francaise d'Ornithologie 49, 31–37.

Jouventin, P., Weimerskirch, H., 1991. Changes in the population size and demography of southern seabirds: management implications. In: Perrins, C.M., Lebreton, J.D., Hirons, G.J.M. (Eds.), Bird Populations Studies, Relevance to Conservation and Management. Oxford University Press, Oxford, pp. 297–314.

Jouventin, P., Weimerskirch, H., 1990. Satellite tracking of wandering albatrosses. Nature 343, 746–748.

Jouventin, P., Guillotin, M., Cornet, A., 1979. Calls of the Emperor penguin and their adaptive significance. Behaviour 70, 230–250.

Jouventin, P., Le Maho, Y., Mougin, J.-L., 1980. Les manchots. Pour La Science 105, 1058–1066.

Jouventin, P., Capdeville, D., Cuenot-Chaillet, F., Boiteau, C., 1994. Exploitation of pelagic resources by a nonflying seabird – satellite tracking of the King penguin throughout the breeding cycle. Mar. Ecol. Prog. Ser. 106, 11–19.

Jouventin, P., Barbraud, C., Rubin, M., 1995. Adoption in the Emperor penguin, *Aptenodytes forsteri*. Anim. Behav. 50, 1023–1029.

Jouventin, P., Aubin, T., Lengagne, T., 1999a. Finding a parent in a King penguin colony: the acoustic system of individual recognition. Anim. Behav. 57, 1175–1183.

Jouventin, P., Lequette, B., Dobson, F.S., 1999b. Age-related mate choice in the wandering albatross. Anim. Behav. 57, 1099–1106.

Jouventin, P., Nolan, P.M., Ornborg, J., Dobson, F.S., 2005. Ultraviolet beak spots in King and Emperor penguins. Condor 107, 144–150.

Jouventin, P., Cuthbert, R.J., Ottvall, R., 2006. Genetic isolation and divergence in sexual traits: evidence for the northern rockhopper penguin *Eudyptes moseleyi* being a sibling species. Mol. Ecol. 15, 3413–3423.

Jouventin, P., McGraw, K.J., Morel, M., Celerier, A., 2007. Dietary carotenoid supplementation affects orange beak but not foot coloration in Gentoo penguins *Pygoscelis papua*. Waterbirds 30, 573–578.

Jouventin, P., Nolan, P.M., Dobson, F.S., Nicolaus, M., 2008. Coloured patches influence pairing rate in King penguins. Ibis 150, 193–196.

Jouventin, P., Christen, Y., Dobson, F.S., 2016. Altruism in wolves explains the coevolution of dogs and humans. Ideas Ecol. Evol. 9, 4–11.

Kearton, C., 1930. The Island of Penguins. Longmans, London.

Keddar, I., Andris, M., Bonadonna, F., Dobson, F.S., 2013. Male-biased matecompetition in King penguin trio parades. Ethology 119, 389–396.

Keddar, I., Couchoux, C., Jouventin, P., Dobson, F.S., 2015a. Variation of mutual colour ornaments of King penguins in response to winter resource availability. Behaviour 152, 1684–1705.

Keddar, I., Jouventin, P., Dobson, F.S., 2015b. Color ornaments and territory position in King penguins. Behav. Process. 119, 32–37.

Keddar, I., Altmeyer, S., Couchoux, C., Jouventin, P., Dobson, F.S., 2015c. Mate choice and colored beak spots of King penguins. Ethology 121, 1048–1058.

Kinsky, F.C., 1960. The yearly cycle of the Norther blue penguin (*Eudyptula minor* novaehollandiae) in the Wellington Harbour area. Rec. Dom. Mus. 3, 145–218.

Kinsky, F.C., Falla, F.A., 1976. A subspecific revision of the Australiasian blue penguin (*Eudyptula minor*) in the New Zealand area. Nat. Mus. N. Z. Rec. 1, 105–126.

Kodric-Brown, A., Brown, J., 1984. Truth in advertising – the kinds of traits favored by sexual selection. Am. Nat. 124, 309–323.

Kokko, H., 1997. Evolutionarily stable strategies of age-dependent sexual advertisement. Behav. Ecol. Sociobiol. 41, 99–107.

Kokko, H., 1998. Good genes, old age and life-history trade-offs. Evol. Ecol. 12, 739–750.

Kokko, H., Johnstone, R.A., 2002. Why is mutual mate choice not the norm? Operational sex ratios, sex roles and the evolution of sexually dimorphic and monomorphic signalling. Philos. Trans. R. Soc. Lond. Ser. B Biol. Sci. 357, 319–330.

Kraaijeveld, K., Kraaijeveld-Smit, F.J.L., Komdeur, J., 2007. The evolution of mutual ornamentation. Anim. Behav. 74, 657–677.

Lack, D., 1956. Swifts in a Tower. Methuen, London.

Lack, D., 1968. Ecological Adaptations for Breeding in Birds. Methuen, London.

LaCock, G.D., Duffy, D.C., Cooper, J., 1987. Population dynamics of the African penguin *Spheniscus demersus* at Marcus Island in the Benguela upwelling ecosystem: 1979–1985. Biol. Conserv 40, 117–126.

Lande, R., 1980. Sexual dimorphism, sexual selection, and adaptation in polygenic characters. Evolution 34, 292–305.

Le Bohec, C., Durant, J.M., Gauthier-Clerc, M., Stenseth, N.C., Park, Y.-H., Pradel, R., Gremillet, D., Gendner, J.-P., Le Maho, Y., 2008. King penguin population threatened by Southern Ocean warming. Proc. Natl. Acad. Sci. U.S.A. 105, 2493–2497.

Le Maho, Y., Whittington, J.D., Hanuise, N., Pereira, L., Boureau, M., Brucker, M., Chatelain, N., Courtecuisse, J., Crenner, F., Friess, B., et al., 2014. Rovers minimize human disturbance in research on wild animals. Nat. Methods 11, 1242–1244.

Lengagne, T., Lauga, J., Jouventin, P., 1997. A method of independent time and frequency decomposition of bioacoustic signals: inter-individual recognition in four species of penguins. C. R. Acad. Sci. III 320, 885–891.

Lengagne, T., Jouventin, P., Aubin, T., 1999a. Finding one's mate in a King penguin colony: efficiency of acoustic communication. Behaviour 136, 833–846.

Lengagne, T., Aubin, T., Lauga, J., Jouventin, P., 1999b. How do King penguins (*Aptenodytes patagonicus*) apply the mathematical theory of information to communicate in windy conditions? Proc. R. Soc. B Biol. Sci. 266, 1623–1628.

Lengagne, T., Aubin, T., Jouventin, P., Lauga, J., 1999c. Acoustic communication in a King penguin colony: importance of bird location within the colony and of the body position of the listener. Polar Biol. 21, 262–268.

Lengagne, T., Aubin, T., Jouventin, P., Lauga, J., 2000. Perceptual salience of individually distinctive features in the calls of adult King penguins. J. Acoust. Soc. Am. 107, 508–516.

Lengagne, T., Lauga, J., Aubin, T., 2001. Intra-syllabic acoustic signatures used by the King penguin in parent-chick recognition: an experimental approach. J. Exp. Biol. 204, 663–672.

Lequette, B., Verheyden, C., Jouventin, P., 1989. Olfaction in Sub-Antarctic seabirds – its phylogenetic and ecological significance. Condor 91, 732–735.

Levick, G.M., 1915. Natural history of the Adelie penguin British "Terra Nova" expedition 1910. Zoology 1, 55–84.

Lorenz, K., 1951. Comparative studies on the behaviour of the Anatinae. Avic. Mag. 57, 157–182.

Lormée, H., Jouventin, P., Chastel, O., Mauget, R., 1999. Endocrine correlates of parental care in an Antarctic winter breeding seabird, the Emperor penguin, *Aptenodytes forsteri*. Horm. Behav. 35, 9–17.

Lovell, S.F., Lein, M.R., 2005. Individual recognition of neighbors by song in a suboscine bird, the alder, flycatcher *Empidonax alnorum*. Behav. Ecol. Sociobiol. 57, 623–630.

Lowe, P.R., 1933. On the primitive characters of the penguins, and their bearing on the phylogeny of brids. Proc. Zool. Soc. Lond. 483–538.

MacFarlane, G.R., Blomberg, S.P., Kaplan, G., Rogers, L.J., 2007. Same-sex sexual behavior in birds: expression is related to social mating system and state of development at hatching. Behav. Ecol. 18, 21–33.

Mackintosh, N., 1960. The pattern of distribution of the Antarctic fauna. Proc. R. Soc. Ser. B Biol. Sci. 152, 624–631.

Marchant, S., Higgins, P.J., 1990. Handbook of Australian, New Zealand and Antarctic birds, vol. 1. Oxford University Press, Melbourne.

Marler, P., 1955. Characteristics of some animal calls. Nature 176, 6–8.

Marler, P., 1957. Specific distinctiveness in the communication signals of birds. Behaviour 11, 13–39.

Martinez, A., Barbosa, A., 2010. Are pterins able to modulate oxidative stress? Theor. Chem. Acc. 127, 485–492.

Massaro, M., Davis, L.S., Darby, J.T., 2003. Carotenoid-derived ornaments reflect parental quality in male and female yellow-eyed penguins (*Megadyptes antipodes*). Behav. Ecol. Sociobiol. 55, 169–175.

Mathevon, M., Casey, C., Reichmuth, C., Charrier, I., 2017. Northern elephant seals memorize the rhythm and timbre of their rivals' voices. Current Biology. http://dx.doi.org/10.1016/j.cub.2017.06.035.

Mathews, G.M., 1932. The birds of Tristan da Cunha. Nativitates Zoologicae 38, 13–48.

Mauget, R., Jouventin, P., Lacroix, A., Ishii, S., 1994. Plasma-Lh and steroid-hormones in King penguin (*Aptenodytes patagonicus*) during the onset of the breeding cycle. Gen. Comp. Endocrinol. 93, 36–43.

Mauget, R., Garcia, V., Jouventin, P., 1995. Endocrine basis of the reproductive pattern of the Gentoo penguin (*Pygoscelispapua*) – winter breeding and extended laying period in northern populations. Gen. Comp. Endocrinol. 98, 177–184.

Mayr, E., 1963. Animal Species and Evolution. Oxford University Press, Oxford.

Mayr, E., 1970. Population, Species and Evolution. Harvard University Press, Harvard, Massechusetts.

Mayr, G., De Pietri, V.L., Scofield, R.P., 2017. A new fossil from the mid-Paleocene of New Zealand reveals an unexpected diversity of world's oldest penguins. Sci. Nat. 104. http://dx.doi.org/10.1007/s00114-017-1441-0.

Mbu Nyamsi, R.G., Aubin, T., Bremond, J.C., 1994. On the extraction of some time dependent parameters of an acoustic signal by means of the analytic signal concept. Its application to animal sound study. Bioacoustics 5, 187–203.

McComb, K., Reby, D., Baker, L., Moss, C., Sayialel, S., 2003. Long-distance communication of acoustic cues to social identity in African elephants. Anim. Behav. 65, 317–329.

McGraw, K.J., 2005. The antioxidant function of many animal pigments: are there consistent health benefits of sexually selected colourants? Anim. Behav. 69, 757–764.

McGraw, K.J., Wakamatsu, K., Ito, S., Nolan, P.M., Jouventin, P., Dobson, F.S., Austic, R.E., Safran, R.J., Siefferman, L.M., Hill, G.E., et al., 2004. You can't judge a pigment by its color: carotenoid and melanin content of yellow and brown feathers in swallows, bluebirds, penguins, and domestic chickens. Condor 106, 390–395.

McGraw, K.J., Toomey, M.B., Nolan, P.M., Morehouse, N.I., Massaro, M., Jouventin, P., 2007. A description of unique fluorescent yellow pigments in penguin feathers. Pigment Cell Res. 20, 301–304.

McGraw, K.J., Massaro, M., Rivers, T.J., Mattern, T., 2009. Annual, sexual, size- and condition-related variation in the colour and fluorescent pigment content of yellow crest-feathers in Snares penguins (*Eudyptes robustus*). Emu 109, 93–99.

Menzbier, M.A., 1887. Vergleichende Osteologie der Pinguine in Anwendung zur Haupteintheilung der Vogel. Bull. Soc. Imp. Natur. Mouscou. 483–587.

Miyazaki, M., Waas, J.R., 2005. Effects of male call pitch on female behaviour and mate fidelity in little penguins. J. Ethol. 23, 167–171.

Montgomerie, R., 2006. Analyzing colors. In: Bird Coloration, vol. 1. Harvard University Press, Cambridge, Massachusetts, pp. 90–147.

Murphy, R.C., 1936. Oceanic Birds of South America. Macmillan, New York.

Nakagawa, S., Waas, J.R., Miyazaki, M., 2001. Heart rate changes reveal that little blue penguin chicks (*Eudyptula minor*) can use vocal signatures to discriminate familiar from unfamiliar chicks. Behav. Ecol. Sociobiol. 50, 180–188.

Nicolaus, M., Le Bohec, C., Nolan, P.M., Gauthier-Clerc, M., Le Maho, Y., Komdeur, J., Jouventin, P., 2007. Ornamental colors reveal age in the King penguin. Polar Biol. 31, 53–61.

Nolan, P.M., Dobson, F.S., Dresp, B., Jouventin, P., 2006. Immunocompetence is signalled by ornamental colour in King penguins, *Aptenodytes patagonicus*. Evol. Ecol. Res. 8, 1325–1332.

Nolan, P.M., Dobson, F.S., Nicolaus, M., Karels, T.J., McGraw, K.J., Jouventin, P., 2010. Mutual mate choice for colorful traits in King penguins. Ethology 116, 635–644.

Noll, A., 1967. Cepstrum pitch determination. J. Acoust. Soc. Am. 41, 293.

Olsson, O., 1995. Timing and Body Reserve Adjustment in King Penguin Reproduction (Ph.D. thesis). Uppsala University.

Olsson, O., 1996. Seasonal effects of timing and reproduction in the King Penguin: a unique breeding cycle. J. Avian Biol. 27, 7–14.

Olsson, O., 1997. Effects of food availability on fledging condition and post-fledging survival in King penguin chicks. Polar Biol. 18, 161–165.

Olsson, O., 1998. Divorce in King penguins: asynchrony, expensive fat storing and ideal free mate choice. Oikos 83, 574–581.

Olsson, O., van der Jeugd, H.P., 2002. Survival in king penguins *Aptenodytes patagonicus*: temporal and sex-specific effects of environmental variability. Oecologica 132, 509–516.

Olsson, O., Bonnedahl, J., Anker-Nilssen, P., 2001. Mate switching and copulation behaviour in King penguins. J. Avian Biol. 32, 139–145.
Osorio, D., Vorobyev, M., 2008. A review of the evolution of animal colour vision and visual communication signals. Vis. Res. 48, 2042–2051.
Oustalet, E., 1875. Sur differents oiseaux de l'Ile St-Paul. Bull. Soc. Philom. Paris 6, 73.
Owren, M.J., Seyfarth, R.M., Cheney, D.L., 1997. The acoustic features of vowel-like grunt calls in chacma baboons (*Papio cyncephalus ursinus*): implications for production processes and functions. J. Acoust. Soc. Am. 101, 2951–2963.
Papoulis, A., 1977. Signal Analaysis. McGraw-Hill, New York.
Partan, S., Marler, P., 1999. Behavior – communication goes multimodal. Science 283, 1272–1273.
Penney, R.L., 1968. Territorial and social behaviour in the Adelie penguin. Antarct. Res. Ser. 12, 83–131.
Penney, R., Emlen, J., 1967. Further experiments on distance navigation in Adelie penguin *Pygoscelis adeliae*. Ibis 109, 99–109.
Peters, J.L., 1931. Checklist of Birds of the World. Harvard University Press, Harvard.
Petrie, M., Halliday, T., 1994. Experimental and natural changes in the peacocks (*Pave cristatus*) train can affect mating success. Behav. Ecol. Sociobiol. 35, 213–217.
Phillips, A.V., Stirling, I., 2000. Vocal individuality in mother and pup South American fur seals, *Arctocephalus australis*. Mar. Mamm. Sci. 16, 592–616.
Pincemy, G., Dobson, F.S., Jouventin, P., 2009. Experiments on colour ornaments and mate choice in King penguins. Anim. Behav. 78, 1247–1253.
Pincemy, G., Dobson, F.S., Jouventin, P., 2010. Homosexual mating displays in penguins. Ethology 116, 1210–1216.
Press, W.H., Flannery, B.P., Teukolsky, S.A., Vetterling, W.T., 1988. Numerical Recipes in C. Cambridge University Press, New York.
Prévost, J., 1961. Ecologie du Manchot Empereur. Hermann, Paris.
Prévost, J., Bourlière, F., 1957. Vie sociale et thermorégulation chez le Manchot empereur. Alauda 25, 167–173.
Prévost, J., Mougin, J.-L., 1970. Guide des Oiseaux et Mammifkres des Terres Australes et Antarctiques Francaise. Delachaux et Niestle, Paris.
Pryke, S.R., Lawes, M.J., Andersson, S., 2001. Agonistic carotenoid signalling in male red-collared widowbirds: aggression related to the colour signal of both the territory owner and model intruder. Anim. Behav. 62, 695–704.
Pycraft, W.P., 1898. Contribution to the osteology of birds. Part II: impennes. Proc. Zool. Soc. Lond. 958–989.
Randall, R.B., Tech, B., 1987. Frequency Alanysis. Bruel & Kjaer Press, Naerum, Denmark.
Reby, D., Joachim, J., Lauga, J., Lek, S., Aulagnier, S., 1998. Individuality in the groans of fallow deer (*Dama dama*) bucks. J. Zool. 245, 79–84.
Reby, D., Hewison, M., Izquierdo, M., Pepin, D., 2001. Red deer (*Cervus elaphus*) hinds discriminate between the roars of their current harem-holder stag and those of neighbouring stags. Ethology 107, 951–959.
Reiertsen, T.K., Erikstad, K.E., Barrett, R.T., Sandvik, H., Yoccoz, N.G., 2012. Climate fluctuations and differential survival of bridled and non-bridled Common guillemots Uria aalge. Ecosphere 3 (52).
Reilly, P.N., Cullen, J.M., 1981. The little penguin *Eudyptula minor* in Victoria, II: Breeding. Emu 81, 1–19.
Richdale, L.E., 1941. The erect-crested penguin (*Eudyptes sclateri* Buller). Emu 41, 25–53.
Richdale, L.E., 1947. The pair bond in penguins and petrels: a banding study. Bird-Banding 18, 107–117.

Richdale, L.E., 1949. A Study of a Group of Penguins of Known Age. Otago Daily Times and Witness Newspapers Co Ltd., Dunedin.

Richdale, L.E., 1951. Sexual Behaviour in Penguins. University of Kansas Press, Lawrence, Kansas.

Richdale, L.E., 1957. A Population Study of Penguins. Oxford University Press, Oxford.

Richdale, L.E., 1963. Breeding behaviour of the narrow-billed and the broad-billed prion on Whero Island. Transm. Zool. Soc. London 31, 87–155.

Riede, T., Zuberbuhler, K., 2003. The relationship between acoustic structure and semantic information in Diana monkey alarm vocalization. J. Acoust. Soc. Am. 114, 1132–1142.

Roberts, B.A., 1940. The breeding biology of penguins. British Graham Land expedition 1934–1937. Sci. Rep. 1, 195–254.

Robisson, P., 1992. Vocalizations in *Aptenodytes* penguins – application of the 2-voice theory. Auk 109, 654–658.

Robisson, P., Aubin, T., Bremond, J., 1989. Individual recognition in the Emperor penguin (*Aptenodytes-forsteri*) – respective parts of the temporal pattern and the sound structure of the courtship song. C. R. Acad. Sci. III 309, 383–388.

Rowley, I., 1983. Re-mating in birds. In: Bateson, P. (Ed.), Mate Choice. Cambridge University Press, London, pp. 331–360.

Sapin-Jaloustre, J., 1960. Ecologie du Manchot Adélie. Hermann, Paris.

Sapin-Jaloustre, J., Bourliére, F., 1952. Parades et attitudes caracteristiques de Pygoscelis adeliae. Alauda 20, 39–53.

Saraux, C., Le Bohec, C., Durant, J.M., Viblanc, V.A., Gauthier-Clerc, M., Beaune, D., Park, Y.-H., Yoccoz, N.G., Stenseth, N.C., Le Maho, Y., 2011. Reliability of flipper-banded penguins as indicators of climate change. Nature 469, 203–206.

Schull, Q., Dobson, F.S., Stier, A., Robin, J.-P., Bize, P., Viblanc, V.A., 2016. Beak color dynamically signals changes in fasting status and parasite loads in King penguins. Behav. Ecol. 27, 1684–1693.

Scolaro, J.A., 1990. Effects of nest density on breeding success in a colony of Magellanic penguins (*Spheniscus magellanicus*). Colon. Waterbirds 13, 41–49.

Searby, A., Jouventin, P., 2003. Mother-lamb acoustic recognition in sheep: a frequency coding. Proc. R. Soc. B Biol. Sci. 270, 1765–1771.

Searby, A., Jouventin, P., 2004. How to measure information carried by a modulated vocal signature? J. Acoust. Soc. Am. 116, 3192–3198.

Searby, A., Jouventin, P., 2005. The double vocal signature of crested penguins: is the identity coding system of rockhopper penguins *Eudyptes chrysocome* due to phylogeny or ecology? J. Avian Biol. 36, 449–460.

Searby, A., Jouventin, P., Aubin, T., 2004. Acoustic recognition in macaroni penguins: an original signature system. Anim. Behav. 67, 615–625.

Segonzac, M., Ms. note. La distinction entre les Gorfous Sauteurs des iles St-Paul-Amsterdam et ceux de l'archipel Crozet (missions 1969-70-71).

Shaughnessy, P.D., 1975. Variation in facial colour of the Royal penguin. Emu 75, 147–152.

Shawkey, M.D., Hill, G.E., 2006. Significance of a basal melanin layer to production of non-iridescent structural plumage color: evidence from an amelanotic Steller's jay (*Cyanocitta stelleri*). J. Exp. Biol. 209, 1245–1250.

Shawkey, M.D., Estes, A.M., Siefferman, L.M., Hill, G.E., 2003. Nanostructure predicts intraspecific variation in ultraviolet-blue plumage colours. Proc. R. Soc. B Biol. Sci. 270, 1455–1460.

Sieber, O., 1985. Individual recognition of parental calls by bank swallow chicks (*Riparia-riparia*). Anim. Behav. 33, 107–116.

Simpson, G.G., 1976. Penguins – Past and Present, Here and There. Yale University Press, New Haven and London.

Sivak, J., 1976. Role of a flat cornea in amphibious behavior of blackfoot penguin (*Spheniscus-demersus*). Can. J. Zool. 54, 1341–1345.

Sladen, W.J.L., 1958. The Pygoscelis Penguins. Falkland Island Departmental Surveys in Science Report 17, pp. 1–97.

Sparks, J.H., 1962. Clumping and social preening in the red advadavat. Birds Illus. 8, 48–49.

Sparks, J.H., 1965. On the role of allopreening invitation behaviour in reducing aggression among red avadavats, with comments on its evolution in the Spermestidae. Proc. Zool. Soc. Lond. 145, 387–403.

Sparks, J., Soper, T., 1967. Penguins. David and Charles, Newton Abbot, Devon.

Speirs, E., Davis, L., 1991. Discrimination by Adelie penguins, *Pygoscelis-adeliae*, between the loud mutual calls of mates, neighbors and strangers. Anim. Behav. 41, 937–944.

Spurr, E.B., 1975. Communication in the Adelie penguin. In: Stonehouse, B. (Ed.), The Biology of Penguins. Macmillan, London, pp. 449–501.

Stoddard, M.C., Prum, R.O., 2008. Evolution of avian plumage color in a tetrahedral color space: a phylogenetic analysis of new world buntings. Am. Nat. 171, 755–776.

Stonehouse, B., 1953. The Emperor penguin (*Aptenodytes forsteri* Gray). Falkl. Isl. Dep. Surv. Sci. Rep. 17, 1–33.

Stonehouse, B., 1960. The King penguin (*Aptenodytes patagonicus*) of South Georgia. Falkl. Isl. Dep. Surv. Sci. Rep. 23, 1–81.

Suthers, R., 1994. Variable asymmetry and resonance in the avian vocal-tract – a structural basis for individually distinct vocalizations. J. Comp. Physiol. A 175, 457–466.

Thiebot, J.-B., Cherel, Y., Crawford, R.J.M., Makhado, A.B., Trathan, P.N., Pinaud, D., Bost, C.-A., 2013. A space oddity: geographic and specific modulation of migration in Eudyptes penguins. PLoS One 8, e71429. UNSP.

Thomas, D.B., McGoverin, C.M., McGraw, K.J., James, H.F., Madden, O., 2013. Vibrational spectroscopic analyses of unique yellow feather pigments (spheniscins) in penguins. J. R. Soc. Interface 10, 20121065.

Thompson, D.H., Emlen, J.T., 1968. Parent-chick individual recognition in the Adelie penguin. Antarct. J. U. S. 3, 132.

Thorpe, W., North, M., 1965. Origin and significance of power of vocal imitation – with special reference to antiphonal singing of birds. Nature 208, 219–222.

Thorpe, W., North, M., 1966. Vocal imitation in tropical Bou-Bou shrike *Laniarius aethiopicus* major. Ibis 108, 432.

Tollu, B., 1988. Les Manchots, Écologie et Vie Sociale. Le Rocher, Paris.

Trivelpiece, W.Z., Trivelpiece, S.G., 1990. Courtship period of Adélie, Gentoo, and Chinstrap penguins. In: Davis, L.S., Darby, J.T. (Eds.), Penguin Biology. Academic Press, San Diego, pp. 113–127.

Tschanz, B., 1968. Trottellummen: die Entstehung der persoenlichen Beziehungen zwischen Jungvogel und Eltern. Zeitschrift fur Tierpsychologie (Beiheft 4), Berlin, Germany.

van Tets, G.P., 1965. A comparative study of some social communication patterns in the Pelecaniformes. Ornithol. Monogr. 2, 1–88.

van Zinderen Bakker Jr., E.M., 1971. A behavioural analysis of the Gentoo penguin *Pygoscelis papua* Forster. In: van Zinderen Bakker Jr., E.M., Winterbottom, J.M., Dyer, R.A. (Eds.), Marion and Prince Edward Islands, Report on the South Africa Biological and Geological Expeditino 1965–1966. Balkema, Cape Town, South Africa, pp. 251–272.

van Heezik, Y., Seddon, P., Cooper, J., Plos, A., 1994. Interrelationships between breeding frequency, timing and outcome in King penguins *Aptenodytes-patagonicus* – are King penguins biennial breeders. Ibis 136, 279–284.

Verheyden, C., Jouventin, P., 1994. Olfactory behavior of foraging *Procellariiformes*. Auk 111, 285–291.

Vianna, J.A., Noll, D., Dantas, G.P.M., Petry, M.V., Barbosa, A., Gonzalez-Acuna, D., Le Bohec, C., Bonadonna, F., Poulin, E., 2017. Marked phylogeographic structure of Gentoo penguin reveals an ongoing diversification process along the Southern Ocean. Mol. Phyl. Evol. 107, 486–498.

Viblanc, V.A., Dobson, F.S., Stier, A., Schull, Q., Saraux, C., Gineste, B., Pardonnet, S., Kauffmann, M., Robin, J.-P., Bize, P., 2016. Mutually honest? Physiological 'qualities' signalled by colour ornaments in monomorphic King penguins. Biol. J. Linn. Soc. 118, 200–214.

Viera, V.M., Nolan, P.M., Cote, S.D., Jouventin, P., Groscolas, R., 2008. Is territory defence related to plumage ornaments in the King penguin *Aptenodytes patagonicus*? Ethology 114, 146–153.

Vorobyev, M., Osorio, D., Bennett, A.T.D., Marshall, N.J., Cuthill, I.C., 1998. Tetrachromacy, oil droplets and bird plumage colours. J. Comp. Physiol. A 183, 621–633.

Waas, J., 1988. Acoustic displays facilitate courtship in little blue penguins, *Eudyptula-minor*. Anim. Behav. 36, 366–371.

Wang, X., Clarke, J.A., 2014. Phylogeny and forelimb disparity in waterbirds. Evolution 68, 2847–2860.

Warham, J., 1958. The nesting of the little penguin *Eudyptula minor*. Ibis 100, 605–616.

Warham, J., 1963. The rockhopper penguin, *Eudoptes chrysocome*, at Macquarie Island. Auk 80, 229–256.

Warham, J., 1971. Aspects of breeding behaviour in the Royal penguin *Eudyptes chrysolophus schlegeli*. Notornis 18, 91–115.

Warham, J., 1972. Aspects of the biology of the erect-crested penguin, *Eudyptes sclateri*. Ardea 60, 145–184.

Warham, J., 1974. The Fiorland crested penguin. Ibis 116, 1–27.

Warham, J., 1975. The crested penguins. In: Stonehouse, B. (Ed.), The Biology of Penguins. Macmillan, London, pp. 189–269.

Weimerskirch, H., Stahl, J.C., Jouventin, P., 1992. The breeding biology and population dynamics of King penguins *Aptenodytes patagonica* on the Crozet Islands. Ibis 134, 107–117.

Williams, T.D., 1995. The Penguins. Oxford University Press, Oxford.

Williams, T.D., Rodwell, S., 1992. Annual variation in return rate, mate and nest-site fidelity in breeding gentoo and macaroni penguins. Condor 94, 636–645.

Wilson, E., 1907. Aves. British National Antarctic Expedition 1901–1904. Natural History, pp. 1–121.

Zahavi, A., 1975. Mate selection – selection for a handicap. J. Theor. Biol. 53, 205–214.

Zuk, M., 2006. Family values in black and white. Nature 439, 917.

# Index

'*Note:* Page numbers followed by "f" indicate figures, "t" indicate tables.'

## F



## G



## H



## I



## J



## K

## L



## M

## W



## Y